Mathematics
for Edexcel
IGCSE

Marguerite Appleton
Demetris Demetriou
Derek Huby
Jayne Kranat

OXFORD
UNIVERSITY PRESS

Great Clarendon Street, Oxford OX2 6DP

Oxford University Press is a department of the University of Oxford.
It furthers the University's objective of excellence in research,
scholarship, and education by publishing worldwide in

Oxford   New York

Auckland   Cape Town   Dar es Salaam   Hong Kong   Karachi
Kuala Lumpur   Madrid   Melbourne   Mexico City   Nairobi
New Delhi   Shanghai   Taipei   Toronto

With offices in

Argentina   Austria   Brazil   Chile   Czech Republic   France   Greece
Guatemala   Hungary   Italy   Japan   Poland   Portugal   Singapore
South Korea   Switzerland   Thailand   Turkey   Ukraine   Vietnam

Oxford is a registered trade mark of Oxford University Press
in the UK and in certain other countries

British Library Cataloguing in Publication Data

Data available

ISBN 978-0-19-915262-9

10 9 8 7 6 5 4 3 2 1

Printed in China

Acknowledgements

The publishers would like to thank Edexcel for their kind permission to
reproduce past exam questions. Edexcel Ltd, accepts no responsibility
whatsoever for the accuracy or method of working in the answers given.

The publisher would like to thank the following for permission to
reproduce photographs:
p178l Box Croxford/Atmosphere Picture Library/Alamy; **p178r** Ordnance Survey.

Cover image courtesy of Image DJ

Paper used in the production of this book is a natural,
recyclable product made from wood grown in sustainable forests.
The manufacturing process conforms to the environmental
regulations of the country of origin.

About this book

This book has been specifically written to help you get your best possible grade in your Edexcel IGCSE Mathematics examinations.

The authors are experienced teachers and examiners who have an excellent understanding of the Edexcel two-tier specification and so are well qualified to help you successfully meet your objectives.

The book is made up of units that are based on the Edexcel specification, and provide coverage of the National Curriculum strands at Key Stage 4.

The units are:

Each unit contains double page spreads for each lesson. These are shown on the full contents list.

Problem solving is integrated throughout the material.

How to use this book

This book is made up of units of work that are colour-coded:
Algebra (green), Data (blue), Number (orange) and Shape, space and
measures (pink).

Each unit starts with an overview of the content, so that you know
exactly what you are expected to learn.

This unit will show you how to

- Use powers of 10 to represent large and small numbers
- Calculate with powers of 10 using the index laws
- Understand and use standard index form, in written notation
  and on a calculator display
- Write any integer as a product of its factors
- Find the prime factor decomposition of any integer and the
  highest common factor and least common multiple of any
  two integers

The first page of a unit also provides prior knowledge questions to
help you revise before you start – if refers you to files on the CD
accompanying this book if you need to refresh your memory on a topic.
Then you can apply your knowledge later in the unit:

Before you start ...

You should be able to answer these questions.

1 Write all the factors of each number.

    a 12   **b** 30

    c 120   **d** 360

Review

CD – N1

Inside each unit, the content develops in double page spreads that all
follow the same structure.

Each spread starts with a list of the learning outcomes and a summary of
the keywords:

S1.5   Volume and surface area

This spread will show you how to:

- Solve problems involving surface areas and volumes of prisms

Keywords

Pythagoras's theorem
Surface Area
Volume

Key points are highlighted in the text so you can see the facts you need
to learn:

- Pythagoras's theorem: $a^2 + b^2 = c^2$

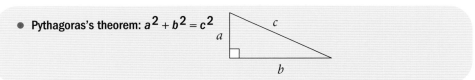

Examples showing the key skills and techniques you need to develop are shown in boxes. Also, margin notes show tips and reminders you may find useful:

Find the probability of rolling three consecutive sixes on a fair dice.

$P(6) = \frac{1}{6}$

$P(6 \text{ and } 6 \text{ and } 6) = P(6) \times P(6) \times P(6)$

$\qquad = \frac{1}{6} \times \frac{1}{6} \times \frac{1}{6}$

$\qquad = \frac{1}{216}$

The outcomes of the rolls of the dice are independent.

Each exercise is carefully graded, set at three levels of difficulty:
- The first few questions are mainly repetitive to give you confidence, and simplify the content of the spread
- The questions in the middle of the exercise consolidate the topic, focusing on the main techniques of the spread
- Later questions extend the content of the spread – some of these questions may be problem-solving in nature, and may involve different approaches.

At the end of the unit is an exam review page so that you can revise the learning of the unit before moving on. The key Edexcel objectives are identified:

A5 Exam review

Key objectives

Use formulae from mathematics and other subjects.

Change the subject of a formula including cases where the subject occurs twice.

Summary questions, including past exam questions, are provided to help you check your understanding of the key concepts covered and your ability to apply the key techniques.

An Edexcel IGCSE formula page is provided near the end of the book so you can see what information you will be given in the exam.

You will find the answers to all exercises at the back of the book so that you can check your own progress and identify any areas that need work.

Contents

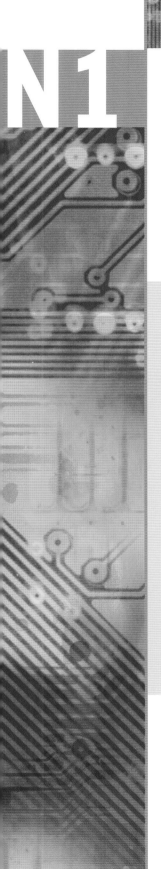

This unit will show you how to

- Use powers of 10 to represent large and small numbers
- Calculate with powers of 10 using the index laws
- Understand and use standard index form, in written notation and on a calculator display
- Write any integer as a product of its factors
- Find the prime factor decomposition of any integer and the highest common factor and least common multiple of any two integers

Before you start ...

You should be able to answer these questions.

Review

1 Write all the factors of each number.

    a 12   **b** 30

    c 120   **d** 360

CD – N1

2 Write a list of all of the prime numbers up to 100. (There are 25 of them.)

CD – N1

3 **a** Write the highest number that is a factor of both 12 and 30.

   b Write the smallest number that is a multiple of both 12 and 30.

CD – N1

4 Write each number as a power of 10.

    a 100   **b** 1000

    c 10 000   **d** 1 000 000

CD – N1

N1.1 Powers of 10

This spread will show you how to:

- Use powers of 10 to represent large and small numbers
- Calculate with powers of 10 using the index laws

Keywords
Index form
Index laws
Power
Reciprocal

You can use **powers** of 10 to represent large and small numbers. Numbers written with powers are in **index form**.

Example

Write these numbers as powers of 10.

a 100   **b** 1000   **c** 1 000 000

a $100 = 10 \times 10 = 10^2$ **b** $1000 = 10^3$ **c** $1\,000\,000 = 10^6$

The number of zeros tells you the power of 10.

You can calculate with powers of 10, using the **index laws**.

- To multiply powers of 10 you add the indices: $10^a \times 10^b = 10^{a+b}$
- To divide powers of 10 you subtract the indices: $10^a \div 10^b = 10^{a-b}$

Example

Work out these calculations. Give your answers in index form.

a $10^2 \times 10^3$ **b** $10^2 \times 10^2$ **c** $10^3 \div 10^2$

a $10^2 \times 10^3 = 100 \times 1000 = 100\,000 = 10^5$
b $10^2 \times 10^2 = 10^{2+2} = 10^4$
c $10^3 \div 10^2 = 10^{3-2} = 10^1 = 10$

$10^1 = 10$

- Any number to the power 0 is equal to 1:
  $3^0 = 1$ $12^0 = 1$ $10^0 = 1$

$x^0 = 1$ for any value of x except zero.

- A negative index represents a **reciprocal**:

  $10^{-2} = \dfrac{1}{10^2} = 0.01$ $10^{-4} = \dfrac{1}{10^4} = 0.0001$

$x^{-n} = 1 \div x^n$ or $\frac{1}{x^n}$

Example

a Write 0.000 001 as a power of 10.
b Work out $10^3 \div 10^3$
c Work out $10^0 \div 10^1$

a Count the number of digits from the decimal point to the digit 1, starting at the decimal point
  $0.000\,001 = 10^{-6}$
b $10^3 \div 10^3 = 10^{3-3} = 10^0 = 1$
c $10^0 \div 10^1 = 10^{0-1} = 10^{-1} = \frac{1}{10} = 0.1$

1 Write these numbers as powers of 10.

   a 100   **b** 10   **c** 1000   **d** 1

   e 10 000   **f** 1 000 000   **g** 100 000   **h** 100 000 000

2 Write these numbers as powers of 10.

   a 0.01   **b** 0.1   **c** 0.001   **d** 0.000 01

   e 0.0001   **f** 0.000 000 1   **g** 0.000 001   **h** 1.0

3 Write these as ordinary numbers.

   a 10^3 **b** 10^6 **c** 10^5 **d** 10^9

   e 10^4 **f** 10^1 **g** 10^2 **h** 10^7

4 Write these as ordinary numbers.

   a 10^0 **b** 10^{-2} **c** 10^{-5} **d** 10^{-3}

   e 10^{-7} **f** 10^{-1} **g** 10^{-4} **h** 10^{-6}

5 Work out these calculations, and give your answers in index form.

   a $10^2 \times 10^3$ **b** $10^4 \times 10^5$ **c** $10^5 \times 10^3$

   d $10^6 \div 10^3$ **e** $10^8 \div 10^4$ **f** $10^6 \div 10^2$

6 Work out these calculations, and give your answers in index form.

   a $10^4 \div 10^6$ **b** $10^3 \div 10^7$ **c** $10^2 \div 10^{10}$

   d $10 \div 10^9$ **e** $1 \div 10^8$ **f** $10^7 \div 10$

7 Work out these calculations, and give your answers in index form.

   a $10^4 \times 10^3 \div 10^5$ **b** $10^6 \div 10^{-6}$

   c $10^3 \times 10^{-4} \div 10^5$ **d** $10^{-5} \div 10^{-2} \times 10^3$

8 Find the value of the letters in these equations.

   a $10^4 \times 10^w \times 10^2 = 10^9$ **b** $10^x \div 10^2 \times 10^5 = 10^8$

   c $10^{-2} \times 10^y \div 10^3 = 10^{-2}$ **d** $10^z \times 10^{-4} \times 10^z = 10$

9 Find the value of the letters in these equations.

   a $300 = 3 \times 10^a$ **b** $0.000\ 005 = b \times 10^{-6}$

   c $0.15 = 1.5 \times 10^c$ **d** $130 = d \times 10^2$

10 Find the value of the letters in these equations.

   a $0.004\ 005 = m \times 10^{-3}$ **b** $10^n \times 10^{-5} \times 10^n = 10^n$

   c $1 \div 10^{-7} = 10^p$ **d** $10^q \times 10^q \times 10^q = 10^{-12}$

This spread will show you how to:

● Understand and use standard index form, in written notation and on a calculator display

Keywords

Index
Standard form

You can write the number 44 506 as 4.4506×10^4.
This is called **standard form**.

> ● You can write a large number in standard form as $A \times 10^n$, where n is an integer and $1 \leqslant A < 10$

Standard form makes it easier to compare and calculate with large numbers.

● Write the number with the place values adjusted so that it is between 1 and 10.

For 3667 write 3.667.

● Work out the value of the **index**, n, the number of columns the digits have moved.

In 3.667 the digits have moved 3 columns right so $n = 3$.
$3667 = 3.667 \times 10^3$

Example

Express these numbers in standard form.

a 435   **b** 4067
c 15.5   **d** 23 517

a $435 = 4.35 \times 10^2$ **b** $4067 = 4.067 \times 10^3$
c $15.5 = 1.55 \times 10^1$ **d** $23\ 517 = 2.3517 \times 10^4$

Example

Write these numbers in order, starting with the smallest.

6.35×10^4, 5.44×10^4, 6.95×10^3, 7.075×10^2

The correct order is:

7.075×10^2, 6.95×10^3, 5.44×10^4, 6.35×10^4

First order the powers of 10:
$10^2 < 10^3 < 10^4$.
Then compare numbers with the same powers of 10:
$5.44 < 6.35$

You can use numbers written in standard form in calculations.

Exam question

The Andromeda Galaxy is 21 900 000 000 000 000 000 km from the Earth.

a Write 21 900 000 000 000 000 000 in standard form.
b Light travels 9.46×10^{12} km in one year.
Calculate the number of years that light takes to travel from the Andromeda Galaxy to Earth. Give your answer in standard form correct to 2 significant figures. *(Edexcel Ltd., (Spec))*

a $21\ 900\ 000\ 000\ 000\ 000\ 000 = 2.19 \times 10^{19}$
b $(2.19 \times 10^{19}) \div (9.46 \times 10^{12}) = 2.3 \times 10^6$ years

Some calculators have an $\boxed{\text{Exp}}$ key.

To enter 3.2×10^4 you press

$\boxed{3}\ \boxed{\cdot}\ \boxed{2}\ \boxed{\text{Exp}}\ \boxed{4}$

The calculator displays

$\boxed{3.2\ {}^{04}_{\times 10}}$

1 Write these numbers in standard form.

    a 1375   **b** 20 554   **c** 231 455   **d** 5.8 billion

> 1 billion
> = 1 000 000 000

2 Rewrite each number from question **1** in standard form correct to two significant figures.

3 Write these numbers in order of size, smallest first.

    4.05×10^4 4.55×10^4 9×10^3 3.898×10^4 1.08×10^4 5×10^4

4 Write these numbers as ordinary numbers.

    a 6.35×10^4 **b** 9.1×10^{17} **c** 1.11×10^2 **d** 2.998×10^8

5 Write these numbers in standard form.

    a 21.5×10^3 **b** 0.7×10^{14} **c** 122.516×10^{18} **d** 0.015×10^9

6 Work out these calculations, giving your answers in standard form.

    a $2 \times 10^3 \times 3 \times 10^4$ **b** $(8 \times 10^{15}) \div (2 \times 10^3)$

    c $7.5 \times 10^3 \times 2 \times 10^5$ **d** $(3.5 \times 10^3) \div (5 \times 10^2)$

    e $5 \times 10^5 \times 3 \times 10^4$ **f** $4 \times 10^3 \times 5 \times 10^2 \times 6 \times 10^4$

7 Use a calculator to find the square of each of these numbers. Give your answers in standard form, correct to 3 significant figures.

    a 2.34×10^8 **b** 9.17×10^{13}

    c 5.49×10^7 **d** 3.33×10^9

8 **a** Use a calculator to work these out.

        i $3.4 \times 10^5 \times 236$ **ii** $6.6 \times 10^7 \div 15$

        iii $292 \times 3.4 \times 10^4$ **iv** $3.145 \times 10^7 \times 17$

    b Write your answers to part **a** in order, smallest to largest.

9 Use a calculator to work these out. Give your answers in standard form correct to 2 significant figures.

    a $6.22 \times 10^5 \times 9.1 \times 10^{11}$ **b** $(9.7 \times 10^4) \div (3.7 \times 10^2)$

    c $(4.15 \times 10^{12}) \div (9.7 \times 10^7)$

10 As the moon orbits Earth the distance between them varies between 4.07×10^5 km and 3.56×10^5 km. Find the difference between these two distances.

> **DID YOU KNOW?**
> The gravitational forces between Earth and the moon cause two high tides a day.

11 The radius of Earth is approximately 6380 km. Find the volume of Earth in cm^3, giving your answer in standard form correct to 3 significant figures. (Use the formula $V = \frac{4}{3}\pi r^3$.)

Standard form for small numbers

This spread will show you how to:

- Understand and use standard index form, in written notation and on a calculator display

You can write small numbers in **standard form**.

For example, $0.000\,004 = 4 \times 10^{-6}$.

> - You can write a small number in standard form as $A \times 10^{-n}$, where n is an integer and $1 \leqslant A < 10$.

- Write the number with the place values adjusted so that it is between 1 and 10.

    For 0.0876 write 8.76.

To get from 8.76 to 0.0876 you divide by 100, or multiply by $\frac{1}{100} = 10^{-2}$.

- Work out the value of the **index**, n, the number of columns the digits have moved. Remember the negative sign for n.

    In 8.76 the digits have moved 2 columns left so $n = -2$.

    $0.0876 = 8.76 \times 10^{-2}$

Example

Express these numbers in standard form.

a 0.000 483   **b** 0.001 003   **c** 0.000 000 007

a $0.000\,483 = 4.83 \times 10^{-4}$
b $0.001\,003 = 1.003 \times 10^{-3}$
c $0.000\,000\,007 = 7 \times 10^{-9}$

In part **a**, the digits have moved 4 places:

0 . 0 0 0 4 8 3
      4 3 2 1

Example

Calculate $0.000\,000\,42 \div 0.0036$, giving your answer in standard form to 3 significant figures.

$(4.2 \times 10^{-7}) \div (3.6 \times 10^{-3})$
   $= (4.2 \div 3.6) \times (10^{-7} \div 10^{-3})$
   $= 1.166\,666 \ldots \times 10^{-7 - -3}$
   $= 1.17 \times 10^{-4}$ (to 3 sf)

Example

Rosie wrote

$(7.2 \times 10^{-4}) \div (3.6 \times 10^{-8}) = 2 \times 10^{-12}$

Is she correct?

No. She has multiplied 10^{-4} and 10^{-8}, instead of dividing.
The correct answer is:

$(7.2 \div 3.6) \times (10^{-4} \div 10^{-8}) = 2 \times 10^{4}$

Examiner's tip
A quick check shows Rosie's answer is incorrect. Since you are dividing one number by another smaller number, the answer must be **greater** than 1.

1 Write these numbers in standard form.

   a 0.000 34   **b** 0.1067   **c** 0.000 0091

   d 0.315   **e** 0.000 0505   **f** 0.0182

   g 0.008 45   **h** 0.000 000 000 306

2 Use a calculator to find the reciprocal of each of these numbers. Give your answers in standard form to 3 significant figures.

Reciprocal of $x = 1 \div x$ or $\frac{1}{x}$

   a 3499   **b** 121 878   **c** 42 510   **d** 4.23×10^5

   e 648   **f** 9.16×10^{-3} **g** 10 002   **h** 4 237 551

3 Write these as ordinary numbers.

   a 4.5×10^{-3} **b** 3.17×10^{-5} **c** 1.09×10^{-6} **d** 9.79×10^{-7}

4 Work out these without using a calculator.

   a $(4 \times 10^6) \div (2 \times 10^8)$ **b** $5 \times 10^4 \times 2 \times 10^{-6}$

   c $(3 \times 10^{-3}) \div (2 \times 10^5)$ **d** $(4 \times 10^5) \div (2 \times 10^2)$

   e $5 \times 10^{-3} \times 5 \times 10^{-4}$ **f** $(9.3 \times 10^{-2}) \div (3 \times 10^{-6})$

Use the standard form function on your calculator to check each of your answers.

5 Work these out using a calculator. Give your answers in standard form, correct to 3 significant figures.

   a $(9.35 \times 10^3) \div (2.72 \times 10^5)$ **b** $(2.16 \times 10^{-7}) \times (3.33 \times 10^{-6})$

   c $(6.12 \times 10^{-11}) \div (2.12 \times 10^{-6})$ **d** $(7.88 \times 10^4) \times (8.26 \times 10^{-7})$

   e $(5.92 \times 10^{-3}) \times (3.87 \times 10^{-4})$ **f** $(4.06 \times 10^{-5}) \div (1.73 \times 10^{-8})$

6 Check your answers to question **5** by rounding the numbers in each question to one significant figure, and then working out an estimate without using a calculator.

7 Light travels about 3×10^8 metres per second. Find the time it takes for light to travel 1 metre. Give your answer in standard form.

8 A mass of 12 grams of carbon contains about 6.023×10^{23} carbon atoms.

   a Write 12 grams as a mass in kilograms using standard form.

   b Estimate the mass of one carbon atom, giving your answer in kilograms in standard form, correct to 2 significant figures.

9 Cosmologists currently think that the universe was created in a 'Big Bang' about 14 thousand million years ago. How many days is it since the Big Bang? Take one year as $365\frac{1}{4}$ days and write your answer in standard form.

Prime numbers and factorisation

This spread will show you how to:
- Write any integer as a product of its factors
- Find the prime factor decomposition of any integer

Keywords
Factor
Integer
Prime
Prime factor
  decomposition

You can write any **integer** as the product of its **factors**.
Prime factor decomposition means writing a number as the product of **prime** factors only.

- The prime factor decomposition of a number can only be written in one way.

Example

Write 30 as the product of its prime factors.

$30 = 3 \times 10 = 3 \times 2 \times 5$
$ = 2 \times 3 \times 5$

Write the prime factors in ascending order.

- Use index notation to show any repeated factors.
  $3 \times 2 \times 2 \times 2 = 2^3 \times 3$

With large numbers, work systematically through possible prime factors.

Example

Write 4095 as a product of its prime factors.

2 is not a factor
3 is a factor $\rightarrow 4095 = 3 \times 1365$
3 is a factor of 1365 $\rightarrow 4095 = 3 \times 3 \times 455$
5 is a factor of 455 $\rightarrow 4095 = 3 \times 3 \times 5 \times 91$
7 is a factor of 91 $\rightarrow 4095 = 3 \times 3 \times 5 \times 7 \times 13$

$4095 = 3^2 \times 5 \times 7 \times 13$

Examiner's tip
Working systematically and writing each line in full will help you keep track of the factors.
Check (by multiplying) that your prime factor decomposition is correct.

Example

A classroom has an area of 144 m^2.
a Write the prime factor decomposition of 144.
b Find the dimensions of the classroom if it is
  i square   **ii** 16 m long   **iii** 18 m long.

a $144 = 2 \times 2 \times 2 \times 2 \times 3 \times 3$
  $ = 2^4 \times 3^2$
b i For a square room each side is $\sqrt{2^4 \times 3^2} = 2^2 \times 3$
  $\phantom{For a square room each side is \sqrt{2^4 \times 3^2}} = 12$ m

  ii $144 \div 16 = (2^4 \times 3^2) \div 2^4$
    $ = 3^2$
    So the classroom is 16 m $\times$ 9 m.

  iii $144 \div 18 = (2^4 \times 3^2) \div (2^4 \times 3^2)$
    $ = 2^3$
    So the classroom is 18 m $\times$ 8 m.

1 Sam thinks that each of these numbers is prime. Explain why he is wrong every time.

   a 201   **b** 995   **c** 777   **d** 441   **e** 71 536

2 Find the prime factor decomposition of each number.

   a 21   **b** 8   **c** 15   **d** 90   **e** 124

3 Write each number as a product of its prime factors. Check your answers by multiplying.

   a 900   **b** 630   **c** 1001   **d** 2205   **e** 1371

   f 891   **g** 2788   **h** 1431   **i** 3377   **j** 2 460 510

4 Copy and complete the table to show which numbers from 2000 to 2010 are prime. Give the prime factor decomposition of those that are not prime.

Number	Prime? (Yes/No)	Prime factor decomposition for non-primes
2000		
2001		
⋮		
2009		
2010		

5 **a** Write each number as a product of its prime factors.

   i 1728   **ii** 423   **iii** 812   **iv** 23

   b For each number in part **a**, find the smallest number you need to multiply by to give a square number.

6 A rectangular carpet has an area of 84 square metres. Its length and width are both whole numbers of metres greater than 1 m.

   > The prime factor decomposition of 84 is $2^2 \times 3 \times 7$.
   > One way of grouping the prime factors of 84 is $2^2 \times (3 \times 7) = 4 \times 21$.
   > The dimensions of the rectangle could be 4 m by 21 m.

   By grouping the prime factors of 84 in different ways, find all the possible dimensions of the carpet.

7 A metal cuboid has a volume of 1815 cm^3. Each side of the cuboid is a whole number of centimetres, and each edge is longer than 1 cm.

   a Find the prime factor decomposition of 1815.

   b Use your answer to part **a** to find all the possible dimensions of the cuboid.

Using prime factors: HCF and LCM

This spread will show you how to:

- Find the prime factor decomposition of any integer and the highest common factor and least common multiple of any two integers

Keywords

Co-prime
Highest common factor
Least common multiple
Prime factor

- The **highest common** factor (HCF) of two numbers is the largest number that is a factor of them both.
- The **least common multiple** (LCM) of two numbers is the smallest number that they both divide into.

You can find the HCF and LCM of two numbers by writing their **prime factors** in a Venn diagram.

You do not need to use a Venn diagram method in your exam.

- The HCF is the product of the numbers in the intersection.
- The LCM is the product of all the numbers in the diagram.

Example

Find the HCF of 54 and 84.

$54 = 2 \times 3^3$ $84 = 2^2 \times 3 \times 7$

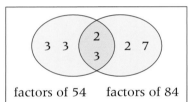

factors of 54   factors of 84

An alternative is to write:
$54 = (2) \times 3 \times (3) \times 3$
$84 = (2) \times 2 \times (3) \times 7$
Identify the common factors.
$HCF = 2 \times 3 = 6$

The prime factors are written in the diagram.
Common factors are in the intersection.
The HCF is the product of the numbers in the intersection:

$2 \times 3 = 6$, so the HCF of 54 and 84 is 6.

Example

Find the LCM of 60 and 280.

$60 = 2^2 \times 3 \times 5$ $280 = 2^3 \times 5 \times 7$

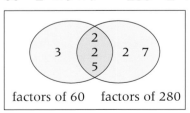

factors of 60   factors of 280

An alternative is to write:
$60 = 2^2 \times 3 \times 5$
$280 = 2^3 \times 5 \times 7$
Identify the highest power of each factor.
$LCM = 2^3 \times 3 \times 5 \times 7 = 840$

The LCM is the product of all the numbers in the diagram.

$2^3 \times 3 \times 5 \times 7 = 840$, so the LCM of 60 and 280 is 840.

1 Find the highest common factor (HCF) of each pair of numbers, by drawing a Venn diagram or otherwise.

    a 35 and 20   **b** 48 and 16   **c** 21 and 24

    d 25 and 80   **e** 28 and 42

2 Find the least common multiple (LCM) of each pair of numbers, by drawing a Venn diagram or otherwise.

    a 24 and 16   **b** 32 and 100   **c** 22 and 33

    d 104 and 32   **e** 56 and 35

3 Find the LCM and HCF of each pair of numbers.

    a 180 and 420   **b** 77 and 735   **c** 240 and 336

    d 1024 and 18   **e** 762 and 826

4 Write the prime factor decomposition of each number.

    a 288   **b** 4725   **c** 67 375

5 **a** Find the LCM and HCF of each pair of numbers.

      i 19 and 25   **ii** 36 and 23   **iii** 17 and 19   **iv** 11 and 9

    b What can you say about the HCF of two numbers, if one (or both) of the numbers is prime?

    c What can you say about the LCM of two numbers, if one (or both) of the numbers is prime?

6 Two numbers are **co-prime** if they have no common factors (other than 1). By drawing Venn diagrams or otherwise, decide whether each pair of numbers is co-prime.

    a 105 and 429   **b** 63 and 715   **c** 121 and 175   **d** 455 and 693

7 **a** Find the HCF and LCM of each pair of numbers in question **6**.

    b Use your answers to part **a** to explain how you can easily find the HCF and LCM of two numbers that are co-prime.

8 Look at this statement.

> The product of any two numbers is equal to the product of their HCF and their LCM.

    a Test the statement. Do you think it is true?

    b Use a Venn diagram to justify your answer to part **a**.

9 Find the HCF and LCM of each set of three numbers.

    a 30, 42, 54   **b** 90, 350 and 462   **c** 462, 510 and 1105

Key objectives

- Use standard index form expressed in conventional notation and on a calculator display
- Understand highest common factor, least common multiple, prime number and prime factor decomposition

1 a Express these integers as products of their prime factors: (4)

    i 84

    ii 63.

  b Hence find the highest common factor (HCF) of 84 and 63. (2)

2 A spaceship travelled for 6×10^2 hours at a speed of 8×10^4 km/h.

  a Calculate the distance travelled by the spaceship. Give your answer in standard form. (3)

One month an aircraft travelled 2×10^5 km.

The next month the aircraft travelled 3×10^4 km.

  b Calculate the total distance travelled by the aircraft in the two months.

    Give your answer as an ordinary number. (2)

(Edexcel Ltd., 2003)

This unit will show you how to

- Calculate the area of common 2-D shapes including composite shapes
- Calculate the area and arc length of a sector of a circle
- Use Pythagoras's theorem to find missing lengths in right-angled triangles
- Calculate the volume and surface area of prisms, pyramids, cones and spheres
- Solve problems involving volumes and surface areas of pyramids, cones and spheres

Before you start ...

You should be able to answer these questions.

Review

1 Work out the

   i area   **ii** circumference

   of each circle.

CD – S1

a 5 cm   **b** 7.4 cm

2 Find the volume of these prisms. State the units of your answers.

CD – S1

a  5 cm 20 cm 6 cm   **b** 4 cm 7 cm

c 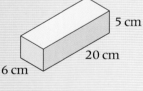 2.8 cm 8.2 cm 4 cm

3 Work out the surface area of the prisms in question 2.

CD – S1

Area of a parallelogram and a trapezium

This spread will show you how to:

● Calculate the area of parallelograms and trapeziums

Keywords

Area
Parallelogram
Trapezium

You can cut a triangle from one side and stick it on the other.

The height is at right angles to the base.

You can use the formula for the **area** of a rectangle to find the formula for the area of a **parallelogram**.

This parallelogram can be made into a rectangle.

● **Area of parallelogram = base × height.**

Two congruent **trapeziums** make a parallelogram.

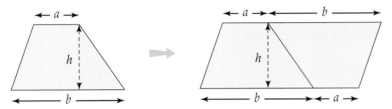

Area of the parallelogram = base × height

$$= (a + b) \times h$$

● **Area of trapezium $= \frac{1}{2} \times (a + b) \times h$**

The area of a trapezium is half the sum of the parallel sides times the distance between them.

Example

Find the areas of these shapes.

a
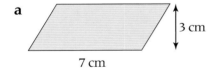

b
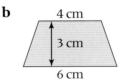

a Area of parallelogram

$$= \text{base} \times \text{height}$$

$$= 7 \times 3$$

$$= 21 \text{ cm}^2$$

b Area of trapezium

$$= \frac{1}{2} \times (a + b) \times h$$

$$= \frac{1}{2} \times (4 + 6) \times 3$$

$$= 5 \times 3$$

$$= 15 \text{ cm}^2$$

1 Find the area of each parallelogram.

a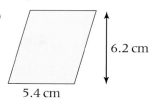
2.5 cm
6 cm

b
6.2 cm
5.4 cm

c 3.8 cm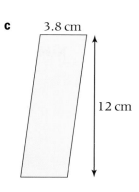
12 cm

d 4.2 cm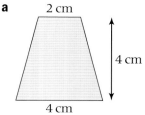
7.5 cm

e 2.9 cm
4.6 cm

2 Find the area of each trapezium.

a 2 cm
4 cm
4 cm

b 3 cm
4 cm
7 cm

c 5 cm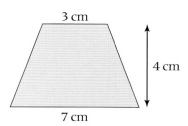
5 cm
3.2 cm

d 24 mm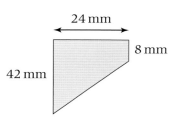
8 mm
42 mm

e 28 mm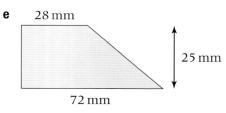
25 mm
72 mm

3 Caroline has drawn a sandcastle.
What is the area of her castle and flag?
Start by dividing the shape into parts.

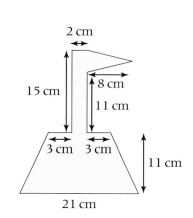
2 cm
15 cm
8 cm
11 cm
3 cm   3 cm
11 cm
21 cm

Arc length and sector area

This spread will show you how to:

- Calculate the area and arc length of a sector of a circle

Keywords
Arc
Circle
Diameter
Perimeter
Radius
Sector

The circumference of a **circle** is in proportion to its **diameter**.

$$c \propto d \qquad\qquad C \approx 3 \times d$$

The proportion is an irrational number. You can represent it with the symbol π (pi).

$\pi = 3.14159\ldots$

- $C = \pi \times d = \pi d$ or $C = 2\pi r$

$d = 2 \times r$

The area of a circle is in proportion to the square of its **radius**.

- $A = \pi \times r^2 = \pi r^2$

An **arc** is a fraction of the circumference of a circle.

A **sector** is a fraction of a circle, shaped like a slice of pie.

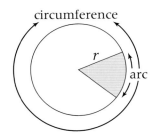

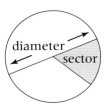

To calculate the area and arc length of a sector, you use the angle at the centre of the sector.

- **Arc length** $= \frac{\theta}{360} \times$ **circumference of whole circle**
  $= \frac{\theta}{360} \times 2\pi r$

- **Sector area** $= \frac{\theta}{360} \times$ **area of whole circle**
  $= \frac{\theta}{360} \times \pi r^2$

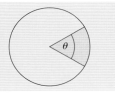

θ is the Greek letter 'theta'.

Example

This fan is a sector. Find
a the arc length
b the area
c the perimeter of the fan.
Give your answers to 1 decimal place.

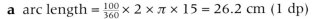

a arc length $= \frac{100}{360} \times 2 \times \pi \times 15 = 26.2$ cm (1 dp)

b sector area $= \frac{100}{360} \times \pi \times 15^2 = 196.3$ cm^2 (1 dp)

c perimeter $=$ arc length $+ 2 \times 15$

$\qquad\qquad = 26.2 + 30$

$\qquad\qquad = 56.2$ cm

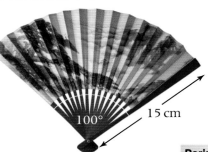

Perimeter of sector = arc length + 2 × radius

1 Find the area of each sector.

 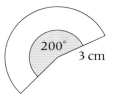

a 108° 4 cm   b 200° 3 cm   c 6 cm 36°

 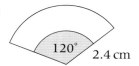

d 72° 5 cm   e 120° 2.4 cm   f 115° 3.5 cm

In this exercise give your answers to 1 decimal place where appropriate.

2 Find the arc length of each sector in question **1**.

3 Find the perimeter of each sector.

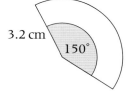 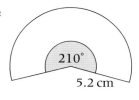

a 48° 16 mm   b 3.2 cm 150°   c 210° 5.2 cm

4 Find the area of this incomplete annulus.

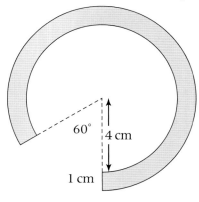

60°   4 cm

1 cm

An annulus is the region between two concentric circles, shown here as the purple ring.

5 Jan wants to lay a patio in a corner of her garden. The patio is shaped like the sector of a circle.

a What is the area of the patio?

She wants to lay a brick path around the curved edge of the patio. Each brick is 30 cm long.

b Calculate the number of bricks she needs.

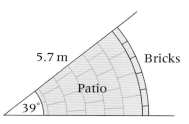

5.7 m   Bricks   Patio   39°

17

Surface area of 3-D shapes

This spread will show you how to:

● Find the surface area of simple 3-D shapes

Keywords

Cuboid
Cylinder
Prism
Surface area

● **Surface area** is the total area of all the faces of a 3-D shape.

To find the surface area, first imagine the net of the shape.

The faces of this **cuboid** are in pairs.
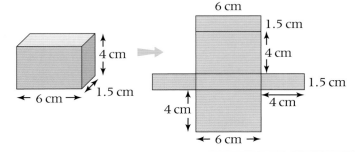

  Front/back: Area = $6 \times 4 = 24$
  Side: Area = $4 \times 1.5 = 6$
  Top/bottom: Area = $6 \times 1.5 = 9$

Surface area $= 2 \times (24 + 6 + 9)$
             $= 2 \times 39$
             $= 78$ cm^2

Example

Find the surface area of this triangular **prism**.

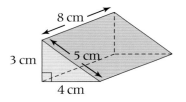

The two end faces of a prism are identical.

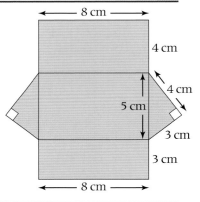

Triangle: Area $= \frac{1}{2} \times 4 \times 3 = 6$
Triangle: Area $= \frac{1}{2} \times 4 \times 3 = 6$
Side: Area $= 3 \times 8 = 24$
Bottom: Area $= 4 \times 8 = 32$
Sloping side: Area $= 5 \times 8 = 40$

Surface area
$= 6 + 6 + 24 + 32 + 40 = 108$ cm^2

Example

Calculate the surface area of this **cylinder**.

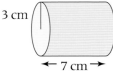

The curved surface of a cylinder is a rectangle.
Length of rectangle = circumference of circle.
Width of rectangle = height of cylinder.

Area of top circle
    $\pi \times 3^2 = 28.27$
Area of bottom circle
    $\pi \times 3^2 = 28.27$
Area of curved surface
    $2 \times \pi \times 3 \times 7 = 131.95$

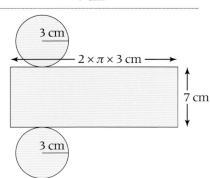

Surface area of cylinder $= 28.27 + 28.27 + 131.95$
                                  $= 188.5$ cm^2 (to 1 dp)

1 Work out the surface areas of these cuboids.

Give your answers to 1 dp.

a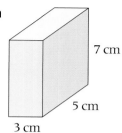
7 cm
5 cm
3 cm

b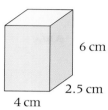
6 cm
2.5 cm
4 cm

c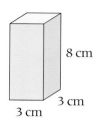
8 cm
3 cm
3 cm

d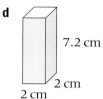
7.2 cm
2 cm
2 cm

e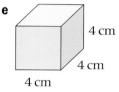
4 cm
4 cm
4 cm

f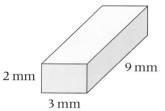
9 mm
2 mm
3 mm

2 Work out the surface areas of these cylinders.

a 2 cm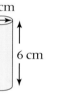
6 cm

b 5 cm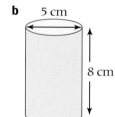
8 cm

c 4 cm
4 cm

d 3.2 cm
5 cm

The top measurements give the diameters.

3 Work out the surface areas of these prisms.

a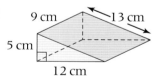
9 cm 13 cm
5 cm
12 cm

b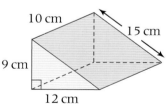
10 cm 15 cm
9 cm
12 cm

4 A scout troop is making a tent out of canvas and a groundsheet out of PVC.

  a What area of canvas do they need for the cover (including front and back flaps)?

  b What area of PVC do they need for the groundsheet?

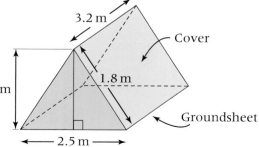
3.2 m
Cover
1.8 m
1.3 m
Groundsheet
2.5 m

Volume of prisms

This spread will show you how to:

- Calculate volumes of right prisms

Keywords

Cross-section
Cylinder
Prism
Volume

- A **prism** is an object with constant **cross-section**.

- **Volume** of a prism = area of cross-section × length.
  = **A** × **l**

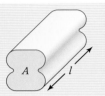

In a right prism there is a right angle between the length and the base.

Example

a Work out the volume of this cuboid.

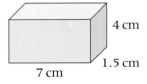
4 cm
1.5 cm
7 cm

b Work out the volume of this prism.

8 cm
3 cm
4 cm

a Area of cross-section =
$4 × 1.5 = 6$ cm^2
Volume $= 6 × 7 = 42$ cm^3

b Area of cross-section =
$\frac{1}{2} × 4 × 3 = 6$ cm^2
Volume $= 6 × 8 = 48$ cm^3

Area of triangle $= \frac{1}{2}bh$

- A **cylinder** is a prism with circular cross-section.

- **Volume** of a **cylinder** = area of circle × height

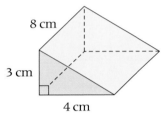

Example

Find the volume of this cylinder.

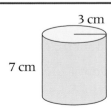
3 cm
7 cm

Do not round intermediate steps of the calculation.

Area of circle $= \pi × 3^2 = 28.274 \ldots$ cm^2
Volume $= 28.274 \ldots × 7 = 198$ cm^3

Give answers to a sensible degree of accuracy.

1 Find the volume of each cuboid.

a
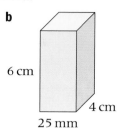
3 cm
5 cm
7 cm

b
6 cm
4 cm
25 mm

c
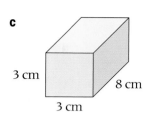
3 cm
8 cm
3 cm

d
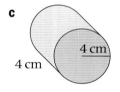
9 cm
2cm
2 cm

e
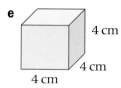
4 cm
4 cm
4 cm

f
2 mm
7 mm
2 mm

2 Find the volume of each cylinder.

a
2 cm
6 cm

b
5 cm
8 cm

c
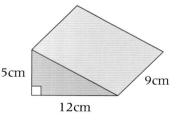
4 cm
4 cm

d
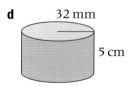
32 mm
5 cm

Be careful with units in part **d**.

3 Find the volume of each prism.

a
5cm
9cm
12cm

b
10mm
8mm
15mm

c
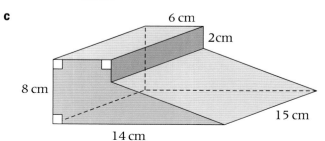
6 cm
2cm
8 cm
15 cm
14 cm

Volume and surface area

This spread will show you how to:

- Solve problems involving surface areas and volumes of prisms

Keywords
Pythagoras's theorem
Surface area
Volume

- **Surface area** is the total area of all the faces of a solid.

Example

This cube has surface area 150 cm². Find the volume of the cube.

A cube has six faces.
Each face has the same area.

So area of one face is $150 \div 6 = 25$ cm²

Each face is a square, so length of each side $\sqrt{25} = 5$ cm

Volume of cube = area of cross-section × length
$$= 25 \times 5$$
$$= 125 \text{ cm}^3$$

Each face is a cross-section.

You can use **Pythagoras's theorem** to find the perpendicular height of a triangle.

- **Pythagoras's theorem:** $a^2 + b^2 = c^2$

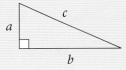

Example

This triangular prism has length 9 cm and its end faces are equilateral triangles with side length 4 cm.
Work out its volume.

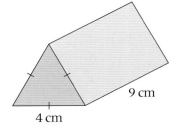

9 cm

4 cm

Using Pythagoras's theorem to find the height, h, of the triangle.
$$h^2 = 4^2 - 2^2$$
$$h = \sqrt{12} = 3.464 \dots$$
Area of cross-section $= \frac{1}{2} \times 4 \times 3.464 \dots$
$$= 6.928 \dots$$
Volume of prism: $6.928 \dots \times 9 = 62.4$ cm³

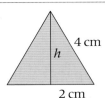

4 cm

h

2 cm

To work out the area of the cross-section you need the perpendicular height of the triangle.

1 Find the volumes of the cubes with these surface areas.

   a Surface area 54 cm^2

   b Surface area 294 cm^2

   c Surface area 96 cm^2

   d Surface area 1.5 m^2

2 Find the volumes of these triangular prisms which have equilateral triangles as cross-sections.

a

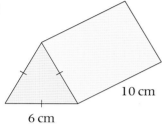

6 cm · 10 cm

b

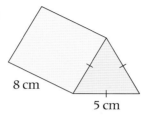

8 cm · 5 cm

c 8 cm

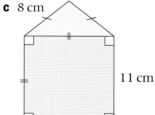

11 cm

d

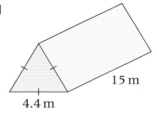

4.4 m · 15 m

3 Find the surface areas of the cubes with these volumes.

   a Volume 512 cm^3

   b Volume 1000 cm^3

   c Volume 216 cm^3

   d Volume 1 m^3

4 The volume of a cuboid is 80 cm^3.
The cuboid has square ends.
The sides of the cuboid are whole
numbers of centimetres.

   a Find the possible dimensions of the cuboid.

   b Find

     i the smallest possible surface area

     ii the largest possible surface area of this cuboid.

5 Repeat question **4** for cuboids with volume

   a 24 cm^3

   b 64 cm^3

S1.6 Volume of a pyramid and a cone

This spread will show you how to:

● Calculate the volume of pyramids and cones

Keywords
Cone
Pyramid
Tetrahedron
Volume

Pyramids and **cones** are 3-D shapes with sides that taper to a point.

Regular tetrahedron	Square-based pyramid	Cone	Irregular pyramid
All four faces are identical equilateral triangles.	Base is a square, all four sides are identical isosceles triangles.	Base is a circle.	Base is an irregular polygon.

● Volume of a pyramid = $\frac{1}{3}$ of base area × vertical height

Volume is the amount of space taken up by an object.

This formula also holds true for a cone.

Find the volume of each solid.

a

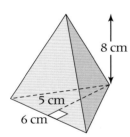

a Volume of pyramid
$= \frac{1}{3}$ base area × height
$= \frac{1}{3} \times (\frac{1}{2} \times 6 \times 5) \times 8$
$= 40 \text{ cm}^3$

Area of triangle
$= \frac{1}{2} \times$ base × height
$= \frac{1}{2} bh$

b

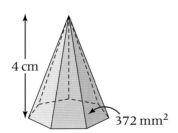

b Volume of pyramid
$= \frac{1}{3}$ base area × height
$= \frac{1}{3} \times 372 \times 40$
$= 4960 \text{ mm}^3$

Dimensions need to be in the same units.

c

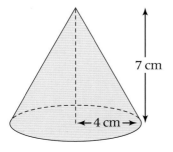

c Volume of cone
$= \frac{1}{3}$ base area × height
$= \frac{1}{3} \times \pi \times 4^2 \times 7$
$= 117.3 \text{ cm}^3$ (1 dp)

Area of base = πr^2
Volume of cone
$= \frac{1}{3}\pi r^2 h$

Round your answer to a sensible degree of accuracy.

Example

24

1 Find the volume of each solid.

a

3 cm
42 cm²

b
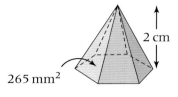
2 cm
265 mm²

c
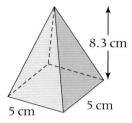
8.3 cm
5 cm
5 cm

d
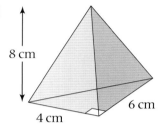
8 cm
6 cm
4 cm

e
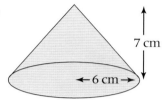
7 cm
←6 cm→

f
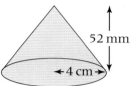
52 mm
←4 cm→

2 James is going to fill a paper cone with sweets.
He can choose between cone X and cone Y.
Cone X has top radius 4 cm and height 8 cm.
Cone Y has top radius 8 cm and height 4 cm.

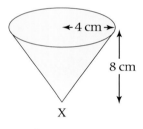
←4 cm→
8 cm
X

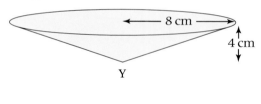
8 cm
4 cm
Y

a Which cone has the greater volume?

b What is the difference in volume between cone X and cone Y?

3 A square-based pyramid has
base side length x cm and
vertical height $2x$ cm.
The volume of the pyramid is
18 cm³.
Work out the value of x.

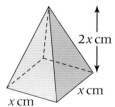
$2x$ cm
x cm
x cm

4 Laura made a model rocket
from a cylinder with height
6 cm and a cone with vertical
height 2.4 cm.
The radius of the cylinder and
the cone is 1.8 cm.

Find the volume of the rocket.

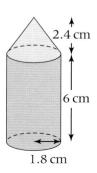
2.4 cm
6 cm
1.8 cm

Volume of
cylinder = area of
base × height

Surface area of a pyramid

This spread will show you how to:

● Calculate the surface area of pyramids

Keywords
Pythagoras's theorem
Surface area

Surface area is the total area of all the surfaces of a 3-D solid.

● For 3-D solids with flat surfaces, you work out the area of each surface and add them together.

You use **Pythagoras's theorem** to find missing lengths in right-angled triangles.

● $a^2 + b^2 = c^2$

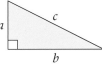

Example

Find the surface areas of these 3-D solids. Give your answer to 1 dp.

a

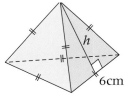

b

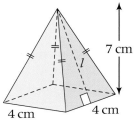

a Each face is an equilateral triangle.

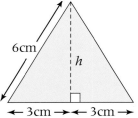

Using Pythagoras: $6^2 = 3^2 + h^2$

$h = \sqrt{6^2 - 3^2} = 5.196$ cm

Area of one face
$= \frac{1}{2} \times 6 \times 5.196 = 15.588$ cm^2

Surface area
$= 4 \times 15.588 = 62.4$ cm^2 (1 dp)

b Base area $4 \times 4 = 16$ cm^2

Each face is an isosceles triangle.

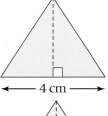

The sloping side of this triangle is l, the height of the triangular face of the pyramid.

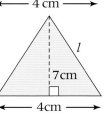

Using Pythagoras: $7^2 + 2^2 = l^2$

$l = \sqrt{7^2 + 2^2} = \sqrt{53} = 7.28$ cm

Area of each triangular face
$= \frac{1}{2} \times 4 \times 7.28 = 14.56$ cm^2

Surface area
$=$ base area $+$ area of 4 triangular faces
$= 16 + (4 \times 14.56) = 74.2$ cm^2 (1 dp)

You are only given the base measurement. To find the height of each side triangle, imagine a vertical slice through the pyramid:

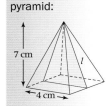

1 Find the surface area of each 3-D solid.

a

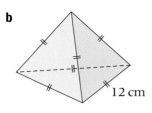

16 mm

b

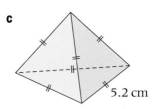

12 cm

c

5.2 cm

d

5 cm

6 cm   6 cm

e

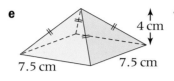

4 cm

7.5 cm   7.5 cm

f

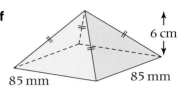

6 cm

85 mm   85 mm

g

6 cm

7 cm   3cm

h

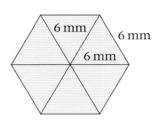

8 cm

5 cm   4 cm

i

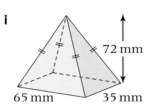

72 mm

65 mm   35 mm

2 A pencil is in the shape of a regular hexagonal prism.
Each side of the hexagon is 6 mm.

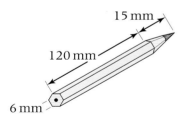

15 mm

120 mm

6 mm

6 mm

6 mm

6 mm

a Find the area of the base of the prism.

The pencil is sharpened at one end to form a pyramid.

The pyramid has height 15 mm and sides 6 mm.

The prism that makes the remainder of the pencil is 120 mm long.

b Find the total surface area of the pencil.

Curved surface area of a cone

This spread will show you how to:

● Solve problems involving surface areas of cones

Keywords
Circumference
Cone
Radius
Surface area

A **cone** is a solid with a circular base.
You form the curved surface by folding a sector of a circle.

 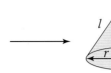

Cones were
intoduced in S1.2
on page 16.

The **radius** of the sector, *l*, is the sloping side of the cone.

Arc length of the sector = **circumference** of the base of the cone
so $\frac{\theta}{360} \times 2\pi l = 2\pi r$

● **Radius of cone** $r = \frac{\theta}{360} \times l$

Remember this
formula.
$r = \frac{\theta}{360} \times l$

Area of sector = curved **surface area** of cone
$$= \frac{\theta}{360} \times \pi l^2$$
$$= \frac{\theta}{360} \times l \times \pi \times l$$
$$= r \times \pi \times l$$

● **Curved surface area of cone** $= \pi r l$

● **Surface area of a cone** = curved surface area + area of base
$$= \pi r l + \pi r^2$$

This sector is folded to form a cone.
Find, giving your answers to 1 dp:

a the radius of the cone
b the curved surface area of the cone
c the total surface area of the cone
d the vertical height of the cone.

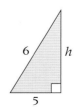

a $\theta = 300°$, $l = 6$ cm
$r = \frac{\theta}{360} \times l = \frac{300}{360} \times 6 = 5$ so $r = 5$

$r = 4$
$l = 6$

b Curved surface area $= \pi r l = \pi \times 5 \times 6 = 94.2$ cm^2

c Base area $= \pi r^2 = \pi \times 5^2 = 78.5$ cm^2
Total surface area $94.24 + 78.53 = 172.8$ cm^2

d Vertical height
$$h^2 = 6^2 - 5^2$$
$$h = \sqrt{6^2 - 5^2} = 3.3 \text{ cm}$$

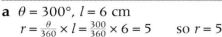

Base radius,
vertical height and
slant side make a
right-angled
triangle, so use
Pythagoras.

Example

1 Find the curved surface areas of cones with

   a base radius 5 cm
      slant height 8.2 cm
   b base radius 3 cm
      slant height 6 cm

   c base radius 45 mm
      slant height 30 mm
   d base radius 2.5 cm
      slant height 30 mm

   e base radius 6.7 cm
      slant height 10.5 cm
   f base radius 135 mm
      slant height 18.5 cm

In this exercise, give your answers to 1 dp where appropriate.

2 Find the curved surface area of each cone.

   a
   8 cm
   6 cm

   b
   5.2 cm
   4 cm

   c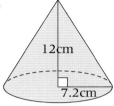
   12 cm
   7.2 cm

Remember to find the slant height first.

3 These sectors are folded to form cones.
   Find the curved surface area of each cone.

   a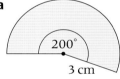
   200°
   3 cm

   b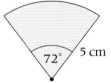
   72°
   5 cm

   c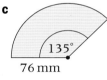
   135°
   76 mm

   d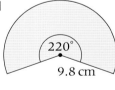
   220°
   9.8 cm

   e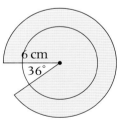
   6 cm
   36°

   f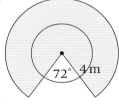
   72°
   4 m

4 Charlie made a model rocket from a
   cylinder with height 12 cm and a
   cone with vertical height 3.2 cm.
   The radius of the cylinder and the
   cone is 1.7 cm.

   Find the total surface area of the
   rocket.

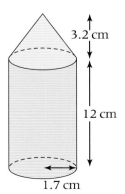
3.2 cm
12 cm
1.7 cm

Volume and surface area of a sphere

This spread will show you how to:

● Solve problems involving volumes and surface areas of spheres

Keywords
Sphere
Surface area
Volume

A **sphere** is like a ball. It has one curved surface, no edges and no vertices.

For a sphere with radius = r cm

● **Volume** $V = \frac{4}{3}\pi r^3$

● **Surface area** $SA = 4\pi r^2$

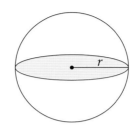

Example

A tennis ball has radius 4.2 cm. Find, giving your answers to 1 dp

a its volume
b its surface area.

a Volume of tennis ball $= \frac{4}{3}\pi r^3$
  $= \frac{4}{3}\pi \times 4.2^3 = 310.3$ cm^3

b Surface area $= 4\pi r^2$
  $= 4\pi \times 4.2^2 = 221.7$ cm^2

The earth is split into two hemispheres.

● **A hemisphere is half a sphere.**

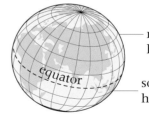

northern hemisphere

equator

southern hemisphere

Example

A paperweight is in the shape of a hemisphere.
It has radius 25 mm.
Find, giving your answer to the nearest whole unit

a its volume
b its surface area.

25 mm

a Volume of paperweight $= \frac{1}{2} \times \frac{4}{3}\pi \times 25^3 = 32\ 725$ mm^3

b Curved surface area $= \frac{1}{2} \times 4\pi \times 25^2$
                $= 3926.99$ mm^2

   Area of base $= \pi \times 25^2$
                $= 1963.49$ mm^2

   Total surface area $= 3926.99 + 1963.49$
                $= 5890$ mm^2 (to the nearest mm^2)

$V = \frac{1}{2} \times \frac{4}{3}\pi r^3$
for a hemisphere.

Curved SA =
$\frac{1}{2} \times 4\pi r^2$

1 Find the volumes of these spheres.

a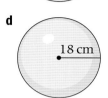
7 cm

b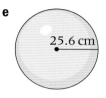
24 mm

c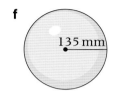
6.3 cm

d
18 cm

e
25.6 cm

f
135 mm

In these questions, give your answers to an appropriate accuracy.

2 Find the surface area of each sphere in question **1**.

3 The surface area of a sphere is 616 cm². Work out the radius of the sphere.

4 Find
  a the volume
  b the curved surface area
    of a hemisphere with radius 4 cm.

5 Rachel bought a cylindrical tube containing 3 power balls.
Each power ball is a sphere of radius 5 cm.
The power balls touch the sides of the tube.
The balls touch the top and bottom of the tube.

Work out the volume of empty space in the tube.

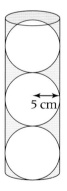
5 cm

6 **a** Which has the greater volume, a sphere with diameter 3 cm or a cube with side length 3 cm?

  b Which has the greater surface area, a sphere with diameter 3 cm or a cube with side length 3 cm?

7 A sphere with radius 6 cm fits exactly inside a cylinder.
  a Write **i** the radius
      ii the height of the cylinder.

  b Work out the surface area of the sphere.

  c Work out the curved surface area of the cylinder.

  d For a sphere radius r that fits exactly inside a cylinder, find an expression for:
    i the surface area of the sphere
    ii the curved surface area of the cylinder.

  e Explain why your answers to **b** and **c** must be the same. You may want to use your answer to part **d** to help you.

Exam review

Key objectives

- Calculate the area of common 2-D shapes, including composite shapes
- Calculate the lengths of arcs and the areas of sectors of circles
- Solve problems involving surface areas and volumes of pyramids, cones and spheres
- Understand, recall and use Pythagoras's theorem in 2-D and 3-D problems

1 The diagram shows a square-based pyramid.

Calculate:

a the surface area   (2)
b the volume   (3)

of the pyramid.

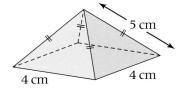

2 The diagram represents a large cone of height 6 cm and base diameter 18 cm.

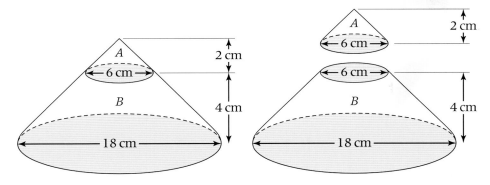

Diagrams NOT accurately drawn

The large cone is made by placing a small cone *A* of height 2 cm and base diameter 6 cm on top of a frustum *B*.
Calculate the volume of the frustum *B*.
Give your answer in terms of π.   (4)

(Edexcel Ltd., 2003)

This unit will show you how to

- Use index notation and simple laws of indices
- Expand single and double brackets
- Manipulate algebraic expressions by taking out common factors
- Factorise quadratic expressions

Before you start ...

You should be able to answer these questions.

1 Evaluate these without a calculator.

   a 14^2 **b** 1^{50}

   c $(-2)^5$ **d** $\left(\frac{2}{3}\right)^3$

   e 0.02^3 **f** 10^{10}

   g $(-5)^1$ **h** $\sqrt{\frac{4}{25}}$

2 Given that $x = -2$, put these expressions in ascending order:

   x^2 $\sqrt{12 - 2x}$ $\left(\frac{1}{x}\right)^2$ $(-x)^4$ $(1 + 2x)^3$

3 Write these statements using the rules of algebra.

   a I think of a number, multiply it by 4 and add 7.

   b I think of a number, subtract 6 and then multiply by 3.

   c I think of a number, multiply it by itself then subtract this from 10.

   d I think of a number, treble it, take away 6 then divide this by two.

   e I think of a number and multiply it by itself five times.

4 a Find the highest common factor of these sets of numbers.

     i   24 and 36   **ii**   30 and 75

     iii   56 and 72   **iv**   90, 180 and 225

     v   17, 28 and 93

   b What do you think the highest common factor of $2x$, x^2 and $5x^3$ is?

Review

CD – A1

CD – A1

CD – A1

Unit N1

This spread will show you how to:

● Use index notation and simple laws of indices

Keywords
Base
Coefficient
Index
Indices

An **index** is a power. The **base** is the number which is raised to this power.

$$3^4 = 3 \times 3 \times 3 \times 3 = 81$$

base   index

The plural of index is **indices**.

You can simplify expressions using the three index laws.

● when multiplying, add the indices $3^2 \times 3^3 = (3 \times 3) \times (3 \times 3 \times 3) = 3^5$

● when dividing, subtract the indices $5^4 \div 5^2 = (5 \times 5 \times 5 \times 5) \div (5 \times 5) = 5^2$

● with brackets, multiply the indices $(4^2)^3 = (4 \times 4) \times (4 \times 4) \times (4 \times 4) = 4^6$

To use the index laws, the bases must be the same. The first two index laws are shown on page 2 for base 10.

When terms have number **coefficients**, deal with these first.

$$5p^7 \times 8p^{-3} = 40p^4$$

$7 + (-3)$

5×8

Example

Simplify each of these using the index laws.

a $p^9 \lozenge p^{-7}$ **b** $\frac{q^{-3}}{q^{-5}}$

c $(w^6)^{-4}$ **d** $p^5 \times q^3$

e $\frac{(k^3 \lozenge k^2)^7}{k}$ **f** $(5p^2)^3 \times 2p^{-7}$

a $p^9 \times p^{-7} = p^2$ **b** $\frac{q^{-3}}{q^{-5}} = q^2$

c $(w^6)^{-4} = w^{-24}$ **d** $p^5 \times q^3 = p^5 q^3$

e $\frac{(k^3 \lozenge k^2)^7}{k} = \frac{(k^5)^7}{k}$ **f** $(5p^2)^3 \times 2p^{-7} = 125p^6 \times 2p^{-7}$

$\quad = \frac{k^{35}}{k}$ $\quad = 250p^{-1}$

$\quad = k^{34}$ $\quad = \frac{250}{p}$

Take care with negatives:
$9 + (-7) = 2$
$(-3) - (-5) = 2$
$6 \times (-4) = -24$

Remember k is really k^1.

$\frac{1}{p} = \frac{p^0}{p} = p^{0-1} = p^{-1}$

Remember $p^0 = 1$.

Example

Expand $(2p^2 q)^3$.

$(2p^2 q)^3 = 8p^6 q^3$

Everything inside the bracket is cubed:
$2^3 = 8$, $(p^2)^3 = p^6$,
$(q^1)^3 = q^3$

1 Simplify each of these.

a $y^3 \times y^9$ **b** $k^9 \div k^5$ **c** $(m^3)^4$ **d** $g^8 \times g^{-5}$

e $\frac{h^{-2}}{h^4}$ **f** $(b^{-4})^3$ **g** $j^{-4} \times j^{-2}$ **h** $(t^{-5})^{-2}$

i $n^{-8} \div n^{-6}$

2 Simplify each of these fully.

a $5h^7 \times 3h^6$ **b** $\frac{15p^3}{5p}$ **c** $(2p^8)^2$ **d** $10r^3 \times 6r^{-4}$

e $(3h^{-3})^3$ **f** $9b^3 \div 3b^{-5}$ **g** $(3m^3 \times 2m^{-7})^2 \div 18m$

h $18(f^{-4})^4 \div 9f^{-16}$

3 Write a simplified expression for the area of this triangle.

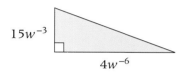

$15w^{-3}$

$4w^{-6}$

4 Show that the expression $(4p^4)^3 \div (8p^7)^2$ simplifies to $\frac{1}{p^2}$.

5 True or false? $3^x \times 3^y$ simplifies to give 9^{x+y}.
Explain your answer.

6 If $x = 3$ and $y = 4$, evaluate these expressions.

a $x^2 y$ **b** $3(x-y)^2$ **c** y^x

d $(x+y)(x-y)$ **e** $(x+y)^2$

7 Which expression is the odd one out?
Explain your answers.

$$5t^2 \times 10t^{-4} \div (5t^{+3})^2 \qquad\qquad \frac{2}{t^6} \qquad\qquad \left(\frac{4t^2}{16t}\right) \times 8t^{-7}$$

8 Find the value of x in this equation.

$$(2^2)^x \times 2^{3x} = 32$$

9 **a** If $u = 3^x$, show that $9^x + 3^{x+1}$ can be written as $u^2 + 3u$.

b Write an expression in terms of x for $u^3 + 9u$.

c Write an expression in terms of x for $u^2 - \frac{1}{u}$.

d Write $81^x - 9^{x-1}$ in terms of u.

A1.2 Expanding single and double brackets

This spread will show you how to:

● Expand single and double brackets

Keywords

Expand
F.O.I.L.

To **expand** a single bracket, you multiply all terms in the bracket by the term outside.

$$3(2x - 9) = 6x - 27$$

To expand double brackets, you multiply each term in the second bracket by each term in the first bracket.

F ... **F**irsts
O ... **O**uters
I ... **I**nners
L ... **L**asts

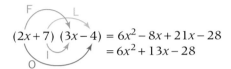

$$(2x + 7)\ (3x - 4) = 6x^2 - 8x + 21x - 28$$
$$= 6x^2 + 13x - 28$$

Example

Expand and simplify each of these.

a $6(2y - 5) - 3(2 - 2y)$ **b** $(2m - 7)(4m - 2)$

a $6(2y - 5) - 3(2 - 2y) = 12y - 30 - 6 + 6y$
$$= 18y - 36$$
b $(2m - 7)(4m - 2) = 8m^2 - 4m - 28m + 14$
$$= 8m^2 - 32m + 14$$

$-3 \times 2 = -6$
$-3 \times -2y = +6y$

Use the index laws:
$2m^1 \times 4m^1 = 8m^2$

Example

Expand $(3x - 1)^2$.

$$(3x - 1)^2 = (3x - 1)(3x - 1)$$
$$= 9x^2 - 3x - 3x + 1$$
$$= 9x^2 - 6x + 1$$

This question is a double bracket in disguise!

Example

Given that the length and width of a rectangle are $x + 5$ and $x - 2$ respectively and its area is 15, show that $x^2 + 3x - 25 = 0$.

Sketch a diagram:

$x - 2$
$x + 5$

Area of rectangle = length × width
$$15 = (x + 5)(x - 2)$$
$$15 = x^2 + 5x - 2x - 10$$
$$15 = x^2 + 3x - 10$$
Hence, $x^2 + 3x - 25 = 0$

Collect like terms
Subtract 15 from each side.

36

1 Expand and simplify each of these.

   a $3(5x + 9)$ **b** $2p(4p - 8)$

   c $3m(5 - 2m)$ **d** $3(2y + 9) + 5(3y - 2)$

   e $5x(2x + 2y - 9)$ **f** $4(t + 9) - 3(2t - 7)$

   g $(7h + 9) - (3h - 7)$ **h** $x(3x^2 + x^3)$

2 Expand and simplify each of these.

   a $(x + 7)(x + 6)$ **b** $(2x - 8)(x + 3)$

   c $(3p + 2)(4p + 5)$ **d** $(3m - 7)(2m - 6)$

   e $(5y - 9)(2y + 7)$ **f** $(3t - 2)^2$

   g $(x + 4)(x - 6) + (x + 3)^2$ **h** $(4 + 8b)(2 - 3b) - (3 - b)^2$

3 Write an expression for the area of each of these.

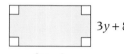 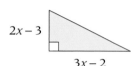

4 Write an expression, in terms of p, for the length of the hypotenuse of this right-angled triangle.

5 Given that the perimeter of this rectangle is equal to 24 cm, find an expression, in terms of x, for its area.

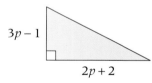

6 Given that two consecutive odd numbers can be written as $2n - 1$ and $2n + 1$ respectively, show that the difference between their squares is $8n$.

7 Expand each of these.

   a $(x + 2)^3$

   b $(2y - 1)(y + 5)(y - 3)$

> To expand a triple bracket,
> - first multiply two of the brackets together
> - multiply each term in your answer with each term in the remaining bracket.

Factorisation

This spread will show you how to:

- Manipulate algebraic expressions by taking out common factors and factorising quadratic expressions

Keywords

Common factors
Factorise
Quadratic

- Factorising is the 'reverse' of expanding.

To **factorise** into single brackets, remove the **common factors** of the terms.
Find the HCF of the coefficients first, then the HCF of the algebraic terms.

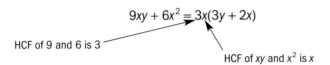

$$9xy + 6x^2 = 3x(3y + 2x)$$

HCF of 9 and 6 is 3

HCF of xy and x^2 is x

HCF of 12 and 18 is 6.
HCF of mn and m is m.

$5p \div 5p$ is 1.
A common mistake is to write $5p(0 - 2p^2)$

Quadratic expressions often factorise into double brackets

Example

Factorise.

a $12mn + 18m$　　**b** $5p - 10p^3$

---

a $12mn + 18m = 6m(2n + 3)$

b $5p - 10p^3 = 5p(1 - 2p^2)$

To factorise into double brackets, look for two numbers that add to give the coefficient of x and multiply to give the constant.

⟹ EXPAND ⟹

$(X + 4)(X + 7)$ ⟹ $x^2 + 11x + 28$

$4 + 7$　4×7

⟸ FACTORISE ⟸

Example

Factorise.

a $x^2 + 9x + 18$　　　　**b** $x^2 - 5x - 36$　　　　**c** $3(a - b) + (a - b)^3$

---

a $x^2 + 9x + 18 = (x + 3)(x + 6)$

Two numbers that multiply to 18 and add to 9 are +6 and +3.

b $x^2 - 5x - 36 = (x - 9)(x + 4)$　　List the factor pairs of −36, then check if they add to −5:
−1 and 36　−2 and 18　−3 and 12　−4 and 9　−6 and 6
$-9 \times 4 = -36$; $-9 + 4 = -5$

c $3(a - b) + (a - b)^2 = (a - b)(3 + (a - b))$
　　　　　　　　　　　$= (a - b)(3 + a - b)$

The terms have $(a - b)$ in common.

1 Factorise each of these by removing common factors.

   a $3x + 6y + 9z$ **b** $10p - 15$ **c** $5xy + 7x$

   d $6mn + 9mt$ **e** $16x^2 - 12xy$ **f** $3p + 9pq$

   g $7xy - 56x^2$ **h** $3x^2 + 12x^3 - 6x$ **i** $3(m + n) + (m + n)^2$

   j $4(p - q) + (p - q)^3$ **k** $ax + bx + ay + by$

2 Factorise each of these using double brackets.

   a $x^2 + 7x + 10$ **b** $x^2 + 8x + 15$ **c** $x^2 + 8x + 12$

   d $x^2 + 12x + 35$ **e** $x^2 - 3x - 10$ **f** $x^2 - 2x - 35$

   g $x^2 - 8x + 15$ **h** $x^2 - x - 20$ **i** $x^2 - 8x - 240$

   j $x^2 + 3x - 108$ **k** $x^2 - 25$ **l** $x^2 - 6 - x$

3 **a** Factorise $x^2 + 2xy + y^2$.

   b Use your answer to **a** to find, without a calculator, the value of $12.3^2 + 2 \times 12.3 \times 7.7 + 7.7^2$.

4 **a** Given that the length of a rectangle is $x + 4$ and its width is $2x - 3$, write a factorised expression for its perimeter.

   b Repeat for a rectangle with length $x^2 + 7x$ and width $24 - 17x$.

5 Given that $x(x + 10) = -21$, write a fully factorised expression with a value of zero.

6 Some expressions can be factorised 'twice' by first removing common factors and then using double brackets.

   For example

   $3x^2 + 15x + 18$ ⟹ $3(x^2 + 5x + 6)$ ⟹ $3(x + 2)(x + 3)$

             Common factors   Double brackets

   Factorise each of these twice.

   a $2x^2 + 16x + 24$ **b** $3y^2 + 45y + 108$ **c** $4m^2 - 4m - 80$

   d $x^3 + 8x^2 + 15x$ **e** $xy^2 - 3xy - 108x$ **f** $x^2y - 16y$

7 Factorise these using common factors, double brackets or both.

   a $p^2 - p - 12$ **b** $3p^2 + 6p$ **c** $10x^2 + 70x + 120$

   d $x^2y - 63y - 2xy$ **e** $a^3 + ab^2 - 2a^2b$ **f** $3am + 3an + 3ab + 3ac$

More factorisation

This spread will show you how to:

- Manipulate algebraic expressions by factorising quadratic expressions

Keywords
Coefficient
Factorise
Quadratic

To **factorise** quadratics with more than one x^2, you need to adapt your method.

$$2x^2 + 5x + 6 = (2x + \ )(x + \ )$$

Two numbers that multiply to give 6 and add to give 5 are +2 and +3

$$(2x + 3)(x + 2) = 2x^2 + 7x + 6 \quad \times$$

or

$$(2x + 2)(x + 3) = 2x^2 + 8x + 6 \quad \times$$

Not all quadratics will factorise easily.

To factorise a **quadratic** where the coefficient of x^2 is not 1:

- Multiply the **coefficient** of x^2 and the constant.

$$2x^2 + 11x + 12 \rightarrow 2 \times 12 = 24$$

- Find two numbers that multiply to give this value and add to give the coefficient of x.

Find two numbers that multiply to give +24 and add to give +11 $\rightarrow$ +3, +8

- Write the quadratic with the x-term split into two x-terms, using these numbers.

$$2x^2 + 3x + 8x + 12$$

This is a suggested method but you may find a method that works better for you.

- Factorise the pairs of terms.

$$x(2x + 3) + 4(2x + 3)$$

- Factorise again, taking the bracket as the common factor.

$$(2x + 3)(x + 4)$$

Check by expanding:
$$(2x + 3)(x + 4) = 2x^2 + 8x + 3x + 12$$
$$= 2x^2 + 11x + 12 \quad \checkmark$$

Example

Factorise.

a $3x^2 + 26x + 16$ **b** $5x^2 + 34x - 7$

---

a $3x^2 + 26x + 16$
$3 \times 16 = 48$ so find two numbers that multiply to +48 and add to +26.
These are +2 and +24
Splitting the x term $3x^2 + 26x + 16 = 3x^2 + 2x + 24x + 16$
$$= x(3x + 2) + 8(3x + 2)$$
$$= (3x + 2)(x + 8)$$

Try the factor pairs systematically.

b $5x^2 + 34x - 7$
$5 \times -7 = -35$
Two numbers that add to +34 and multiply to -35 are +35, -1.
Splitting the x term $5x^2 + 35x - 1x - 7 = 5x(x + 7) - 1(x + 7)$
$$= (5x - 1)(x + 7)$$

Be careful with signs as you take out the factor -1.

1 Factorise fully.

   a $2x^2 + 5x + 3$ **b** $3x^2 + 8x + 4$

   c $2x^2 + 7x + 5$ **d** $2x^2 + 11x + 12$

   e $3x^2 + 7x + 2$ **f** $2x^2 + 7x + 3$

   g $2x^2 + x - 21$ **h** $3x^2 - 5x - 2$

   i $4x^2 - 23x + 15$ **j** $6x^2 - 19x + 3$

   k $12x^2 + 23x + 10$ **l** $8x^2 - 10x - 3$

   m $6x^2 - 27x + 30$ **n** $4x^2 - 9$

   o $6x^2 + 7x - 3$ **p** $18x^2 + 21x - 4$

2 Explain why you cannot factorise $2x^2 + 4x + 3$.

3 Factorise these expressions, using common factors (single brackets) and double brackets, or both.

   a $18x^2 - 9x$ **b** $4ab - 16ab^3$

   c $3mn + 8m - m^3$ **d** $x^2 - 7x - 18$

   e $2x^2 - x - 15$ **f** $x^3 + 7x^2 + 12x$

   g $2px^2 + 11px + 12p$ **h** $50x^2 - 50x - 1000$

   i $40p^2 - 230p + 150$

> Sometimes you may need to factorise twice.

4 Show that the mean average of these expressions is $(2x + 3)(x + 5)$.

$6x^2 + 10x + 23$		$2x^2 + 20x - 11$

	$4x^2 + 13x + 21$		$9x - 4x^2 + 27$

5 If a quadratic expression has a negative x^2 term, you can factorise it by first taking out a common factor of '-1'.

   Use this method to factorise these expressions.

   a $3 - 7y - 6y^2$

   b $10p + 3 - 8p^2$

   c $11y - 3y^2 - 10$

   d $27m - 6m^2 - 30$

   e $5xy + 2y - 3x^2y$

> $21 - x - 2x^2$
> $= -(-21 + x + 2x^2)$
> $= -(2x^2 + x - 21)$
> $= -(x - 3)(2x + 7)$
> $= (3 - x)(2x + 7)$

The difference of two squares

This spread will show you how to:

● Manipulate algebraic expressions by factorising quadratic expressions, including the difference of two squares

Keywords
Factorise
Factors
Quadratic

When you expand brackets, sometimes the 'x' terms cancel each other out.

$$(x + 3)(x - 3) = x^2 - 9$$
$$(x + 5)(x - 5) = x^2 - 25$$
$$(2x - 7)(2x + 7) = 4x^2 - 49$$
$$\uparrow \qquad \uparrow$$
$$(2x)^2 \quad 7^2$$

You call quadratic expressions of the form $x^2 - 16 = (x + 4)(x - 4)$ the 'difference of two squares'.

You can use what you know about expanding brackets to factorise DOTS expressions.

$$x^2 - 64 = (x + 8)(x - 8)$$
$$\sqrt{x^2} = x, \ \sqrt{64} = 8$$

DOTS is shorthand for '**d**ifference **o**f **t**wo **s**quares'.

● When you factorise an expression, check for:
  – **common factors**
  – **double brackets**
  – **difference of two squares.**

Example

Factorise.

a $x^2 - 81$ **b** $16y^2 - 49$ **c** $25a^2 - 36b^2$ **d** $2x^2 - 50$.

a $x^2 - 81 = (x + 9)(x - 9)$ using DOTS
b $16y^2 - 49 = (4y - 7)(4y + 7)$
c $25a^2 - 36b^2 = (5a - 6b)(5a + 6b)$
d $2x^2 - 50 = 2(x^2 - 25)$
$\qquad\qquad = 2(x - 5)(x + 5)$

$\sqrt{16y^2} = 4y,$
$\sqrt{49} = 7.$

Take out the common factor 2.
$x^2 - 25$ is DOTS.

Example

By writing 2491 as $50^2 - 3^2$, find the prime factors of 2491.

$$50^2 - 3^2 = (50 - 3)(50 + 3)$$
$$= 47 \times 53$$

47 and 53 are both prime.
The prime factors of 2491 are 47 and 53.

1 Factorise these expressions.

a $x^2 - 100$ **b** $y^2 - 16$ **c** $m^2 - 144$ **d** $p^2 - 64$

e $x^2 - \frac{1}{4}$ **f** $k^2 - \frac{25}{36}$ **g** $w^2 - 2500$ **h** $49 - b^2$

i $4x^2 - 25$ **j** $9y^2 - 121$ **k** $16m^2 - \frac{1}{4}$ **l** $400p^2 - 169$

m $x^2 - y^2$ **n** $4a^2 - 25b^2$ **o** $9w^2 - 100v^2$ **p** $25c^2 - \frac{1}{4}d^2$

q $x^3 - 16x$ **r** $50y - 2y^3$ **s** $\left(\frac{16}{49}\right)x^2 - \left(\frac{64}{81}\right)y^2$

2 Use factorisation to work out these, without a calculator.

a $101^2 - 99^2$ **b** $10\,006^2 - 9994^2$

c $100^2 - 99^2$ **d** $407^2 - 93^2$

3 Rewrite each of these numbers as the difference of two squares in order to find their prime factors.

a 851 **b** 9991 **c** 627 **d** 319

4 Without using a calculator, find the missing side in this right-angled triangle in surd form.

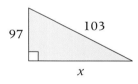

5 Factorise these algebraic expressions.

a $6x^2 - 15xy + 9y^2$ **b** $16a^2 - 9b^2$ **c** $x^2 - 11x + 28$

d $2x^2 + 11x - 21$ **e** $x^3 + 3x^2 - 18x$ **f** $5ab + 10(ab)^2$

g $10 - 3x - x^2$ **h** $10 - 10x^2$ **i** $2y + y^2 - 63$

j $2x^3 - 132x$ **k** $6x^2 + 6 - 13x$ **l** $x^4 - y^4$

6 Show that the shaded area in the diagram is 6160 cm².
Do not use a calculator.

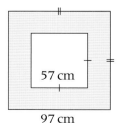

Exam review

Key objectives

- Use index notation for simple positive integer powers, and simple instances of index laws
- Manipulate algebraic expressions, factorise quadratic expressions, including the difference of two squares and cancel common factors in rational expressions

1 A rectangle has width $(5 - x)$ cm and length $(2 + 2x)$ cm.

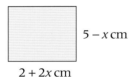

$5 - x$ cm

$2 + 2x$ cm

Show that the area of the rectangle is $(8x - 2x^2 + 10)$ cm^2.   (2)

2 a Simplify $k^5 \div k^2$.   (1)

  b Expand and simplify:

    i $4(x + 5) + 3(x - 7)$

    ii $(x + 3y)(x + 2y)$ (4)

  c Factorise $(p + q)^2 + 5(p + q)$ (1)

  d Simplify $(m^{-4})^{-2}$ (1)

  e Simplify $2t^2 \times 3r^3 t^4$ (2)

(Edexcel Ltd., 2004)

This unit will show you how to

- Round to a given number of significant figures
- Estimate answers to calculations
- Use appropriate degrees of accuracy for solutions
- Understand that data and measurements are not always exact
- Calculate the upper and lower bounds of data
- Multiply and divide by a number between 0 and 1, using the associated language
- Use mental and written methods for calculations involving fractions, decimals and percentages

Before you start ...

You should be able to answer these questions.

Review

1 Work out these additions and subtractions, showing your methods clearly for each.

   a 145 + 298   **b** 775 − 402

   c 16.9 + 4.4   **d** 973 − 708

   e 83.78 + 48.39   **f** 315.7 − 48.89

   g 66.8 − 39.77

   h 102.65 + 35.88 + 17.7

   CD – N2

2 Work out these multiplications and divisions, showing your methods clearly for each.

   a 116 × 6   **b** 7728 ÷ 7

   c 38 × 19   **d** 104 ÷ 8

   e 3.7 × 4.2   **f** 238 × 8.5

   g 4552 ÷ 9   **h** 809 ÷ 11

   CD – N2

3 Round each number to the decimal place shown.

   a 4.45 to 1 dp   **b** 3.666 to 2 dp

   c 1.07 to 1 dp   **d** 3.382 to 1 dp

   e 1.055 to 2 dp   **e** 17.8249 to 2 dp

   CD – N2

This spread will show you how to:
- Round to a given number of significant figures
- Estimate answers to calculations
- Use appropriate degrees of accuracy for solutions

Keywords
Digit
Round
Significant
   figures

Numbers are rounded to make them easier to handle.

- Numbers **round** up if the 'next' **digit** is a 5 or more.

Example

Round 104.458 to

a 2 dp   **b** 1 dp   **c** the nearest 10

d the nearest 100   **e** the nearest 1000

a 104.46   **b** 104.5   **c** 100

d 100   **e** 0

- When rounding to a given number of **significant figures**, start counting at the first non-zero digit.

Example

Round to 2 significant figures

a 16.668   **b** 4.923   **c** 14 559

d 105   **e** 0.000 675

a 16.668 = 17   **b** 4.923 = 4.9   **c** 14 559 = 15 000
   to 2 sf   to 2 sf   to 2 sf

d 105 = 110   **e** 0.000 675 = 0.000 68
   to 2 sf   to 2 sf

You should round measurements to a realistic degree of accuracy. For example, if you are using a metre rule your measurements will be accurate to the nearest half centimetre.

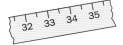

Example

A block of glass weighs 46 grams, and has a volume of 18.3 cm^3. Find the density of the glass in g/cm^3, giving your answer to a suitable degree of accuracy.

Density = mass ÷ volume
First estimate: $46 \div 18.3 \approx 50 \div 20 = 2.5$
Density = $46 \div 18.3 \approx 2.513\ 661\ 2$ g/cm^3
The mass, 46 g, is given to 2 sf so correct your answer to 2 sf.
Density of glass = 2.5 g/cm^3 to 2 sf.

To estimate the result of a calculation, round the numbers to 1 significant figure.

It is not sensible to have 7 dp in your final answer.

1 Round each of these numbers to the nearest 10.

   a 306   **b** 445   **c** 534.5   **d** 2174.9   **e** 56 685

2 Round each of these numbers to the nearest whole number.

   a 43.475   **b** 0.508   **c** 23.486   **d** 31.503   **e** 44.499

3 Round each of these numbers
  i to the nearest 100   **ii** to the nearest 1000.

   a 1286   **b** 1094   **c** 49   **d** 508   **e** 41 450

4 Round each of these numbers to the accuracy shown.

   a 0.31 (1 dp)   **b** 0.735 (2 dp)   **c** 0.1505 (3 dp) **d** 0.675 (2 dp)

5 Round each of these numbers to two significant figures.

   a 0.0564   **b** 3.175   **c** 14.67   **d** 948

6 Round these numbers to the accuracy shown.

   a 0.518 (2 sf)   **b** 34 591 (3 sf)   **c** 72 736 (3 sf)   **d** 0.004 49 (2 sf)

7 Round these numbers to one significant figure.

   a 352   **b** 0.632   **c** 0.0045   **d** 651 330

8 Use a calculator to work out these. Write your answers correct to two significant figures.

   a $5 \div 12$ **b** $6 \div 7$ **c** 37×49 **d** 239×18

9 **a** Round all the numbers to one significant figure and write a calculation that you could do in your head to estimate the answer to each of these calculations.

     i $451 \div 18$ **ii** $17 + 38 \div 4$ **iii** $1736 + 33 \times 48$ **iv** $1.9173 + 3.2013$

   b Now write the answers to the calculations you wrote in part **a**.

   c Use a calculator to find an exact answer for each calculation in part **a**. For each one, write a sentence to say how well the calculator result agrees with the estimate that you wrote in part **b**.

10 **a** Write a calculation that you can do in your head to estimate the answers to these calculations.

     i $258 + 362$ **ii** $64 \div 27$ **iii** 62.7×211.8 **iv** $96.7 - 64.8$

   b Explain carefully whether each of the calculations that you wrote in part **a** will produce an overestimate or an underestimate of the actual result.

   c Use a calculator to find the exact answers, and check your answers to part **b**.

This spread will show you how to:

- Understand that data and measurements are not always exact
- Calculate the upper and lower bounds of data

Keywords

Lower bound
Upper bound

● Measurements are not exact. Their accuracy depends on the precision of the measuring instrument and the skill of the person making the measurement.

The height of a tree is given as 5 metres, correct to the nearest metre.

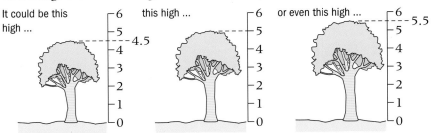

It could be this high ...   this high ...   or even this high ...

- 4.5 m is the **lower bound**.
- 5.5 m is the **upper bound**.

Compare this to data obtained by **counting**, which are exact.

The upper bound is not actually included in the range of possible values.

You always use the lower bound and the upper bound when stating the range of possible values of measurements or data.

Example

The mass of a meteorite is given as 235.6 g.
Find the lower and upper bounds of the mass.

The mass is given correct to the nearest 0.1 g.
Lower bound of the actual mass is 235.55 g.
Upper bound is 235.65 g.
If the mass of the meteorite is m then

$$235.55 \leqslant m < 235.65$$

Example

A student was asked to give the upper and lower bounds for a measurement given as 4.5 seconds. He gave the lower bound as 4.45 seconds, and the upper bound as 4.5499999 seconds.
What is wrong with his answer?

The lower bound is correct, but the upper bound should be 4.55 seconds. The student was trying to show that the maximum possible value is 'a bit less than 4.55 seconds'. However, this is never necessary; the upper bound is not a possible value of the measurement.

1 Each of these measurements was made correct to one decimal place. Write the upper and lower bounds for each measurement.

a 5.8 m   b 16.5 litres   c 0.9 kg   d 6.3 N

e 10.1 s   f 104.7 cm   g 16.0 km   h 9.3 m/s

2 Find the upper and lower bounds of these measurements, which were made to varying degrees of accuracy.

a 6.7 m   b 7.74 litres   c 0.813 kg   d 6 N

e 0.001 s   f 2.54 cm   g 1.162 km   h 15 m/s

3 Find the maximum and minimum possible total weight of

a 12 boxes, each of which weighs 14 kg, to the nearest kilogram

b 8 parcels, each weighing 3.5 kg.

4 Find the upper and lower bounds of these measurements, which are correct to the nearest 5 mm.

a 35 mm   b 40 mm   c 110 mm   d 4.5 cm

5 A box contains 38 nails, each weighing 12 g, to the nearest gram. Alex calculated the maximum and minimum possible total weight of the nails. He wrote

- Maximum = 38.5×12.5 g = 481.25 g
- Minimum = 37.5×11.5 g = 431.25 g

These calculations are incorrect. Explain why, and correct them.

6 A lift can hold up to six people, with a maximum safe load of 460 kg. A group of six people are waiting for the lift. Their weights are 85 kg, 96 kg, 63 kg, 73 kg, 68 kg and 73 kg (all measured to the nearest kilogram). Is it possible that the total weight of the group exceeds the limit for the lift? Show your working.

7 A completed jigsaw puzzle measures 24 cm by 21 cm (both to the nearest centimetre). Find the maximum and minimum possible area of the puzzle. Show your working.

8 A bag contains 98 g of sugar, then 34 g of sugar are taken out. Find the upper and lower bounds for the amount of sugar remaining in the bag, if the measurements are all correct to the nearest gram.

9 Unladen, a lorry weighs 2.3 tonnes, measured to the nearest 100 kg. The lorry is loaded with crates, each weighing 250 kg, correct to the nearest 10 kg. On its journey the lorry crosses a bridge with a maximum safe load of 15 tonnes. What is the maximum number of crates that the driver can load onto the lorry?

This spread will show you how to:

● Multiply and divide by a number between 0 and 1

Keywords
Divide
Multiply
Reciprocal

Multiplying doesn't always make numbers bigger.
Similarly, **dividing** doesn't always make them smaller.

Multiplying by 1
No change

Multiplying by **numbers < 1**
make positive numbers smaller

Multiplying by **numbers > 1**
make positive numbers bigger

0

1

Dividing by **numbers < 1**
make positive numbers bigger

Dividing by **numbers > 1**
make positive numbers smaller

Dividing by 1
No change

Dividing by a number is the same as multiplying by its **reciprocal**, for example

$$9 \div 4 = 9 \times \frac{1}{4} = 2\frac{1}{4}$$

Multiplying by a number is the same as dividing by its reciprocal, for example

$$2 \times 4 = 2 \div \frac{1}{4} = 8$$

A number multiplied by its reciprocal is equal to 1.

There are 8 quarters in 2.

Example

State whether a positive number will be larger, smaller or the same size when it is

a multiplied by 1.3   **b** divided by 0.3   **c** multiplied by 0.4

d divided by 1.5   **e** multiplied by 1.

a Larger   **b** Larger   **c** Smaller

d Smaller   **e** The same

Example

Use reciprocals to give an equivalent division for each multiplication.

a 7×2 **b** 1.8×5 **c** 29×1 **d** 15×0.5 **e** 46×0.2

a $7 \div 0.5$ **b** $1.8 \div 0.2$ **c** $29 \div 1$ **d** $15 \div 2$ **e** $46 \div 5$

Example

Use reciprocals to give an equivalent multiplication for each division.

a $17 \div 4$ **b** $5.9 \div 8$ **c** $44 \div 1$ **d** $2.9 \div 0.25$ **e** $78 \div 0.4$

a 17×0.25 **b** 5.9×0.125 **c** 44×1 **d** 2.9×4 **e** 78×2.5

$0.4 = \frac{2}{5}$, so the reciprocal of $0.4 = \frac{5}{2} = 2.5$

1 Explain whether a positive number will give you an answer that is larger, smaller or the same size as the original number, when you
   a multiply by 8   **b** divide by 7   **c** multiply by $\frac{1}{4}$
   d multiply by 0.01   **e** divide by 0.75   **f** divide by $\frac{3}{5}$
   g multiply by 0   **h** divide by 1.

2 Use reciprocals to write each of these multiplications as a division.
   a 8×0.5 **b** 10×0.2 **c** 12×0.25 **d** 18×0.1

3 Write an equivalent division for each of these multiplications.
   a $15 \times \frac{1}{5}$ **b** $28 \times \frac{1}{4}$ **c** $10 \times \frac{2}{5}$ **d** $72 \times \frac{1}{8}$

4 Write an equivalent multiplication for each of these divisions.
   a $18 \div \frac{1}{2}$ **b** $24 \div 0.25$ **c** $8 \div \frac{2}{3}$ **d** $5.5 \div 0.1$
   e $5.9 \div 1$ **f** $66 \div \frac{3}{5}$ **g** $7 \div \frac{7}{10}$ **h** $8 \div 1.25$

5 Calculate these, without using a calculator. Show your method.
   a $12 \div 0.5$ **b** $36 \div 0.25$ **c** $13 \div 0.2$ **d** $7.2 \div 0.1$
   e 8×0.25 **f** 45×0.2 **g** $24 \times \frac{1}{3}$ **h** $20 \div \frac{2}{3}$

6 Evaluate these without using a calculator. Show your method.
   a $9 \div \frac{1}{3}$ **b** $12 \div \frac{1}{5}$ **c** $8 \div 0.5$ **d** 84×0.25
   e 35×0.2 **f** $60 \div \frac{2}{3}$ **g** $16 \div 0.4$ **h** $15 \div 2.5$

7 Find the value of each expression. Do not use a calculator.
   a $24 \div \frac{3}{2}$ **b** $30 \div \frac{5}{4}$ **c** $15 \div 1.25$ **d** $24 \div 1.5$
   e $35 \div \frac{7}{6}$ **f** $28 \div 1\frac{1}{6}$ **g** $45 \div 1\frac{1}{4}$ **h** $32 \div 1\frac{1}{7}$

8 For any pair of numbers, the greater number is the one further to the right on the number line. Write the greater number in each of these.
   a 2 and 2.5   **b** 2 and −2   **c** −3 and 3.5   **d** 0 and −4.2
   e 1 and −25   **f** −7 and −23   **g** 4.8 and −38   **h** −3.9 and −3.85

9 Explain whether each statement is true or false, giving examples to support your argument.
   a When you divide a number by 2, the answer is always smaller than the number you started with.
   b Dividing by $\frac{1}{2}$ is always the same as doubling a number.
   c When you multiply by 5, the answer is always bigger than the number you started with.
   d When you multiply by 10, the answer will always be different from the original number.

10 **a** Explain the effect of repeatedly multiplying a number by +0.9.
   b Explain the effect of repeatedly multiplying a number by −0.9.

This spread will show you how to:

● Use mental methods for calculations involving fractions, decimals and percentages

Keywords
Equivalent
    calculation
Estimate
Product

You can use strategies such as finding **estimates** and **equivalent calculations** to help with mental calculations.

Example

Work out   **a** $\frac{2}{3} + \frac{3}{4}$ **b** $\frac{2}{5} \times \frac{3}{8}$ **c** $0.3 - 0.23$

d 0.4×0.3 **e** $0.8 \div 0.02$ **f** 28% of 40.

---

a Start with an estimate. Both fractions are between $\frac{1}{2}$ and 1, so the total is between 1 and 2.
The lowest common denominator is 12, and the equivalent fractions are $\frac{8}{12}$ and $\frac{9}{12}$. The total is $\frac{17}{12} = 1\frac{5}{12}$.

b Both fractions are less than $\frac{1}{2}$, so the **product** is less than $\frac{1}{4}$.
Multiplying numerators and denominators gives $\frac{6}{40}$, which cancels down to $\frac{3}{20}$.

c $0.3 - 0.2 = 0.1$, so the answer must be a little smaller than 0.1.
$30 - 23 = 7$, so the answer is 0.07.

d Half of 0.3 is 0.15, so 0.4×0.3 must be a little smaller than 0.15.
$4 \times 3 = 12$, so the answer is 0.12.

e Multiply both numbers by 100, to give the equivalent calculation $80 \div 2 = 40$.

f The answer is a bit more than a quarter of 40, which is 10.
Work out $28 \times 4 = 56 \times 2 = 112$; the answer is 11.2.

Or you could work out 40% of 28.

Example

Work out $1\frac{1}{4} - \frac{3}{7}$.

Estimate first.
$\frac{3}{7}$ is nearly $\frac{1}{2}$, so the answer must be a little more than $\frac{3}{4}$.
The common denominator is 28 so work out

$$1\frac{7}{28} - \frac{12}{28} = 1 - \frac{5}{28}$$
$$= \frac{23}{28}$$

All of the calculations in this exercise should be done mentally, with jottings where necessary.

1 Write a fraction equivalent to each of these decimals.

    a 0.5   **b** 0.3   **c** 0.25   **d** 0.4   **e** 0.6   **f** 0.05   **g** 0.45   **h** 0.375

2 Write the decimal equivalents of these fractions.

    a $\frac{3}{4}$ **b** $\frac{2}{5}$ **c** $\frac{5}{8}$ **d** $\frac{3}{20}$ **e** $\frac{1}{25}$ **f** $\frac{4}{25}$ **g** $\frac{7}{20}$ **h** $\frac{3}{50}$

3 Add these fractions.

    a $\frac{1}{2}+\frac{1}{4}$ **b** $\frac{1}{5}+\frac{1}{10}$ **c** $\frac{1}{3}+\frac{1}{6}$ **d** $\frac{1}{4}+\frac{1}{5}$ **e** $\frac{1}{8}+\frac{1}{2}$ **f** $\frac{1}{6}+\frac{1}{4}$ **g** $\frac{1}{3}+\frac{1}{5}$ **h** $\frac{3}{8}+\frac{1}{4}$

4 Find the the answers to these subtractions.

    a $\frac{3}{4}-\frac{1}{2}$ **b** $\frac{5}{8}-\frac{1}{4}$ **c** $\frac{4}{5}-\frac{1}{10}$ **d** $\frac{3}{4}-\frac{1}{6}$ **e** $\frac{4}{9}-\frac{1}{3}$ **f** $\frac{5}{6}-\frac{2}{3}$ **g** $\frac{3}{8}-\frac{1}{6}$ **h** $\frac{5}{8}-\frac{2}{7}$

5 Evaluate these multiplications, giving your answers as fractions in their simplest form.

    a $\frac{2}{3}\times\frac{1}{5}$ **b** $\frac{3}{4}\times\frac{1}{2}$ **c** $\frac{2}{9}\times\frac{3}{4}$ **d** $\frac{4}{5}\times\frac{5}{8}$ **e** $\frac{3}{8}\times\frac{4}{9}$ **f** $\frac{5}{12}\times\frac{3}{10}$ **g** $\frac{6}{7}\times\frac{1}{2}$ **h** $\frac{7}{24}\times\frac{3}{14}$

6 Work out these divisions, and simplify your answers.

    a $\frac{1}{2}\div\frac{1}{8}$ **b** $\frac{2}{3}\div\frac{1}{6}$ **c** $\frac{2}{5}\div\frac{1}{10}$ **d** $\frac{2}{3}\div\frac{3}{4}$ **e** $\frac{5}{8}\div\frac{1}{4}$ **f** $\frac{7}{10}\div\frac{2}{4}$ **g** $\frac{4}{9}\div\frac{2}{3}$ **h** $\frac{7}{20}\div\frac{1}{4}$

7 Do these additions and subtractions.

    a $0.3+0.4$ **b** $0.5+0.7$ **c** $1.2+0.9$ **d** $1.0-0.8$

    e $2.4-1.5$ **f** $13.3-6.8$ **g** $0.4+0.32$ **h** $0.5-0.17$

    i $6.7-4.85$ **j** $19-2.77$ **k** $4.27+5.944$ **l** $16.4-8.517$

8 Do these multiplications and divisions.

    a 4.8×0.2 **b** $6.2\div0.2$ **c** $25.4\div0.25$ **d** 4.3×0.3

    e 2.8×0.4 **f** $2.8\div0.4$ **g** 0.75×0.6 **h** $0.75\div0.05$

9 Find these percentages. Show your method in each case.

    a 10% of 480   **b** 15% of 340   **c** 19% of 800   **d** 41% of 600

    e 32% of 75   **f** 53% of 740   **g** 65% of 90   **h** 79% of 4000

10 Work out the result of these percentage changes. Show your method in each case.

    a £350 is increased by 12%   **b** £840 is decreased by 12.5%

    c £50 is increased by 17%   **d** £390 is increased by 21%

    e £780 is decreased by 5%   **f** £245 is decreased by 15%

Written calculations

This spread will show you how to:

- Use written methods for calculations involving fractions, decimals and percentages

Keywords

Decimal
  equivalent
Estimate
Place value

You need to know a range of efficient written techniques for calculating with fractions, decimals and percentages.

To divide by a fraction, multiply by its multiplicative inverse (invert the fraction).

Multiply fractions by multiplying numerators and denominators.

Notice that you can still use mental methods to estimate the answers, which gives you a check on the written calculations.

Example

Calculate

a 2.6×3.28 **b** $\frac{2}{3} \div \frac{3}{5}$ **c** 48% of 73 **d** 37 as a percentage of 120.

a
$$2.6 \times 3.28$$
$$\text{Estimate} \approx 3 \times 3 = 9$$
$$\begin{array}{r} 328 \\ \times\ 26 \\ \hline 1968 \\ 6560 \\ \hline 8528 \end{array} \Rightarrow Ans = 8.528$$

Do the calculation with whole numbers, then use the **estimate** to adjust the **place value**.

b
$$\frac{2}{3} \div \frac{3}{5}$$
(Estimate ≈ a bit more than 1)
$$\frac{2}{3} \div \frac{3}{5} = \frac{2}{3} \times \frac{5}{3}$$
$$= \frac{10}{9} = 1\frac{1}{9}$$

Note the initial estimate, $\frac{2}{3}$, is slightly larger than $\frac{3}{5}$ (think of their **decimal equivalents**).

c
$$48\% \text{ of } 73$$
$$\text{Estimate} \approx \frac{1}{2} \times 72 = 36$$
$$\begin{array}{r} 73 \\ \times\ 48 \\ \hline 584 \\ 2920 \\ \hline 3504 \end{array} \Rightarrow Ans = 35.04$$

Here you need to work out $73 \times 48 \div 100$. You can just do the whole-number multiplication, and adjust the place value.

d
$$\text{Estimate } \frac{37}{120} \approx 30\%$$
$$\begin{array}{r} 30833 \\ 12\overline{)370000} \\ \underline{36} \\ 100 \\ \underline{96} \\ 40 \\ \underline{36} \\ 4 \end{array} \quad Ans = 30.8\dot{3}\%$$

Use long division. Notice the recurring digit, and how the place value is adjusted.

Remember that on the calculator exam paper, you must write down the calculation to indicate your method, even if you use a calculator for the actual calculation.

Unless you are told otherwise, you should use a written method for the calculations in this exercise.

1 First write a mental estimate, and then use a written method to calculate an exact answer for each of these.

   a $134.6 - 7.859$ b 2.47×7 c 64.8×0.9

   d $434.28 \div 7$ e $23.862 + 38.779$ f 5.7×8.9

   g $84.66 \div 0.3$ h 13.2×0.74 i $234.2 - 166.89$

2 Use a calculator to check your answers to question **1**.

3 Evaluate these.

   a $\frac{2}{7} + \frac{3}{5}$ b $\frac{4}{9} \div \frac{3}{5}$ c $\frac{3}{8} \times \frac{5}{6}$

   d $\frac{7}{8} - \frac{3}{7}$ e $\frac{2}{3} + \frac{1}{6} + \frac{2}{5}$ f $\frac{4}{9} + \frac{2}{3} - \frac{5}{6}$

   g $\frac{2}{5} \times \frac{5}{6} \times \frac{1}{2}$ h $\frac{4}{5} - \frac{3}{8}$ i $\frac{1}{2} \div \frac{3}{8}$

4 Use the fraction facility on a scientific calculator to check your answers for question **3**.

On some calculators this looks like

5 Calculate these percentages, giving your answers to 3 significant figures.

   a 47% of 93   b 32% of 49   c 17% of 604

   d 64% of 123   e 23% of 290   f 56% of 324

   g 43.9% of 600   h 88.3% of 17   i 26.4% of 428

6 Use a calculator to check your answers to question **5**.

7 Find the results of these percentage changes.

   a 480 is decreased by 4.75%   b 725 is increased by 10.6%

   c 285 is increased by 67.5%   d 48.49 is decreased by 18.75%

8 Use a calculator to check your answers to question **7**.

9 For each pair of numbers, use long division (or an equivalent written method) to express one number as a percentage of the other. Give your answers to a suitable degree of accuracy.

   a 15 as a percentage of 240   b 28.8 as a percentage of 120

   c 81.9 as a percentage of 450   d 1.7 as a percentage of 130

   e 48.75 as a percentage of 90   f 267.5 as a percentage of 1800

10 Use a calculator to check your answers to question **9**.

N2 Exam review

Key objectives

- Estimate answers to problems involving decimals, round to a given number of significant figures
- Recognise limitations on the accuracy of data and measurements
- Develop a range of strategies for mental calculation.
- Understand 'reciprocal' as multiplicative inverse, knowing that any non-zero number multiplied by its reciprocal is 1

1 a Write a mental estimate for:

    i $54.32 + 26.16$

    ii $2.75 \div 0.4$ (2)

  b Use a written method to calculate the exact answers to the calculations in part **a**.   (2)

  c Calculate these, representing your answer in decimal form to 2 significant figures:

    i 75% of 30

    ii $4 \times \frac{3}{7}$ (4)

2 A boy has a pen that is of length 10 cm, measured to the nearest centimetre.

His pen case is of length 10.1 cm, measured to the nearest millimetre.

Explain why it might not be possible for him to fit the pen in the pen case.   (3)

(Edexcel Ltd., (Spec.))

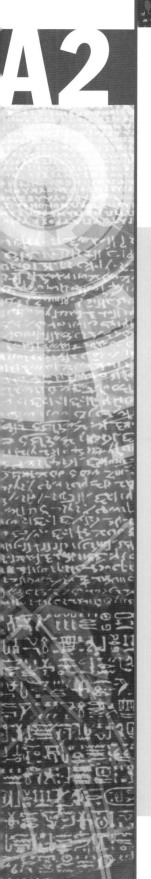
This unit will show you how to

- Set up simple equations
- Simplify equations involving fractions by using common denominators
- Cancel common fractions in algebraic expressions
- Solve linear equations, including equations in which the term including the unknown is negative or fractional

Before you start ...

You should be able to answer these questions.

Review

1 Given that $x = 4$, find the value of these expressions.

   a $3x - 4$ **b** $10 - 4x$ **c** $(3x - 1)^2$

   d $\frac{x}{2} - 9$ **e** $5x^2 - 1$

CD – A2

2 Simplify these expressions, by collecting like terms where possible.

   a $5x + 7y - 4x + 9y$ **b** $11x + 9x^2 - 8x$

   c $2ab + 5ab$ **d** $7p - 9$

   e $11x^3 - 2x^3 + 5x^3$ **f** $3(2x + y) - x(x + 4)$

Unit A1

3 Expand these sets of brackets, simplifying your answer where possible.

   a $3(2x - 1) + 6(3x - 4)$

   b $2(4y - 8) - 3(y - 9)$

   c $x(3x - 2y)$ **d** $(x + 9)(x - 7)$

   e $(2w - 8)(3w - 4)$ **f** $(p - q)^2$

Unit A1

4 Evaluate these, without a calculator, expressing your answers in their simplest form.

   a $\frac{12}{30}$ **b** $\frac{1}{5} + \frac{2}{9}$ **c** $\frac{3}{4} - \frac{1}{4}$

   d $2\frac{1}{6} + 3\frac{2}{5}$ **e** $\frac{5}{6} \times \frac{7}{10}$ **f** $\frac{5}{8} \div \frac{1}{5}$

CD – A2

Solving equations

This spread will show you how to:

- Set up simple equations
- Solve linear equations, including equations in which the term including the unknown is negative

Keywords
Equation
Inverse
Solution

- To solve an **equation** with the unknown on one side only, you use **inverse** operations.

$$3x - 4 = 16 \qquad \text{Add 4 to both sides}$$
$$3x = 20 \qquad \text{Divide both sides by 3}$$
$$x = \frac{20}{3} \text{ or } 6\frac{2}{3}$$

You can 'read' the equation as: 'I think of a number, multiply it by 3 and subtract 4: this gives 16'. You can 'undo' the operations using the inverse operations 'add 4' then 'divide by 3'.

- To solve an equation when the unknown is in a negative term, first add this term to both sides to give a simpler equation.

$$20 - 3x = 16 \qquad \text{Add } 3x \text{ to both sides}$$
$$20 = 16 + 3x$$

- To solve an equation with the unknown on both sides, first subtract the smallest algebraic term from both sides to give a simpler equation.

$$2x - 17 = 5x + 8 \xrightarrow{\substack{\text{Subtract } 2x \text{ from} \\ \text{both sides}}} -17 = 3x + 8 \qquad \text{Smallest algebraic term is } 2x$$

$$10 - 3x = 19 - 7x \xrightarrow{\substack{\text{Subtract } -7x \text{ from} \\ \text{both sides}}} 10 + 4x = 19$$

Notice that subtracting $(-7x)$ is the same as adding $7x$, as $-(-7x) = +7x$.

Example

Solve **a** $25 - 8x = 5$ **b** $2(3 - 9x) = 4(10 - 2x)$

a $25 - 8x = 5$ Add $8x$ to both sides
     $25 = 5 + 8x$ Subtract 5 from both sides
     $20 = 8x$ Divide both sides by 8
       $x = \frac{20}{8} \text{ or } 2\frac{1}{2}$ Change improper fractions to mixed numbers

Inverse of $+5$ is -5
Inverse of $\times 8$ is $\div 8$

b $2(3 - 9x) = 4(10 - 2x)$ Expand the brackets
     $6 - 18x = 40 - 8x$ Subtract $-18x$ from both sides: $-(-18x) = +18x$
         $6 = 40 + 10x$ Subtract 40 from both sides
     $-34 = 10x$ Divide both sides by 10
       $x = -3.4$ Only use decimals if they are exact, otherwise, use fractions

---

Example

Solve $6p + 3 = 2(p - 1)$

$6p + 3 = 2(p - 1)$
$6p + 3 = 2p - 2$
$4p + 3 = -2$
   $4p = -5$
    $p = -\frac{5}{4} \text{ or } -1.25$

1 Solve these equations.

   a $5x - 4 = 17$

   b $3(2x - 4) = 17$

   c $4x - 9 = 27$

   d $10 - 4x = 11$

   e $3(2 - x) = 19$

   f $15 = 9 - 5y$

   g $x + 2x - 7 - 8 = 12$

   h $3 + 7x - 9 - 15x = 18$

   i $17\frac{1}{4} = 18\frac{3}{4} - \frac{1}{2}x$

2 Solve these equations.

   a $10 + 2x = 7x - 9$ **b** $5x - 4 = 12 - 2x$

   c $2 - 3y = 9 - 8y$ **d** $5 - y = 10 - 2y$

   e $3w + 9 = -2w - 8$ **f** $4(1 - 2x) = 2(3 - x)$

   g $2(3y - 1) = 3(y - 1)$ **h** $3z + 2(4z - 2) = 5(3 - z)$

   i $6(p - 1) + 5(4 - p) = 6p$ **j** $10q - (2q + 4) = 15$

   k $2(r - 7) - (2r - 3) = 9(r - 2)$

3 For each problem, set up an equation and solve it to find the unknown.

   a Find the length of the rectangle:

      $5(7 - 3x)$

      $7x - 9$

   b Expression ① is 10 more than twice expression ②.

      Find p.

      ① $5p - 3$ ② $3p - 4$

4 **a** Find the length of a square of side $2x + 3$ with an area equal to the area of a rectangle measuring $x + 6$ by $4x - 2$.

   b In 10 years' time, my age will be double what it was 11 years ago. How old am I?

Solving equations involving fractions

This spread will show you how to:

- Set up simple equations
- Solve linear equations, including equations in which the term including the unknown is negative or fractional

Keywords
Denominator
Equation
Solve

- To **solve equations** involving fractions, first clear the fractions by multiplying both sides of the equation by the denominator.

$$\frac{3x + 2}{4} = 2x - 1 \qquad \text{Multiply both sides by 4}$$

$$3x + 2 = 4(2x - 1) \qquad \cancel{4} \times \frac{3x + 2}{\cancel{4}} = 3x + 2$$

- For equations with fractions on both sides, multiply both sides of the equation by the product of the denominators.

$$\frac{2x - 1}{5} = \frac{3x - 4}{8} \qquad \text{Multiply both sides by } 8 \times 5.$$

$$8(2x - 1) = 5(3x - 4) \qquad 8 \times \cancel{5} \times \frac{2x - 1}{\cancel{5}} = \cancel{8} \times 5 \times \frac{3x - 4}{\cancel{8}}$$

This method is called 'cross-multiplying'.

- You can simplify equations involving several fractions by multiplying each term by every denominator.

Example

Solve.

a $\dfrac{x}{5} - \dfrac{x}{6} = 10$ **b** $5 - \dfrac{5}{x} = 50$

a $\dfrac{x}{5} - \dfrac{x}{6} = 10$ Multiply each term by 5
Multiply each term by 6

$$x - \frac{5x}{6} = 50$$

$$6x - 5x = 300$$

$$x = 300$$

You could do this in one step by multiplying each term by $5 \times 6 = 30$.

b $5 - \dfrac{5}{x} = 50$ Add $\dfrac{5}{x}$ to both sides

$$5 = 50 + \frac{5}{x} \qquad \text{Subtract 50 from each side}$$

$$-45 = \frac{7}{x} \qquad \text{Multiply each term by } x$$

$$-45x = 5 \Rightarrow x = -\frac{5}{45}$$

Example

Solve $\dfrac{2m + 1}{2} + \dfrac{3m - 1}{3} = 2$

$$\frac{2m + 1}{2} + \frac{3m - 1}{3} = 2$$

$$2m + 1 + \frac{2(3m - 1)}{3} = 4 \qquad \text{Multiply by 2}$$

$$3(2m + 1) + 2(3m - 1) = 12 \qquad \text{Multiply by 3}$$

$$6m + 3 + 6m - 2 = 12$$

$$12m + 1 = 12$$

$$12m = 11 \Rightarrow m = \frac{11}{12}$$

1 Solve these equations. Clear the fractions first.

  a $\dfrac{x+7}{3} = \dfrac{2x-4}{5}$

  b $\dfrac{5y-9}{2} = \dfrac{3-2y}{6}$

  c $\dfrac{3z+1}{5} = \dfrac{2z}{3}$

  d $\dfrac{p}{2} + \dfrac{p}{4} = 7$

  e $\dfrac{4q}{3} - \dfrac{2q}{5} = 9$

  f $\dfrac{3m}{4} + \dfrac{7m}{6} = \dfrac{1}{3}$

2 These equations have the same solution – true or false?

$\dfrac{x}{4} + \dfrac{2x}{5} = 10$

$\dfrac{4x-2}{3} = \dfrac{3-8x}{5}$

3 These two rectangles have equal lengths. Find their areas.

5 mm

Rectangle A
Area = $2x+3$

8 mm

Rectangle B
Area = $5x-1$

4 A number is doubled and divided by 5. Three more than the number is divided by 8. Both calculations give the same answer.
What is the number?

5 The sum of one-fifth of Lucy's age and one-seventh of her age is 12.
Use algebra to work out Lucy's age.

6 These trapezia have equal areas.
What are the lengths of the parallel sides in each?

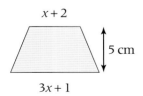

$x+2$

5 cm

$3x+1$

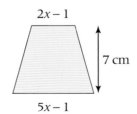

$2x-1$

7 cm

$5x-1$

Dealing with algebraic fractions

This spread will show you how to:

- Cancel common fractions in algebraic expressions

Keywords
Cancel
Denominator
DOTS
Factor
Numerator

- You can **cancel** common **factors** in algebraic expressions.

$$\frac{3p}{3} = \frac{\cancel{3} \times p}{\cancel{3}} = p$$ 3 is a common factor of 3 and $3p$.

- You cannot cancel in the expression $\dfrac{3+p}{3}$ because 3 is not a common factor of $3 + p$ and 3.

- You can factorise an expression and then cancel common factors.

$$\frac{x^2 + 5x + 6}{x + 2} = \frac{(x + 2)(x + 3)}{x + 2} = x + 3$$

- You can cancel common factors in multiplication or division problems to make the calculation easier.

$$\frac{p}{7} \div \frac{3p}{4} = \frac{\cancel{p}}{7} \times \frac{4}{3\cancel{p}} = \frac{4}{21}$$

Use common factors, double brackets or **DOTS** (Difference Of Two Squares).

$x + 2$ is a common factor of the **numerator** and **denominator**:

$$\frac{\cancel{(x+2)}(x + 3)}{\cancel{(x+2)}} = x + 3$$

p is a common factor of $p \times 4$ and $3p \times 7$.

Example

Cancel these fractions fully.

a $\dfrac{5p^3}{10p}$ **b** $\dfrac{3x + 6}{3}$ **c** $\dfrac{2x^2 - 5x - 12}{x^2 - 16}$

---

a $\dfrac{5p^3}{10p} = \dfrac{p^2}{2} = \dfrac{1}{2}p^2$ Divide numerator and denominator by the common factor $5p$.

b $\dfrac{3x + 6}{3} = \dfrac{3(x + 2)}{3} = x + 2$ Factorise first.

c $\dfrac{2x^2 - 5x - 12}{x^2 - 16}$ Numerator: factors of -24 that add to make -5 are 3 and -8.

$$= \frac{(2x + 3)(x - 4)}{(x + 4)(x - 4)}$$

$$= \frac{2x + 3}{x + 4}$$ Denominator: DOTS

$2x^2 - 5x - 12 = 2x^2 - 8x + 3x - 12$
$= 2x(x - 4) + 3(x - 4)$
$= (2x + 3)(x - 4)$

Example

Simplify this expression.

$$\frac{x^2 - 9}{7} \times \frac{5}{x - 3}$$

$$\frac{x^2 - 9}{7} \times \frac{5}{x - 3} = \frac{(x + 3)(x - 3) \times 5}{7(x - 3)}$$

$$= \frac{5(x + 3)}{7}$$

1 Cancel these fractions fully.

a $\dfrac{15w}{5}$ **b** $\dfrac{3b}{9}$ **c** $\dfrac{10c^2}{5c}$ **d** $\dfrac{12bd}{3d^2}$

e $\dfrac{100(bd)^2}{25b}$ **f** $\dfrac{2x+6}{2}$ **g** $\dfrac{x^2+x}{x}$ **h** $\dfrac{5y-10}{15}$

i $\dfrac{x^2+5x+6}{x+3}$ **j** $\dfrac{x^2-3x-28}{x+4}$ **k** $\dfrac{x-5}{x^2-12x+35}$

l $\dfrac{x^2-4}{x+2}$ **m** $\dfrac{4y^2-25}{2y+5}$ **n** $\dfrac{x-9}{x^2-81}$

o $\dfrac{2x^2-7x+5}{x-1}$ **p** $\dfrac{3x^2+10x+8}{x^2-4}$ **q** $\dfrac{x^3-16x}{x^2+4x}$

2 Explain why $\dfrac{x+1}{x-1}$ cannot be simplified, whereas $\dfrac{x+1}{x^2-1}$ can be.

3 Simplify fully $\dfrac{a^3b^2-a}{ab+1}$.

4 By cancelling where possible, simplify these multiplication and division calculations.

a $\dfrac{4p}{3} \times \dfrac{9}{4p}$ **b** $\dfrac{6ab}{7} \times \dfrac{2}{b}$ **c** $\dfrac{4m^2}{8} \times \dfrac{2n}{5m^3}$ **d** $\dfrac{3}{g} \div \dfrac{g}{5}$

e $\dfrac{4w}{3} \div \dfrac{w}{2}$ **f** $\dfrac{2f^2}{p^3} \times \dfrac{p}{4f}$ **g** $\dfrac{y^2}{5} \div \dfrac{y^3}{25}$

h $\dfrac{(x^2+11x+28)}{5} \times \dfrac{15}{(x+4)}$ **i** $\dfrac{x^2-11x+18}{12} \div \dfrac{x^2-17x+18}{24}$

j $\dfrac{2x^2-7x-15}{x^2-36} \times \dfrac{2x+12}{2x^3+3x^2}$

5 A rectangle measures $\dfrac{x+2}{8}$ by $\dfrac{7}{x^2-4}$ and its area is $\frac{1}{4}$ m². Set up an equation and solve it to find the dimensions of this shape.

6 The product of these three expressions is 8.
Use this information to find the value of y.

$$\dfrac{y^2+10y+21}{2y+8} \qquad \dfrac{y^2-16}{15} \qquad \dfrac{60}{y^2-y-12}$$

Adding and subtracting algebraic fractions

This spread will show you how to:

- Simplify equations involving fractions by using common denominators

Keywords
Denominator
Numerator

- To add (or subtract) numerical fractions, you convert them to equivalent fractions with a common **denominator**, and then add (or subtract) the **numerators**.

$$\frac{3}{4} + \frac{1}{6} = \frac{9}{12} + \frac{2}{12} = \frac{11}{12}$$

$$\frac{5}{7} - \frac{2}{3} = \frac{15}{21} - \frac{14}{21} = \frac{1}{21}$$

- You use the same method to add or subtract algebraic fractions.

To convert to an equivalent fraction, multiply numerator and denominator by the same number.

Example

Simplify these expressions.

a $\dfrac{4}{p} - \dfrac{3}{q}$ **b** $\dfrac{x+4}{6} - \dfrac{2x-1}{5}$

a $\dfrac{4}{p} - \dfrac{3}{q}$

$= \dfrac{4q}{pq} - \dfrac{3p}{pq}$

$= \dfrac{4q - 3p}{pq}$

b $\dfrac{x+4}{6} - \dfrac{2x-1}{5}$

$= \dfrac{5(x+4)}{30} - \dfrac{6(2x-1)}{30}$

$= \dfrac{5x+20}{30} - \dfrac{(12x-6)}{30}$

$= \dfrac{5x+20-12x+6}{30}$

$= \dfrac{26-7x}{30}$

In part **a**, convert to equivalent fractions with common denominator pq.

Example

Write each expression as a single fraction.

a $\dfrac{5}{x+2} + \dfrac{3}{x-4}$ **b** $\dfrac{3}{4x} + \dfrac{7}{4x^2}$

a $\dfrac{5}{x+2} + \dfrac{3}{x-4}$

$= \dfrac{5(x-4)}{(x+2)(x-4)} + \dfrac{3(x+2)}{(x+2)(x-4)}$

$= \dfrac{5x-20+3x+6}{(x+2)(x-4)}$

$= \dfrac{8x \,?\, 14}{(x+2)(x-4)} = \dfrac{2(4x-7)}{(x+2)(x-4)}$

b $\dfrac{3}{4x} + \dfrac{7}{4x^2}$

$= \dfrac{3 \times x}{4x \times x} + \dfrac{7}{4x^2}$

$= \dfrac{3x+7}{4x^2}$

In part **a**, the common denominator is $(x+2)(x-4)$.

In part **b**, the common denominator is $4x^2$.

Factorise the numerator and denominator – sometimes you may be able to cancel further.

1 Simplify these expressions.

a $\dfrac{3p}{5}+\dfrac{p}{5}$　　**b** $\dfrac{y}{7}+\dfrac{3y}{7}$　　**c** $\dfrac{1}{3p}+\dfrac{8}{3p}$　　**d** $\dfrac{5y}{4}+\dfrac{y}{8}$

e $\dfrac{2p}{5}-\dfrac{p}{3}$　　**f** $\dfrac{6}{x}-\dfrac{7}{y}$　　**g** $\dfrac{4}{x}+\dfrac{2}{x^2}$

2 Sort these expressions into equivalent pairs.
Which is the odd one out? Create its pair.

$\dfrac{5x}{12}-\dfrac{3x}{12}$　　$\dfrac{2}{3}x-\dfrac{1}{3}x$　　$\dfrac{x}{3}-\dfrac{x}{4}$　　$\dfrac{x}{12}$

$\dfrac{x}{6}+\dfrac{x}{4}$　　$\dfrac{x}{6}$　　$\dfrac{4x^2}{12x}$

3 Write each expression as a single fraction.

a $\dfrac{x+2}{5}+\dfrac{2x-1}{4}$　　　　**b** $\dfrac{3x-2}{7}+\dfrac{5-3x}{11}$

c $\dfrac{2y-5}{3}-\dfrac{3y-8}{5}$　　　　**d** $\dfrac{3(p-2)}{5}-\dfrac{2(7-2p)}{7}$

e $\dfrac{2}{x-7}+\dfrac{3}{x+4}$　　　　**f** $\dfrac{5}{x-2}+\dfrac{3}{x+3}$

g $\dfrac{3}{y-2}-\dfrac{4}{y+1}$　　　　**h** $\dfrac{2}{p+3}-\dfrac{5}{p-1}$

i $\dfrac{3}{w}+\dfrac{9}{w-8}$　　　　**j** $\dfrac{4}{x-2}+\dfrac{5}{(x-2)^2}$

4 Here is a rectangle.

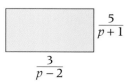

$\dfrac{5}{p+1}$

$\dfrac{3}{p-2}$

Find an expression for the perimeter of the rectangle.

More equations involving fractions

This spread will show you how to:

● Simplify equations involving fractions by using common denominators

Keywords
Cross-multiply
Solve

● You can **solve** equations involving fractions by simplifying first.

For equations of the form 'fraction = fraction' use cross-multiplication.

$$\frac{x+3}{4} \diagdown\kern-1.2em= \kern-1.2em\diagup \frac{2x-1}{7}$$

$$7(x+3) = 4(2x-1)$$

For equations involving several fractions, multiply each term by the product of all the denominators.

$$\frac{x}{3} + \frac{x}{4} = 2$$

$$3 \times 4 \times \frac{x}{3} + 3 \times 4 \times \frac{x}{4} = 3 \times 4 \times 2$$

$$4x + 3x = 24$$

● You can simplify sums (or differences) of fractions to a single fraction and then **cross-multiply**.

Example

Solve the equation $\dfrac{x+3}{6} + \dfrac{x+5}{5} = 4$.

$$\frac{x+3}{6} + \frac{x+5}{5} = 4$$

$$5 \diamond 6 \diamond \frac{(x+3)}{6} + 5 \diamond 6 \diamond \frac{(x+5)}{5} = 5 \diamond 6 \diamond 4$$

$$5x + 15 + 6x + 30 = 120$$

$$11x + 45 = 120 \qquad \Rightarrow x = \frac{75}{11}$$

Example

By simplifying the equation $\dfrac{5}{x+2} - 16 = \dfrac{3}{x-4}$, show that $8x^2 - 17x - 51 = 0$.

$$\frac{5}{x+2} - 16 = \frac{3}{x-4}$$

$$\frac{5}{x+2} - \frac{3}{x-4} = 16$$ Collect the fractions on one side of the equation.

$$\frac{5(x-4)}{(x+2)(x-4)} - \frac{3(x+2)}{(x+2)(x-4)} = 16$$ Convert to fractions with a common denominator.

$$\frac{5x - 20 - 3x - 6}{(x+2)(x-4)} = 16$$ Cross-multiply.

$$\frac{2x - 26}{(x+2)(x-4)} = 16$$

$$2(x-13) = 16(x+2)(x-4)$$

$$x - 13 = 8(x^2 - 2x - 8)$$ Divide both sides by 2.

$$x - 13 = 8x^2 - 16x - 64$$

$$8x^2 - 17x - 51 = 0$$ Rearrange into the required form.

1 Simplify the left-hand side of each original equation in the table and then show that it can be transformed to the new equation.

Original equation ...	Transform to ...
$\dfrac{2}{x} + \dfrac{2}{x+1} = 3$	$(3x+2)(x-1)=0$
$\dfrac{3}{y-1} + \dfrac{3}{y+1} = 4$	$(y-2)(2y+1)=0$
$\dfrac{4}{m+1} + \dfrac{2}{m-2} = 3$	$m^2 = 3m$
$\dfrac{3}{x-4} + \dfrac{5}{(x-4)(x+5)} = 10$	$10x^2 = 220 - 7x$
$\dfrac{6}{(x^2-4)} + \dfrac{3}{(x-2)} = 5$	$3x + 32 = 5x^2$
$\dfrac{5}{x^2+5x+6} + \dfrac{7}{x^2+8x+15} = 1$	$x^3 + 10x + 19x^2 - 9 = 0$

2 Solve these equations in two ways:

i by writing the left-hand side as a single fraction, then using cross-multiplication

ii by multiplying through by each denominator in turn and simplifying.

Check that both methods give the same solution.

> You need to find a **common denominator**.

a $\dfrac{(x+3)}{5} + \dfrac{(2x-1)}{4} = 12$ **b** $\dfrac{2y-3}{4} - \dfrac{3y-8}{2} = 20$

3 Here is an equation:

$$\frac{2}{x+1} - \frac{5}{x+2} = \frac{3}{x+3}$$

Decide which of the following equations are possible as a next step in simplifying the original equation.

$$2(x+2)(x+3) - 5(x+1)(x+3) = 3(x+1)(x+2)$$

$$2x^2 + 10x + 12 - 5x^2 - 20x - 15 = 3x^2 + 9x + 6$$

$$\frac{x+1}{2} - \frac{x-2}{5} = \frac{x+3}{3}$$

Exam review

Key objectives

- Solve linear equations in one unknown, with integer or fractional coefficients, in which the unknown appears on either side or on both sides of the equation

- Manipulate algebraic expressions, factorise quadratic expressions, including the difference of two squares, and cancel common factors in rational expressions

1 Solve this equation.

$$\frac{3}{x} - \frac{3}{2x} = \frac{3}{8}$$ (2)

2 Fully simplify these equations.

a $2(3x + 4) - 3(4x - 5)$ (2)

b $(2xy^3)^5$ (2)

c $\dfrac{n^2 - 1}{n + 1} \times \dfrac{2}{n - 2}$ (3)

(Edexcel Ltd., 2003)

This unit will show you how to

- Find the mean, median, mode, range and interquartile range of a small data set
- Calculate the mean for large and combined sets of data
- Use grouped frequency tables
- Calculate the modal class, class containing the median, estimated mean and range of data sets

Before you start ...

You should be able to answer these questions.

Review

1 For the following numbers

2 1 5 3 3 4 7

find the:

- **a** mode
- **b** median
- **c** range
- **d** mean

CD – D1

2 For the following numbers

3 3 2 2 5 5 5 1 5 3

find the:

- **a** mode
- **b** median
- **c** range
- **d** mean

CD – D1

This spread will show you how to:

● Find an average and a measure of spread for a data set

Keywords
Average
Interquartile range
Lower quartile
Mean
Measure of spread
Median
Mode
Range
Upper quartile

You can summarise data using an **average** and a **measure of spread**.

● An average is a single value.
There are three types of average:
- the **mode** is the value that occurs most often
- the **median** is the middle value when the data are arranged in order
- the **mean** is calculated by adding all the values and dividing by the number of values.

● Spread is a measure of how widely dispersed the data are.
Two measures of spread are:
- the range
- the **interquartile range** (IQR).

If there are one or more extreme values the IQR is a better measure of spread than the range.

An extreme value is a value well outside the range of the rest of the data

● Range = highest value − lowest value
● IQR = upper quartile − lower quartile

Lower quartile
$= \frac{1}{4}(n + 1)$th value

Upper quartile
$= \frac{3}{4}(n + 1)$th value

Example

Louise collected data on the number of times her friends went swimming in one month.

| 4 | 7 | 22 | 1 | 6 | 2 | 1 | 5 | 6 | 6 | 4 |

Work out the: **a** range   **b** mode   **c** mean
              d median   **e** interquartile range.

In order the data are: 1   1   2   4   4   5   6   6   6   7   22

a Range = 22 − 1 = 21

b Mode = 6

c Mean = 5.8
$(4 + 7 + 22 + 1 + 6 + 2 + 1 + 5 + 6 + 6 + 4) \div 11 = 64 \div 11 = 5.8$

The mean does not have to be an integer even if all the data values are integers.

d There are 11 values.
Median $= \frac{11 + 1}{2} = $ 6th value = 5

If there are n values
Median value
$= (\frac{n + 1}{2})$th value.

e Interquartile range = upper quartile − lower quartile

Lower quartile $= (\frac{11 + 1}{4})$th value
           $= (\frac{12}{4})$th value
           = 3rd value
           = 2

Upper quartile $= (\frac{3(11 + 1)}{4})$th value
           = 9th value
           = 6

IQR = 6 − 2 = 4

1 For these sets of numbers work out the
   i range   **ii** mode   **iii** mean
   iv median   **v** interquartile range

   a 5, 9, 7, 8, 2, 3, 6, 6, 7, 6, 5

   b 45, 63, 72, 63, 63, 24, 54, 73, 99, 65, 63, 72, 39, 44, 63

   c 97, 95, 96, 98, 92, 95, 96, 97, 99, 91, 96

   d 13, 76, 22, 54, 37, 22, 21, 19, 59, 37, 84

   e 89, 87, 64, 88, 82, 88, 85, 83, 81, 89, 90

   f 53, 74, 29, 32, 67, 53, 99, 62, 34, 28, 27, 27, 27, 64, 27

   g 101, 106, 108, 102, 108, 105, 106, 109, 103, 105, 107, 104, 104, 105, 105

2 For the set of numbers in question **1e**, explain why the interquartile range is a better measure of spread to use than the range.

3 For the set of numbers in question **1f**, explain why the mode is not the best average to use.

4 **a** Subtract 100 from each of the numbers in question **1g** and write down the set of numbers you get.

   b For your set of numbers in **a**, work out the

   i range   **ii** mode   **iii** mean

   iv median   **v** interquartile range

   c Compare your answers for the measures of spread in part **b i** and **v** and **1g i** and **v**.
   What do you notice?

   d Add 100 to your answers for the measures of average in part **b ii**, **iii** and **iv**.
   Compare these answers to the answers you got in question **1g**.
   What do you notice?

   e Give a reason for what you noticed in parts **c** and **d**.

5 A scientist takes two sets of measurements from her experiment.
   Her results are:

Set A:	0	99	99	100	100	100	100	100	101	101	200
Set B:	0	0	99	99	100	100	100	101	101	200	200

   a For each set of measurements, work out the

   i range   **ii** mode   **iii** mean

   iv median   **v** interquartile range

   b Discuss whay you notice about the measurements and your answers to part **a**.

This spread will show you how to:

● Find the mean, median, mode and range of a small data set

Keywords
Interquartile range
Lower quartile
Mean
Median
Mode
Range
Upper quartile

The three common types of average are

– **mode** – the value that occurs most often
– **median** – the middle value when the data are arranged in order
– **mean** – the total of all the values divided by the number of values.

Two measures of spread are

– **range** = highest – lowest value
– the **interquartile range** is the spread of the middle half of the data
– interquartile range (IQR) = **upper quartile** (UQ) – **lower quartile** (LQ)

For data arranged in ascending order:

● Lower quartile: value $\frac{1}{4}$ of the way along
● Upper quartile: value $\frac{3}{4}$ of the way along

Example

The table shows the lengths of words in a paragraph of writing.

For these data, work out the

a mode **b** median
c mean **d** range
e IQR.

Word length	Frequency
4	4
5	5
6	8
7	7
8	2
9	1

a Mode = 6 letters   Highest frequency = 8, for words with 6 letters.
b Total number of words = 27   Add the values in the frequency column.
Middle value is the $\frac{1}{2}(27 + 1) = 14$th
This occurs in the 'Word length 6' category.   Adding frequencies, 4 + 5 = 9, 9 + 8 = 17, so the 14th is in the 3rd group, that is, 6 letter words.
Median = 6 letters
c Mean

Word length	Frequency	Word length × frequency
4	4	4 × 4 = 16
5	5	5 × 5 = 25
6	8	6 × 8 = 48
7	7	7 × 7 = 49
8	2	8 × 2 = 16
9	1	9 × 1 = 9
Total	**27**	**163**

Total of word length × frequency = total number of letters in paragraph. Total number of words = 27

The mean value does not have to be an integer or a member of the original data set.

Mean = 163 ÷ 27 = 6.03 letters
d Range: Longest – shortest word length   9 – 4 = 5 letters
e IQR = UQ – LQ
LQ = 5 $\frac{1}{4}(27+1)$ = 7th value, in the category 'word length 5'.
UQ = 7 $\frac{3}{4}(27+1)$ = 21st value, in the category 'word length 7'.
IQR = 7 – 5 = 2 letters

You can use the statistical mode of your scientific calculator to calculate the mean of a data set.

1 Anya counted the contents of 12 boxes of paperclips.
Her results are shown in the table.

Number of paperclips	Frequency
44	1
45	5
46	4
47	2

Work out the mean number of paperclips in a box.

2 The tables give information about the length of words in four
different paragraphs.
Copy each table, add an extra working column and find the

i mode **ii** median **iii** mean **iv** range **v** interquartile range.

a

Word length	Frequency
4	8
5	4
6	9
7	5
8	5

b

Word length	Frequency
3	4
4	6
5	9
6	6
7	4

c

Word length	Frequency
4	7
5	3
6	4
7	2
8	3
9	6
10	2

d

Word length	Frequency
3	5
4	5
5	7
6	8
7	6
8	4
9	2

3 A rounders team played 20 matches.
The numbers of rounders scored in
these matches are given in the table.

Number of rounders	Frequency
0	4
1	7
2	5
3	3
4	1

Work out the mean number of rounders scored.

4 Reuben counted the raisins in some mini-boxes.
Five boxes contained 13 raisins, nine boxes contained 14 raisins
and the remainder had 15 raisins.
The mean number of raisins per box was 14.1 (to 1 dp).
Find the number of boxes that contained 15 raisins.

Mean of combined data sets

This spread will show you how to:

- Calculate the mean for large and combined sets of data

- **Mean** = $\dfrac{\text{Total of all values}}{\text{Number of values}}$

If you combine two data sets with known means, you can use these means to calculate the mean for the combined set.

Example

A group of 14 men and 22 women took their driving theory test.
The men's mean mark was 68%.
The women's mean mark was 74%.
Work out the mean mark for the whole group.

Men: Total of all marks: $68 \times 14 = 952$
Women: Total of all marks: $74 \times 22 = 1628$

Total of all marks for men and women: $952 + 1628 = 2580$
Mean mark for whole group $= 2580 \div 36 = 71.66666\ ..$
$= 72\%$

Number in group:
$14 + 22 = 36$

Example

In a survey, 50 bus passengers were asked how long they had had to wait for their bus.
30 passengers were asked on daytime buses, 20 on evening buses.

The mean waiting time for all 50 passengers was 14 minutes.
The mean waiting time for the daytime buses was 12 minutes.

Compare the mean waiting times for daytime and evening buses.

Total waiting time for all 50 passengers $50 \times 14 = 700$ minutes
Total waiting time for daytime buses $30 \times 12 = 360$ minutes
Total waiting time for evening buses $700 - 360 = 340$ minutes
Mean waiting time for evening buses $340 \div 20 = 17$ minutes
The mean waiting time for evening buses (17 minutes) was longer than for daytime buses (12 minutes).

Example

42 scuba divers took a diving exam.
16 divers were under 18 and 26 were adults.
The mean mark for the adults was p.
The mean mark for the under 18s was q.
Find an expression for the mean mark of all 42 divers.

Total of all adult marks $= 26p$
Total of all under 18s marks $= 16q$

Mean for all divers $= \dfrac{26p + 16q}{42}$

1   A swimming club has 180 members of which 110 are male and 70
    are female.
    The mean daily training time for males is 86 minutes.
    The mean daily training time for females is 72 minutes.
    Find the mean daily training time all 180 members.

2   A college has 240 A-level students of which 135 are girls and
    105 are boys.
    A survey of home study time for one week found that the mean time
    for boys was 6.8 hours and the mean time for girls was 8.2 hours.
    Find the mean time spent on home study for all 240
    A-level students.

3   A driving school calculated that 40 of their pupils passed the driving
    test after a mean number of 24.5 lessons. Of the 40 pupils, the six
    that already had a motorcycle driving licence had an average of
    16 lessons.
    Find the mean number of driving lessons that the remaining pupils
    had before passing their driving test.

4   500 people were asked how many times they had visited a museum
    in the past year.
    290 of them lived in a large cities, and the remainder lived in
    small towns.
    The mean number of visits for the whole group was 4.8.
    For the city people, the mean number of visits was 6.3.
    Compare the mean number of museum visits for people in cities and
    small towns.
    Suggest a reason for the difference.

5   From a survey of 800 in store and online customers, a retailer
    calculates the mean amount spent per customer as £74.63.
    For the 575 in store customers, the mean was £71.44.
    Compare the mean amounts spent by in store and online customers.
    Suggest a reason for the difference.

6   36 women and 42 men take a fitness test.
    The mean score for the women is c.
    The mean score for the men is d.
    Write an expression for the mean fitness score for the whole group.

7   A squash club has m male and f female members.
    The average age for the males is x.
    The average age for the females is y.
    Write an expression for the average age for all the members of the
    squash club.

Averages and spread for grouped data

This spread will show you how to:

- Use grouped frequency tables
- Calculate the modal class, class containing the median, and estimated mean and range for large data sets

Keywords

Estimate
Grouped
  frequency
Median
Midpoint
Modal class

You can present a large set of data in a **grouped frequency** table.

Lengths of runner beans in cm

3.9, 5.2, 7.6, 10.6, 12.4, 14.2

- For a grouped frequency table, the **modal class** is the class with the highest frequency.

- You can work out which class contains the median.

- You can **estimate** the mean and range.

Length, x cm	Frequency
$0 < x \leqslant 5$	I
$5 < x \leqslant 10$	II
$10 < x \leqslant 15$	III

The classes must not overlap.

You do not know the actual data values, so you can only estimate the mean and range.

Example

The table shows the times taken, to the nearest minute, for commuters to solve a sudoku puzzle.

Find

a the modal class
b the class containing the median.

Calculate an estimate for
c the mean
d the range.

Time, t, minutes	Frequency
$5 < t \leqslant 10$	4
$10 < t \leqslant 15$	18
$15 < t \leqslant 20$	11
$20 < t \leqslant 25$	5
$25 < t \leqslant 30$	2

a Modal class is $10 < t \leqslant 15$ (10 to 15 minutes).   The class with the highest frequency.

b Total number of commuters = 40   Add the frequency column.

Median is the $\frac{1}{2}(40 + 1) = 20\frac{1}{2}$th value   Look for 20th and 21st values.

which is in the class $10 < t \leqslant 15$ (10 to 15 minutes).

c

Time, t, minutes	Frequency	Midpoint	Midpoint × frequency
$5 < t \leqslant 10$	4	7.5	30
$10 < t \leqslant 15$	18	12.5	225
$15 < t \leqslant 20$	11	17.5	192.5
$20 < t \leqslant 25$	5	22.5	112.5
$25 < t \leqslant 30$	2	27.5	55
Totals	40		615

Add two columns to your table.
Use the **midpoint** as an estimate of the mean value for each class.

Estimate of mean = $615 \div 40 = 15.375 = 15$ minutes

d Range = $30 - 5 = 25$ minutes.   Longest possible time = 30 minutes.
Shortest possible time = 5 minutes.

1 The tables give information about the times taken for visitors to find their way through different mazes.
Copy each table, add extra working columns and find

i the modal class   **ii** the class containing the median

iii an estimate of the mean   **iv** an estimate of the range.

a   Hampton Court Maze

Time, t, minutes	Frequency
$5 < t \leqslant 10$	5
$10 < t \leqslant 15$	8
$15 < t \leqslant 20$	6
$20 < t \leqslant 25$	4
$25 < t \leqslant 30$	2

b   Marlborough Maze

Time, t, minutes	Frequency
$0 < t \leqslant 10$	1
$10 < t \leqslant 20$	9
$20 < t \leqslant 30$	7
$30 < t \leqslant 40$	5
$40 < t \leqslant 50$	2

c   The Maize Maze

Time, t, minutes	Frequency
$5 < t \leqslant 10$	6
$10 < t \leqslant 15$	7
$15 < t \leqslant 20$	5
$20 < t \leqslant 25$	0
$25 < t \leqslant 30$	2
$30 < t \leqslant 35$	1

d   Leeds Castle Maze

Time, t, minutes	Frequency
$5 < t \leqslant 15$	12
$15 < t \leqslant 25$	9
$25 < t \leqslant 35$	8
$35 < t \leqslant 45$	5
$45 < t \leqslant 55$	4
$55 < t \leqslant 65$	2

2 Alfie kept a record of his monthly food bills for one year.

a Find the class interval that contains the median.

b Calculate an estimate for Alfie's mean monthly food bill.

Food bill, B, pounds	Frequency
$100 < B \leqslant 150$	5
$150 < B \leqslant 200$	4
$200 < B \leqslant 250$	2
$250 < B \leqslant 300$	1

3 A computer game and video store gives its staff a discount.
The table shows the amount spent by staff members in one month.

Monies spent, M, pounds	Frequency
$0 < M \leqslant 40$	11
$40 < M \leqslant 80$	8
$80 < M \leqslant 120$	8
$120 < M \leqslant 160$	9
$160 < M \leqslant 200$	3
$200 < M \leqslant 240$	1

a Calculate an estimate for the mean amount of money spent.

b Write down the class interval that contains the median.

c The manager of the shop spent £250, which was not included in the table. If this amount is included, would your answer to **b** change? Explain your answer.

Exam review

Key objectives

- Calculate the mean for (and estimate of) large data sets with grouped data

1 Find an estimate for the mean time.

Time, t, Seconds	Frequency
$0 < t \leqslant 20$	0
$20 < t \leqslant 50$	10
$50 < t \leqslant 60$	5

(3)

2 There are 12 boys and 15 girls in a class.

In a test, the mean mark for the boys was n.

In the same test, the mean mark for the girls was m.

Work out an expression for the mean mark of the whole class of 27 students. (3)

(Edexcel Ltd., (Spec.))

This unit will show you how to

- Add, subtract, multiply and divide with fractions
- Convert between fractions, decimals and percentages
- Recognise terminating and recurring decimals when written as fractions
- Order fractions, decimals and percentages
- Find percentages of quantities
- Calculate percentage increases and decreases
- Solve reverse percentage problems

Before you start ...

You should be able to answer these questions.

Review

1 Find the results of these additions and subtractions involving fractions.

   a $\frac{1}{2} + \frac{1}{4}$ **b** $\frac{1}{2} - \frac{1}{4}$

   c $\frac{3}{10} + \frac{1}{5}$ **d** $\frac{2}{3} + \frac{1}{2}$

Unit N2

2 Find the results of these multiplications and divisions involving fractions.

   a $\frac{1}{2} \times \frac{2}{3}$ **b** $\frac{2}{5} \div \frac{1}{4}$

   c $\frac{3}{7} \times \frac{7}{10}$ **d** $\frac{4}{5} \div \frac{1}{10}$

Unit N2

3 Carry out these conversions.

   a Convert $\frac{3}{5}$ to a percentage.

   b Convert 0.35 to a fraction.

   c Convert 65% to a fraction.

   d Convert $\frac{7}{20}$ to a decimal.

Unit N2

4 Calculate.

   a 30% of £40   **b** 18% of 60 cm

   c 5.5% of 300 m   **d** 65% of 3 hours

Unit N2

Fraction calculations

This spread will show you how to:

- Add, subtract, multiply and divide with fractions

Cancel
Common denominator
Common factors
Multiplicative inverse

You need to be able to:

- Add and subtract fractions.
  You can add or subtract fractions if they have the same denominator.

$$\frac{3}{5} + \frac{1}{4} = \frac{12}{20} + \frac{5}{20}$$
$$= \frac{17}{20}$$

$$3\frac{1}{4} - 2\frac{5}{8} = 1 + \frac{1}{4} - \frac{5}{8}$$ Subtract the whole numbers first.
$$= 1 + \frac{2}{8} - \frac{5}{8}$$
$$= 1 - \frac{3}{8}$$
$$= \frac{5}{8}$$

Use a **common denominator**.

- Multiply fractions.

$$\frac{2}{3} \times \frac{5}{8} = \frac{10}{24}$$
$$= \frac{5}{12}$$

Multiply numerators together and multiply denominators together.

- Divide fractions.
  Dividing by any number is the same as multiplying by its **multiplicative inverse.**

$$\frac{3}{4} \div \frac{2}{5} = \frac{3}{4} \times \frac{5}{2}$$
$$= \frac{15}{8}$$
$$= 1\frac{7}{8}$$

$\frac{5}{2}$ is the inverse of $\frac{2}{5}$.

Example

Work out each of these calculations.

a $\frac{3}{5} + \frac{11}{16}$ **b** $\frac{2}{7} - \frac{1}{4}$ **c** $\frac{3}{5} \div 8$ **d** $\frac{4}{9} \div \frac{1}{3}$ **e** $\frac{3}{8} \times \frac{5}{9}$ **f** $\frac{3}{5} \div \frac{7}{10}$

a $\frac{3}{5} + \frac{11}{16} = \frac{48}{80} + \frac{55}{80}$ The LCM of 5 and 16 is 80. **b** $\frac{2}{7} - \frac{1}{4} = \frac{8}{28} - \frac{7}{28}$ The LCM of 7 and 4 is 28.
$$= \frac{48 + 55}{80}$$ $$= \frac{1}{28}$$
$$= \frac{103}{80}$$
$$= 1\frac{23}{80}$$

c $\frac{3}{5} \div 8 = \frac{3}{5} \times \frac{1}{8}$ The multiplicative inverse **d** $\frac{4}{9} \div \frac{1}{3} = \frac{4}{9} \times \frac{3}{1} = \frac{4 \times 3}{9 \times 1}$ The multiplicative
$$= \frac{3 \times 1}{5 \times 8}$$ of 8 is $\frac{1}{8}$. $$= \frac{12}{9}$$ inverse of $\frac{1}{3}$ is $\frac{3}{1}$
$$= \frac{3}{40}$$ $$= \frac{4}{3}$$ (which is 3).
$$= 1\frac{1}{3}$$

e $\frac{\cancel{3}^{1}}{8} \times \frac{5}{\cancel{9}_{3}} = \frac{1}{8} \times \frac{5}{3}$ Notice how you can **cancel** **f** $\frac{3}{5} \div \frac{7}{10} = \frac{3}{\cancel{5}_{1}} \times \frac{\cancel{10}^{2}}{7}$ Again, cancel
$$= \frac{5}{24}$$ **common factors** before $$= \frac{6}{7}$$ common factors
multiplying.   before multiplying.

1 Add these fractions.

   a $\frac{1}{2}+\frac{1}{2}$ **b** $\frac{1}{2}+\frac{1}{4}$ **c** $\frac{1}{2}+\frac{1}{3}$ **d** $\frac{1}{5}+\frac{1}{10}$ **e** $\frac{1}{3}+\frac{1}{4}$

2 Subtract these fractions.

   a $\frac{2}{3}-\frac{1}{3}$ **b** $\frac{1}{2}-\frac{1}{6}$ **c** $\frac{1}{3}-\frac{1}{4}$ **d** $\frac{2}{5}-\frac{1}{3}$ **e** $\frac{2}{3}-\frac{3}{7}$

3 Do these calculations with mixed numbers.

   a $1\frac{1}{2}+\frac{1}{4}$ **b** $2\frac{1}{3}-\frac{2}{3}$ **c** $1\frac{1}{5}+2\frac{1}{10}$ **d** $3\frac{1}{4}-\frac{1}{8}$

4 Do these multiplications and divisions.

   a $\frac{3}{8}\times 4$ **b** $2\times\frac{2}{5}$ **c** $\frac{7}{8}\div 2$ **d** $\frac{15}{16}\div 3$

5 Calculate.

   a $\frac{2}{3}\times\frac{1}{3}$ **b** $\frac{5}{8}\div\frac{1}{4}$ **c** $\frac{5}{9}\times\frac{1}{5}$ **d** $\frac{9}{20}\div\frac{1}{5}$

6 Calculate.

   a $\frac{2}{5}\div\frac{2}{3}$ **b** $\frac{3}{7}\times\frac{7}{8}$ **c** $\frac{3}{7}\times\frac{2}{5}$ **d** $\frac{4}{9}\div\frac{5}{6}$

7 Evaluate.

   a $2\frac{1}{4}\div\frac{3}{4}$ **b** $3\frac{1}{2}\div\frac{5}{8}$ **c** $2\frac{1}{4}\times 3\frac{2}{3}$ **d** $1\frac{1}{8}\times 2\frac{3}{4}$

8 Calculate.

   a $3\frac{7}{8}+2\frac{1}{4}$ **b** $3\frac{7}{8}-3\frac{1}{4}$ **c** $5\frac{1}{2}\times 1\frac{7}{8}$ **d** $2\frac{1}{2}\div 3\frac{3}{4}$

9 Calculate.

   a $\dfrac{\frac{3}{4}+\frac{1}{2}}{\frac{5}{8}-\frac{1}{4}}$ **b** $\left(\frac{3}{4}-\frac{1}{8}\right)\left(\frac{3}{4}+\frac{1}{8}\right)$

   c $\dfrac{1\frac{1}{2}-\frac{7}{8}}{2\frac{3}{4}-1\frac{1}{6}}$ **d** $\left(2\frac{3}{5}-1\frac{3}{4}\right)\left(1\frac{7}{8}+3\frac{1}{4}\right)$

10 Tom planted 48 daffodils. On Easter Day, $\frac{2}{3}$ of them were in bloom. How many daffodils were blooming?

11 Ed irons his shirts every Sunday, and it takes him $5\frac{3}{4}$ minutes do one shirt.
Last week he washed all his shirts and spent 46 minutes on his ironing. How many shirts does he have?

Fractions and decimals

This spread will show you how to:

* Recognise terminating and recurring decimals when written as fractions
* Convert between fractions, decimals and percentages

* To convert a fraction to a decimal divide the numerator by the denominator.

Example

Write these fractions as decimals.

a $\frac{7}{8}$ **b** $\frac{2}{3}$

a
$$\frac{7}{8} = 7 \div 8 = 8\overline{)7.000} \quad 0.875$$

0.875 is a **terminating** decimal.

b
$$\frac{2}{3} = 2 \div 3 = 3\overline{)2.000...} \quad 0.666...$$

0.666... = $0.\dot{6}$ is a **recurring** decimal.
The dot shows the recurring digit.

To decide if a fraction will be a terminating or a recurring decimal:

* If the only factors of the denominator are 2 and/or 5 or combinations of 2 and 5 then the fraction will be a terminating decimal.
* If the denominator has any factors other than 2 and/or 5 then the fraction will be a recurring decimal.

$\frac{7}{8}$ is a terminating decimal as $8 = 2 \times 2 \times 2$.

$\frac{5}{6}$ is a recurring decimal as $6 = 2 \times \mathbf{3}$.

To convert a terminating decimal to a fraction write the decimal as a fraction with the denominator as a power of ten.

$$0.385 = \frac{385}{1000}$$

Then cancel common factors: $\frac{385}{1000} = \frac{77}{200}$

Example

Write these as fractions.   **a** 0.3   **b** 0.56   **c** 0.625

a $0.3 = \frac{3}{10}$ **b** $0.56 = \frac{56}{100} = \frac{14}{25}$ **c** $0.625 = \frac{625}{1000} = \frac{5}{8}$

To convert a recurring decimal to a fraction use this method.

Write $0.\dot{3}\dot{6}$ as a fraction.   Put $\quad 0.\dot{3}\dot{6} = x$
$$\text{Then } 100x = 36.\dot{3}\dot{6}$$
$$x = 0.\dot{3}\dot{6}$$
$$99x = 36 \quad \text{Subtract one from the other.}$$
$$x = \frac{36}{99} = \frac{4}{11}$$

If there is one recurring digit, find 10x. If there are three, find 1000x.

Example

Write these as fractions.   **a** $0.\dot{3}$ **b** $0.1\dot{6}$ **c** $0.\dot{1}2\dot{3}$

a $10x = 3.\dot{3}$
$\quad\quad x = 0.\dot{3}$
$\quad 9x = 3$
$\quad\quad x = \frac{3}{9} = \frac{1}{3}$

b $100x = 16.1\dot{6}$
$\quad\quad x = 0.1\dot{6}$
$\quad 99x = 16 \quad \text{(subtract)}$
$\quad\quad x = \frac{16}{99}$

c $1000x = 123.\dot{1}2\dot{3}$
$\quad\quad x = 0.\dot{1}2\dot{3}$
$\quad 999x = 123$
$\quad\quad x = \frac{123}{999} = \frac{41}{333}$

The dots tell you that the group of digits, 123, recurs.

1 Write the decimal equivalents of these fractions.

a $\frac{1}{2}$ b $2\frac{2}{5}$ c $\frac{3}{20}$ d $\frac{1}{25}$

2 Use a written method to convert these fractions to decimals.
Show your working.

a $\frac{3}{8}$ b $\frac{4}{5}$ c $\frac{1}{16}$ d $\frac{3}{25}$

3 Use a calculator to check your answers to questions **1** and **2**.

4 Write these fractions as recurring decimals, using the 'dot' notation.
Show your working – do not use a calculator.

a $\frac{1}{3}$ b $\frac{2}{3}$ c $\frac{1}{6}$ d $\frac{1}{9}$ e $\frac{5}{6}$

5 Write each of the fractions $\frac{1}{7}$, $\frac{2}{7}$, $\frac{3}{7}$, $\frac{4}{7}$, $\frac{5}{7}$ and $\frac{6}{7}$ as a recurring decimal,
using 'dot' notation. Describe any patterns that you see in your
results.

6 Convert these decimals to fractions.

a 0.5   b 0.3   c 0.75   d 0.95   e 0.65

7 Convert each of these recurring decimals to a fraction in its simplest
form. Show your working.

a $0.1\dot{1}$ b $0.2\dot{2}$ c $0.\dot{1}\dot{5}$ d $0.\dot{1}2\dot{5}$ e $0.\dot{2}1\dot{6}$

8 Write these decimals as fractions. Show your working.

a $0.2\dot{1}$ b $0.7\dot{2}$ c $0.8\dot{2}\dot{7}$ d $0.6\dot{3}2\dot{1}$ e $0.81\dot{7}\dot{5}$

9 a Which one of these is a recurring decimal?

$\frac{18}{25}$ $\frac{19}{20}$ $\frac{8}{11}$ $\frac{9}{18}$

b Write $\frac{7}{9}$ as a recurring decimal.

c You are told that $\frac{1}{54} = 0.0\dot{1}8\dot{5}$

Write $\frac{4}{54}$ as a recurring decimal.

10 a Prove that $0.\dot{5}\dot{7} = \frac{57}{99}$.

b Hence, or otherwise, write the decimal number $0.3\dot{5}\dot{7}$ as a fraction.

Fractions, decimals and percentages

This spread will show you how to:

- Convert between fractions, decimals and percentages
- Order fractions, decimals and percentages

Keywords
Ascending
Denominator
Numerator
Order

- To convert a decimal to a percentage, multiply by 100%.

  $0.65 = 0.65 \times 100\% = 65\%$

- To convert a percentage to a decimal, divide by 100.

  $18.\dot{3}\% = 18.\dot{3} \div 100 = 0.18\dot{3}$

- To convert a fraction to a percentage, divide the **numerator** by the **denominator**, and then multiply by 100%.

  $\frac{5}{8} = 5 \div 8 \times 100\% = 0.625 \times 100\% = 62.5\%$

Example

a Convert these decimals to percentages.

  i 0.74   **ii** 1.315   **iii** $0.\dot{8}$ **iv** $0.2\dot{8}1\dot{5}$

b Convert these percentages to decimals.

  i 67.5%   **ii** 255%   **iii** 0.1%

c Convert these fractions to percentages.

  i $\frac{3}{7}$ **ii** $\frac{4}{9}$ **iii** $\frac{9}{40}$ **iv** $\frac{13}{25}$

a i $0.74 = 0.74 \times 100\% = 74\%$ **ii** $1.315 = 1.315 \times 100\% = 131.5\%$

  iii $0.\dot{8} = 0.\dot{8} \times 100\% = 88.\dot{8}\% = 88.9\%$ to 1 dp

  iv $0.2\dot{8}1\dot{5} = 0.2815815... \times 100\% = 28.158158...\% = 28.2\%$ to 1 dp

b i $67.5\% = 0.675$ **ii** $255\% = 2.55$ **iii** $0.1\% = 0.001$

c i $\frac{3}{7} = 3 \div 7 = 0.\dot{4}2857\dot{1} = 0.\dot{4}2857\dot{1} \times 100\% = 42.\dot{8}5714\dot{2}\% = 43\%$

  ii $\frac{4}{9} = 4 \div 9 = 0.\dot{4} = 0.\dot{4} \times 100\% = 44.\dot{4}\% = 44\%$

  iii $\frac{9}{40} = 9 \div 40 = 0.225 = 0.225 \times 100\% = 22.5\% = 23\%$

  iv $\frac{13}{25} = 13 \div 25 = 0.52 = 0.52 \times 100\% = 52\%$

In part **c**, answers are given to the nearest whole number.

Example

Write these quantities in **ascending** order.

a $\frac{7}{20}, \frac{3}{8}, \frac{1}{5}, \frac{3}{10}$ **b** $35\%, \frac{2}{7}, \frac{4}{15}, 0.347$

a Rewrite the list as decimals: 0.35, 0.375, 0.2, 0.3

  The original list, in ascending order, is: $\frac{1}{5}, \frac{3}{10}, \frac{7}{20}, \frac{3}{8}$

b The decimal equivalents are 0.35, $0.\dot{2}8571\dot{4}$, $0.2\dot{6}$, 0.347

  The original list, in ascending order, is: $\frac{4}{15}, \frac{2}{7}, 0.347, 35\%$

1 Convert these decimals to percentages.

    a 0.35   **b** 0.607   **c** 0.995   **d** 1.00

    e 2.15   **f** 0.00056   **g** 17   **h** 0.101

2 Convert these recurring decimals to percentages.

    a $0.5\dot{5}$ **b** $0.3\dot{4}$ **c** $0.7\dot{5}$ **d** $0.51\dot{2}\dot{8}$

    e $0.43\dot{7}$ **f** $0.83\dot{8}$ **g** $0.\dot{8}3\dot{8}$ **h** $1.0\dot{5}$

3 Convert these percentages to decimals.

    a 22%   **b** 18.5%   **c** $55.5\dot{5}\%$ **d** $35.\dot{5}\%$

    e $6.\dot{5}\dot{6}\%$ **f** $61.\dot{4}\dot{9}\%$ **g** $54.\dot{4}\dot{6}\%$ **h** $152.\dot{2}\%$

4 Convert these fractions to percentages. You should be able to do these without a calculator.

    a $\frac{4}{5}$ **b** $\frac{3}{20}$ **c** $\frac{7}{8}$ **d** $\frac{3}{4}$

    e $\frac{7}{25}$ **f** $\frac{3}{16}$ **g** $\frac{9}{20}$ **h** $\frac{7}{50}$

5 Convert these fractions to percentages. You may use a calculator.

    a $\frac{2}{3}$ **b** $\frac{1}{15}$ **c** $\frac{3}{7}$ **d** $\frac{5}{6}$

    e $\frac{7}{9}$ **f** $\frac{4}{11}$ **g** $\frac{1}{12}$ **h** $\frac{2}{15}$

6 Rewrite each set of fractions with a common denominator.
    Then use your answers to rewrite each list in ascending order.

    a $\frac{2}{3}, \frac{1}{2}, \frac{3}{5}, \frac{1}{6}$ **b** $\frac{3}{8}, \frac{1}{12}, \frac{7}{24}, \frac{1}{2}, \frac{1}{3}$ **c** $\frac{3}{4}, \frac{1}{20}, \frac{3}{5}, \frac{7}{40}, \frac{5}{8}$ **d** $\frac{7}{36}, \frac{4}{9}, \frac{1}{2}, \frac{2}{3}, \frac{3}{4}, \frac{5}{18}$

7 Rewrite each list of fractions as a list of equivalent decimals.
    Use your answers to rewrite each list in descending order.

    a $\frac{2}{5}, \frac{1}{4}, \frac{3}{8}, \frac{4}{10}$ **b** $\frac{3}{10}, \frac{1}{3}, \frac{2}{9}, \frac{2}{5}, \frac{3}{11}$ **c** $\frac{2}{7}, \frac{7}{20}, \frac{1}{5}, \frac{4}{9}, \frac{1}{3}$ **d** $\frac{4}{13}, \frac{3}{11}, \frac{1}{3}, \frac{5}{9}, \frac{3}{8}, \frac{2}{7}$

8 Rewrite these sets of numbers in ascending order.
    Show your working.

    a 33.3%, 0.33, $33\frac{1}{3}\%$ **b** 0.45, 44.5%, 0.454, $0.\dot{4}$

    c $0.2\dot{3}$, 0.232, 22.3%, 23.22%, 0.233

    d $\frac{2}{3}$, 0.66, $0.6\dot{5}$, 66.6%, 0.6666

    e $\frac{1}{7}$, 14%, 0.142, $\frac{51}{350}$, $14.\dot{1}\%$ **f** 86%, $\frac{5}{6}$, $0.8\dot{6}$, 0.866, $\frac{6}{7}$

9 Jodie says that the recurring decimal $0.\dot{9}$ is a little smaller than 1.
    Abby says that $0.\dot{9}$ is equal to 1. Who is correct? Explain your
    reasoning.

This spread will show you how to:

- Find percentages of quantities
- Calculate percentage increases and decreases

Keywords
Percentage
  decrease
Percentage
  increase

- To work out 72% of a quantity, multiply by 0.72.

  72% of £18 = 0.72 × £18 = £12.96

0.72 is the decimal equivalent of 72%.

- You can find a **percentage increase** in the same way. For example, to find a 2% increase, multiply by 1.02.

  10 million increased by 2% = 1.02 × 10 000 000

  = 10 200 000

The new amount will be 102% of the original amount.

- Use the same method to find a **percentage decrease**. For example, to find a 15% decrease, multiply by 0.85.

  £50 reduced by 15% = 0.85 × £50 = £42.50

The new amount will be 85% of the original amount.

Example

What decimal equivalent would you use to find

a 37.5% of a quantity   b a 12% increase   c a 42% decrease?

a 37.5% = 0.375
b For a 12% increase, multiply by 1.12.
c For a 42% decrease, multiply by 0.58.

$100\% - 42\% = 58\%$

Example

a Increase 500 g by 25%.   b Reduce £265 by $\frac{1}{3}$.

a 500 g × 1.25 = 625 g
b The new price is $\frac{2}{3}$ or $66\frac{2}{3}\%$ of the original price.

  $265 × 0.\dot{6} = £176.67$ to the nearest penny

The new mass is 125% of the original mass.
125% = 1.25

$66\frac{2}{3}\% = 0.\dot{6}$

Example

A 740 ml bottle of shampoo costs £2.45. In a special offer the bottle size is increased by 15% and the price is reduced by 15%.

Find the original and the new cost per litre of the shampoo.

Original cost per litre $= \frac{£2.45}{0.74} = £3.31$ per litre
  Size of new bottle = 740 × 1.15 = 851 ml
  New price = £2.45 × 0.85 = £2.08
New cost per litre $= \frac{2.08}{0.851} = £2.45$ per litre

1 Use a mental method to find these amounts.

  a 25% of £48   **b** 40% of 600 m   **c** 90% of 58 kg

  d 5% of 3.60 km   **e** 110% of €64.00   **f** 30% of 60 seconds

  g 105% of 82 cm   **h** 95% of 36 grams

2 **a** Write a decimal equivalent for each of the percentages in question **1**.

  b Use a calculator to check your answers to part **a**.

3 Write a decimal equivalent that you would multiply by, to find these percentages of a quantity.

  a 30%   **b** 32%   **c** 32.5%

  d 1.25%   **e** 112%   **f** 0.006%

4 Write the decimal multipliers for these percentage increases.

  a 15%   **b** 2.5%   **c** 22.5%

  d 87.5%   **e** 108%   **f** 0.045%

5 Write the decimal multipliers for these percentage decreases.

  a 5%   **b** 8.25%   **c** 22.75%

  d 38.25%   **e** 98%   **f** 100%

6 The price of petrol is increased by 10%.

  a Write the decimal multiplier equivalent to 10%.

  This new price is then increased by a further 10%.

  b Work out the decimal equivalent you would use to find, in one step, the result of the two successive 10% increases.

7 Find the single decimal equivalent for these percentage changes.

  a A 5% increase, followed by a 6% increase.

  b An 8% increase, followed by a 10% decrease.

  c A 22% decrease, followed by a 30% increase.

8 Mandy buys a sweater over the internet. She has to add VAT at 17.5% to the price shown. Her loyalty club card entitles her to a 7.5% discount on the order.

  • Mandy's Dad tells her to deduct the discount before adding the VAT.

  • Her Mum tells her it would be better to add the VAT first, then subtract the discount.

What should Mandy do? Explain your answer.

This spread will show you how to:

● Solve reverse percentage problems

● To find one quantity as a percentage of another, simply divide one quantity by the other, and then multiply by 100%.

> . If 36 people out of 400 wear contact lenses, this represents $(36 \div 400) \times 100\% = 9\%$.

● To find the original amount before a percentage change, first decide what multiplication was needed to find the new amount.

For example

> A train fare is £31.72 after a **22%** increase, so the original fare was multiplied by **1.22**.

New price is 122% of original price.

> If a freshly-baked cake weighs 2200 grams, and this is a **12%** reduction on the uncooked weight, then the uncooked weight was multiplied by **0.88**.

Cooked weight is 88% of uncooked weight.

● To find the original amount before the percentage change, simply 'undo' the multiplication by dividing the new amount by the appropriate number.

For example

This is a **reverse percentage**.

> The original train fare = $31.72 \div 1.22 = £26.00$.

> The uncooked weight of the cake = $2200 \div 0.88 = 2500$ g.

Example

a A shop buys DVD recorders at a wholesale price of £85 each, and sells them at £155. Find the percentage profit.
b A computer costs £998.75 including VAT at 17.5%. Find the cost before VAT.
c The price of a dress is reduced by 30% in a sale. If the sale price is £77, find the original price.

a Profit = £155 − £85 = £70.
Profit as a percentage of the wholesale price = $\frac{70}{85} \times 100\% = 82.4\%$
b Original cost × 1.175 = £998.75
$\Rightarrow$ original cost = £998.75 ÷ 1.175 = £850
c Original price × 0.7 = £77
$\Rightarrow$ original price = £77 ÷ 0.7 = £110

Example

A restaurant sells a bottle of wine for £10.35, making a profit of 130%. How much did the restaurant pay for the wine?

The restaurant adds 130% to the original cost, so the selling price is 230% of the original cost.

Original cost × 2.3 = £10.35
$\Rightarrow$ original cost = £10.35 ÷ 2.3 = £4.50

1 Find

   a 20 as a percentage of 800   **b** 25 as a percentage of 750

   c 16 as a percentage of 29   **d** 22 as a percentage of 1760.

2 A suit is normally sold for £350. In a sale, the price is reduced to £205. What percentage reduction is this?

3 Copy and complete the table to show the percentage reduction for each original price.

Original price	Sale price	Percentage reduction
£25.00	£21.00	
£37.50	£32.75	
£1500.00	£950.50	
£2850.75	£2200.00	

4 A car manufacturer increases the price of a Sunseeker sports car by 6%. The new price is £8957. Calculate the price before the increase.

5 During 2005 the population of Camtown increased by 5%.
At the end of the year the population was 14 280.
What was the population at the beginning of the year?

6 In a clearance sale at a jewellery shop, the price of all unsold items is reduced by 10% at the end of each day.

   a A brooch is on sale for £250 on Monday morning. It is still unsold on Wednesday evening. Find the sale price of the brooch on Thursday morning.

   b A pair of earrings is on sale for £280.67 on Thursday morning. Find the price of the earrings on the previous Monday morning.

7 The Retail Price Index (RPI) is a government statistic that shows the percentage increase in the price of a representative 'basket' of goods and services over the last 12 months. The table shows the RPI for the month of July for selected years.

Year	1975	1976	1977	1978	1979	1980
July RPI	26.3	13.8	17.7	7.4	11.4	21.0

   a The total household expenditure for one family in July 1975 was £104.75. Estimate the same family's expenditure in July 1974.

   b Another family had a monthly expenditure of £216.50 in July 1977. Estimate their monthly expenditure in July 1980. Show your working, and state any assumptions you have made.

   c A third family had a monthly expenditure of £198.45 in July 1979. Estimate their monthly expenditure in July 1976.

Exam review

Key objectives

- Distinguish between fractions with denominators that have only prime factors of 2 and 5, and other fractions
- Draw on knowledge of operations and inverse operations, and of methods of simplification in order to select and use suitable strategies and techniques to solve problems and word problems

1 a Convert $\frac{5}{25}$ to a

    i decimal

    ii percentage. (3)

  b Convert $0.4\dot{5}$ to a

    i fraction in its simplest form

    ii percentage. (4)

2 A garage sells cars.

It offers a discount of 20% off the normal price for cash.

Dave pays £5200 cash for a car.

Calculate the normal price of the car. (3)

(Edexcel Ltd., 2003)

This unit will show you how to

- Recall and use properties of lines and angles including parallel lines
- Calculate and use interior and exterior angles in polygons
- Use the correct vocabulary to describe parts of a circle
- Know and use the theorems for angles in circles
- Know and use the theorems for tangents at a point on a circle

Before you start ...

You should be able to answer these questions.

Review

1 Draw circles to show
  a a sector **b** a segment **c** a chord.

Unit S1

2 Work out the missing angles.

CD – S2

a   **b**

c   **d**

3 Work out the missing angles in these triangles.

CD – S2

a 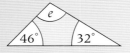   **b**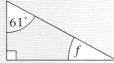

4 Work out the missing angles in these quadrilaterals.

CD – S2

a 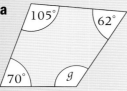   **b**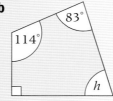

This spread will show you how to:

- Use parallel lines, alternate angles, corresponding angles and interior angles

Keywords

Alternate
Corresponding
Interior
Parallel
Supplementary
Vertically
  opposite

- Angles are formed when two lines cross, as at this crossroads.

$$a = c \quad \text{and} \quad b = d$$

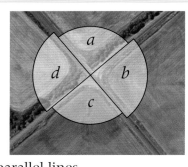

- **Vertically opposite** angles are equal.

Parallel lines are lines that never cross.

You need to remember these angle facts for parallel lines.

- **Alternate** angles are equal.

These are sometimes called Z angles.

- **Corresponding** angles are equal.

These are sometimes called F angles.

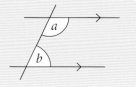

- **Interior** angles are **supplementary**. $a + b = 180°$

Supplementary angles add up to 180°.

When you work out angles, you should always say which angle fact you are using.

Example

Find the missing angles. Give reasons for your answers.

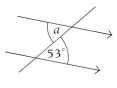

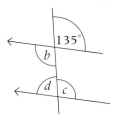

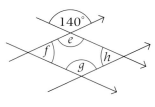

$a = 53°$ alternate angles

$b = 135°$ vertically opposite
$c = 135°$ corresponding
$d = 45°$ interior angles

$e = 140°$ vertically opposite
$f = 40°$ interior angles
$g = 140°$ interior angles
$h = 40°$ interior angles

1 Copy the diagrams.

   i Use colour to show alternate angles in each diagram.

   ii Use another colour to show corresponding angles in each diagram.

You may be able to find more than one pair in each diagram.

a    **b**   **c**

2 Find the missing angles in each diagram.
Write down which angle fact you are using each time.

a 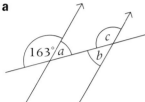   **b** 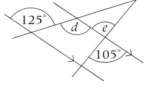   **c**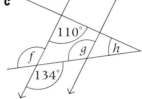

3 a Angles x, y and z are on a straight line.
Write down the value $x + y + z$.

  b Use alternate angles to work out the two missing angles in the triangle.

  c Use your answers to **a** and **b** to show that angles in a triangle add up to $180°$.

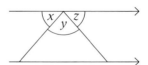

4 a Work out the missing angles in this diagram.

  b Describe the quadrilateral formed between the pairs of parallel lines.

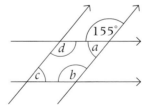

5 Work out the missing angles.

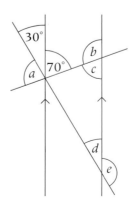

This spread will show you how to:

● Calculate and use the interior and exterior angles of polygons

Keywords
Exterior angle
Interior angle
Polygon
Vertex

A **polygon** is a closed shape with three or more straight sides.

● A regular polygon has all sides the same length and all interior angles equal.

The **interior angles** are inside the polygon.

The **exterior angles** are made by extending each side in the same direction.
Exterior angles are outside the polygon.

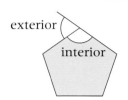

exterior
interior

A square is a regular quadrilateral.

● For any polygon the exterior angle sum = 360°.
● For a regular polygon with n sides, each exterior angle = 360° ÷ n.

You can divide any polygon into triangles by drawing diagonals from a **vertex** (corner).

The number of triangles is always two less than the number of sides.

A pentagon divides into 3 triangles.
Angles in a triangle add up to 180°.
So interior angle sum of a pentagon = $3 \times 180 = 540°$

● For a polygon with n sides the interior angle sum = $(n - 2) \times 180°$.

Example

In a regular octagon find

a an interior angle   **b** an exterior angle.

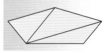

a An octagon has 8 sides and divides into 6 triangles.
Interior angle sum is $6 \times 180° = 1080°$
Each interior angle is $1080° \div 8 = 135°$.

b Each exterior angle is $360° \div 8 = 45°$.

Example

An irregular hexagon has angles 108°, 92°, 120°, 134°, 115° and x.

Find the size of angle x.

Any hexagon has 6 sides and divides into 4 triangles.

Sum of interior angles is $4 \times 180° = 720°$

Sum of the 5 given angles is
$108° + 92° + 120° + 134° + 115° = 569°$

So $x = 720° - 569° = 151°$

1 a Copy the table of regular polygons.

b Draw each polygon and divide it into triangles by drawing diagonals from a vertex.

c Complete the table.

Shape	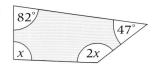					
Number of sides						
Number of triangles the shape splits into						
Sum of the interior angles in the shape						
Size of one interior angle						
Size of one exterior angle						

2 Find the missing angles in these quadrilaterals.

a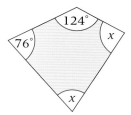

b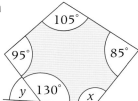

c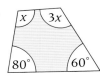

3 Find the missing angles in these irregular polygons.

a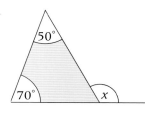

b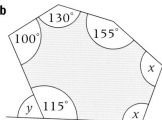

4 a Find the exterior angle marked *x* in each triangle.

i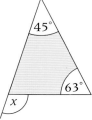

ii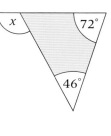

iii

b Use your answers to part **a** to help you copy and complete this statement.
The exterior angle of a triangle is equal to_____

5 Use a diagram to explain why the interior angle sum of a quadrilateral is 360°.

Angle theorems in circles

This spread will show you how to:

- Use the correct vocabulary to describe parts of a circle
- Prove and use the theorems for angles in circles

Keywords

Arc
Chord
Diameter
Radii
Radius
Segment

You need to know parts of a circle.

A straight line joining two points on the circumference is a **chord**.

A chord divides a circle into two **segments**.

The part of the circumference that joins two points is an **arc**.

The longest chord is the **diameter**.

You need to be able to prove and use these circle theorems.

Proof

Prove the angle at the centre is twice the angle at the circumference from the same arc.

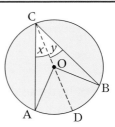

Draw the radius OC and extend it to D.
In △ AOC, AO = OC (both radii)
      ∠OAC = ∠OCA = x (isosceles triangle)
      ∠COA = 180° − 2x
      ∠AOD = 2x (angles on a straight line)

Similarly using △ COB you can prove that
      ∠DOB = 2y

Now ACB = $x + y$ and AOB = 2x + 2y as required.

Proof

Prove angles from the same arc in the same segment are equal.

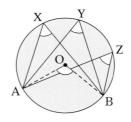

∠AOB = 2 × ∠AXB (angle at centre is twice
∠AOB = 2 × ∠AYB angle at circumference)
∠AOB = 2 × ∠AZB
so ∠AXB = ∠AYB = ∠AZB as required.

Example

Find the missing angles. Give reasons.

a

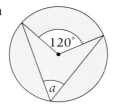

b

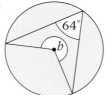

c

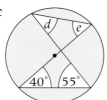

a $a = 60°$
     (angle at centre is
     twice angle at
     circumference)

b   Obtuse angle at
     centre = 128°
     $b = 360° − 128° = 232°$
     (angles at a point)

c $d = 40°$, $e = 55°$
     (angles on same arc
     are equal)

Work out the missing angles.
Give a reason for each answer.

1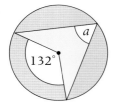

2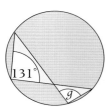

3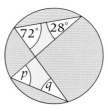

4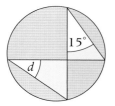

5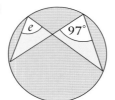

6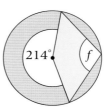

7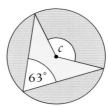

8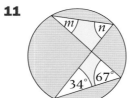

9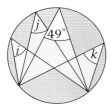

10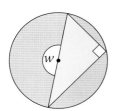

11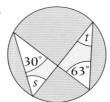

12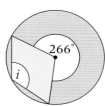

13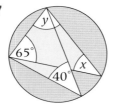

14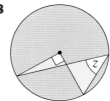

15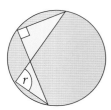

16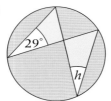

17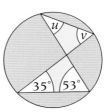

18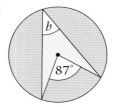

More angle theorems in circles

This spread will show you how to:

- Use the correct vocabulary to describe parts of a circle
- Prove and use the theorems for angles in circles

Keywords

Cyclic
   quadrilateral
Semicircle

You need to be able to prove and use two more circle theorems.

Proof

Prove an angle in a **semicircle** is a right angle.

Angle at the centre = 180°
Angle at centre = 2 × angle at circumference
so angle at circumference = 90°

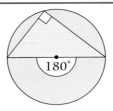

- All four vertices of a **cyclic quadrilateral** lie on the circumference of a circle.

Proof

Prove opposite angles of a cyclic quadrilateral add up to 180°.

Draw radii OA and OC.
Reflex ∠AOC = 2x
Obtuse ∠AOC = 2y

2x + 2y = 360° (angles at a point)
x + y = 180° as required.

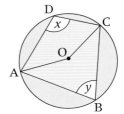

Example

Find the missing angles. Give reasons.

a

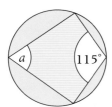

b

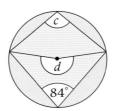

c

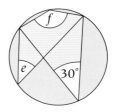

a   a = 65° (opposite angles of cyclic quadrilateral = 180°)

b   c = 96° (opposite angles of cyclic quadrilateral)

d = 2 × 96° = 192° (angle at centre = 2 × angle at circumference)

c   e = 30° (angles on same arc are equal)

f = 150° (opposite angles of cyclic quadrilateral)

Work out the missing angles.
Explain how you worked out each answer.

1
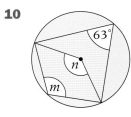

2
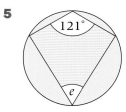

3
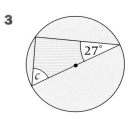

4
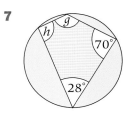

5
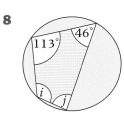

6
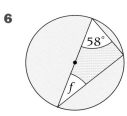

7
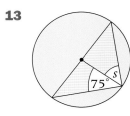

8
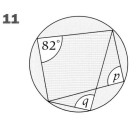

9
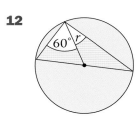

10
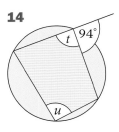

11
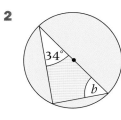

12
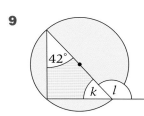

13
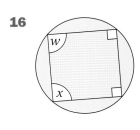

14
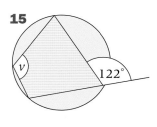

15
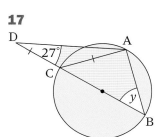

16
17
18

Tangents to circles

This spread will show you how to:

- Use the correct vocabulary to describe parts of a circle
- Know and use the theorems for tangents at a point on a circle

Keywords
Chord
Perpendicular
Radius
Tangent

You need to know some facts about **tangents** to circles.

- The angle between a tangent and the **radius** at the point where the tangent touches the circle is a right angle.

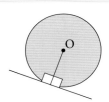

- Two tangents drawn from a point to a circle are equal.

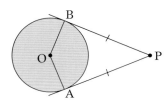

- The **perpendicular** line from the centre of a circle to a **chord** bisects the chord.

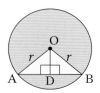

Example

Find the missing angles and lengths. Give reasons.

a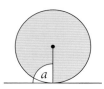

b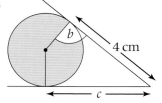

c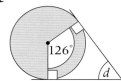

a $a = 90°$ (angle between tangent and radius)

b $b = 90°$ (angle between tangent and radius)

$c = 4$ cm (tangents from a point)

c $d + 90° + 90° + 126° = 360°$ (angles of quadrilateral)

$d = 54°$

Work out the missing angles and lengths.
Give a reason for each answer.

You may need to use other circle theorems or Pythagoras's theorem.

1
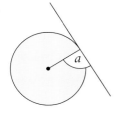

2
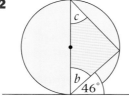

3
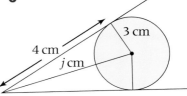

4
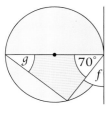

5
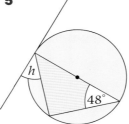

6

7
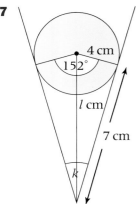

8
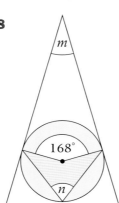

9
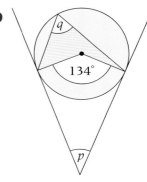

10
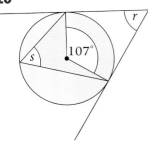

11
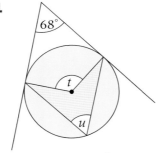

12

13
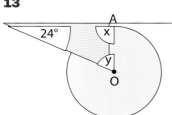

14
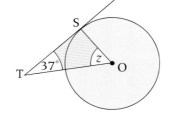

Alternate segment theorem

This spread will show you how to:

- Know and use the theorems for tangents at a point on a circle

Keywords
Alternate
    segment
Chord
Segment
Tangent

You need to know the alternative segment theorem.

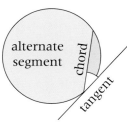

The **chord** divides the circle into two **segments**.
The acute angle between the tangent and the chord is partly in the
minor segment.
The major segment is the **alternate segment**.

Proof

- Prove the angle formed between a **tangent** and a **chord** is equal to
  the angle from that chord in the **alternate segment** of the circle.

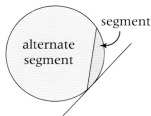 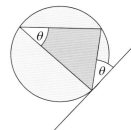

Draw a diameter where the tangent touches the circle (T).
Join this diameter (TX) to the other end of the
chord (P) to form a triangle PTX.
∠XPT = 90° (angle in a semicircle)

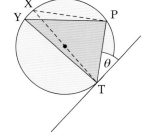

The angle between tangent and radius is a
right-angle, so ∠PTX = 90° − θ.

∠TXP + 90° + (90° − θ) = 180°
(angles in a triangle add up to 180°)
so ∠TXP = 180° − 180° + θ = θ

Angles from the same arc in the same **segment** are equal, so the angle in
the alternate segment, ∠PYT = θ.

Example

Find the missing angles. Give reasons.

a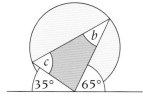

b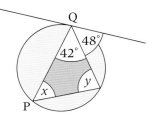

Since $y = 90°$, PQ
must be a diameter.

a $b = 35°$ (alternate segment)
   $c = 65°$ (alternate segment)

b $x = 48°$ (alternate segment)
   $y = 180° − 42° − 48°$ (angles in triangle)
   $= 90°$

Work out the missing angles. Explain how you worked out each answer.
You may need to use angles in circle theorems from the previous pages.

1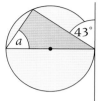

43°
a

2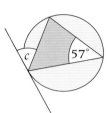

b
72°

3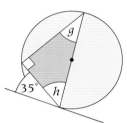

c
57°

4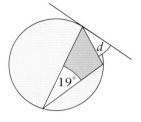

d
19°

5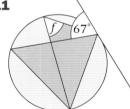

50°
e
f

6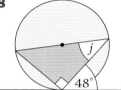

g
35°
h

7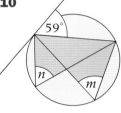

i
64°

8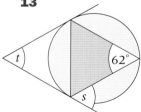

j
48°

9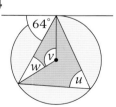

l
k
36°

10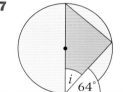

59°
n
m

11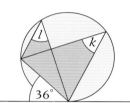

f
67°

12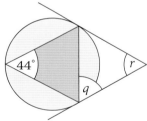

44°
r
q

13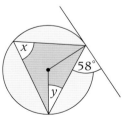

t
62°
s

14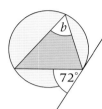

64°
v
w
u

15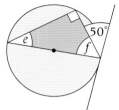

x
58°
y

16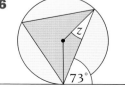

z
73°

Geometrical proof

This spread will show you how to:

● Derive proofs using the circle and tangent theorems

Keywords

Arc
Centre
Circumference
Cyclic
    quadrilateral

You can use circle theorems to explain and prove geometrical facts.

Proof

Prove that ABCD is a **cyclic quadrilateral**.

∠B and ∠D are opposite angles of the
quadrilateral ABCD.
∠B + ∠D = 57° + 123° = 180°
So ABCD is a cyclic quadrilateral.

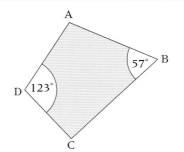

You know that in a
cyclic quadrilateral,
opposite angles
add up to 180°,
so look at the
opposite angles of
ABCD.

Example

PQR is a triangle drawn inside a circle.
Angle PQR = 100°.
Is PR a diameter?

If PR were a diameter, angle PQR would be
an angle in a semicircle, which is 90°.
Angle PQR ≠ 90°, so PR is not a diameter.

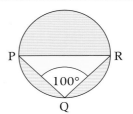

Suppose that PR is
a diameter, and
see if the rest of
the information
fits.
As it does not, your
supposition must
be wrong.

Example

In the diagram AB = ST.

Prove   **a** ∠AOB = ∠SOT.
        b Angles at the circumference from
           equal arcs are equal.

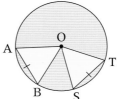

a AO = BO = SO = TO (radii)
   AB = ST (given)
   So △AOB ≡ △SOT (SSS) and ∠AOB = ∠SOT

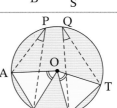

(given) means a
fact you have been
told.

b Arcs AB and ST are equal.
   ∠AOB = 2 × ∠APB   (angle at centre is twice angle at
                         circumference)
   ∠SOT = 2 × ∠SQT
   So ∠APB = ∠SQT

1 In the quadrilateral PQRS ∠PQR = 38° and ∠RSP = 138°.
Explain why PQRS cannot be a cyclic quadrilateral.

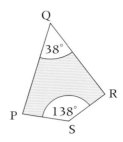

2 ABCD is a parallelogram and it is also a cyclic quadrilateral.
Use circle theorems and properties of quadrilaterals to explain which special type of quadrilateral ABCD could be.

3 D, E and F are three points on the circumference of a circle of radius 68 mm.
DE = 120 mm and EF = 64 mm.

Show that DF is a diameter of the circle.

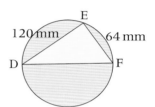

4 Explain why C is not the centre of the circle.

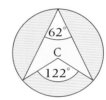

5 PQRS is a cyclic quadrilateral.
X lies outside the circle such that XQR and XPS are straight lines.
XQ and XP are equal in length.

Use the 'angles in cyclic quadrilaterals' theorem to show that PQRS is an isosceles trapezium.

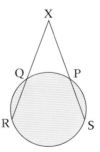

6 X and Y lie on the circumference of a circle with centre O.
P is a point outside the circle.
Angle PXY = 62°
Reflex angle XOY = 256°

Explain why PX is not a tangent to the circle.

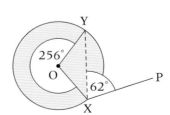

Key objectives

- Recall the definition of a circle and the meaning of related terms, including chord, tangent, arc, sector and segment

- Use the facts that the angle subtended by an arc at the centre of a circle is twice the angle subtended at any point on the circumference, the angle subtended at the circumference by a semicircle is a right angle, angles in the same segment are equal, and opposite angles of a cyclic quadrilateral sum to 180 degrees

- Understand that the tangent at any point on a circle is perpendicular to the radius at that point. Understand and use the fact that tangents from an external point are equal in length

- Use the alternate segment theorem

1 The diagram shows a circle with centre O.
VW is a tangent to the circle.
A, B, and C are points on the circumference of the circle.
Angle ABW is 72°.

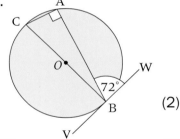

   a Calculate the size of angle ABC.   (2)
     Explain your answer.

   b Hence calculate the size of the angle ACB.   (2)
     Explain your answer.

2 The diagram shows a circle, centre O.
AC is a diameter.
Angle BAC = 35°.
D is the point on AC such that angle BDA is a right angle.

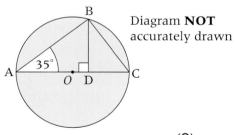

Diagram **NOT** accurately drawn

   a Work out the size of angle BCA.   (2)
     Give reasons for your answer.

   b Calculate the size of angle DBC.   (1)

   c Calculate the size of angle BOA.   (2)

(Edexcel Ltd., 2004)

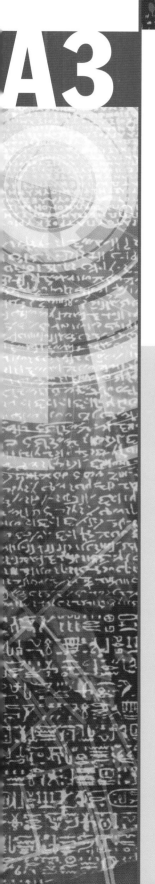

This unit will show you how to

- Generate sequences from a general term
- Describe the behaviour of a sequence, using the relevant vocabulary
- Find the nth term of a linear sequence
- Justify the general term of a linear sequence by looking at the context from which it arises
- Rearrange quadratic expressions
- Solve quadratic equations by factorisation or by using the quadratic formula

Before you start ...

You should be able to answer these questions.

Review

1 Find the next two terms of each sequence.

   a 2, 5, 9, 14, ...   **b** 2, 4, 8, 16, ...

   c $\frac{1}{1}, \frac{1}{4}, \frac{1}{9}, \frac{1}{16}$, ...   **d** 1, 8, 27, 64, ...

   e 5, 7, 12, 19, ...   **f** 3, −9, 27, −81, ...

CD – A3

2 Given that $n = 4$, find the value of each expression.

   a $5n - 4$ **b** $2n^2$

   c $n^3 - 2$ **d** $(5 - n)^3$

   e $(-1)^n$

Unit A2

3 Factorise each of these expressions.

   a $x^2 - 5x$ **b** $x^2 + 10x + 21$

   c $x^2 - 25$ **d** $y^2 - 6y + 9$

   e $p^2 - 100$ **f** $3ab - 9a^2$

   g $16m^2 - 49$ **h** $3x^2 + 7x + 2$

Unit A1

4 Given that $s = ut + \frac{1}{2}at^2$, find:

   a s when $u = 8$, $t = 5$ and $a = -4$

   b u when $s = 100$, $t = 4$ and $a = -2$

Unit A2

This spread will show you how to:

- Generate common sequences and describe how number patterns are formed

Keywords
Sequence
Term

- A **sequence** is a set of numbers that follow a pattern, for example

  5, 9, 13, 17, 21, … are the first five **terms** of a sequence that goes up in 4s

  3, 6, 12, 24, 48, … are the first five terms of a sequence that doubles

  1, 4, 9, 16, 25, … is the sequence of square numbers
      (1 × 1, 2 × 2, 3 × 3 and so on)
  1, 8, 27, 64, 125, … are the cube numbers
      (1 × 1 × 1, 2 × 2 × 2, 3 × 3 × 3 and so on)

Examiner's tip
The techniques learned in this unit will be useful for your Using and Applying Mathematics coursework task.

You can find sequences in patterns, for example,

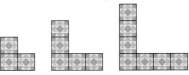

Number of tiles is:
3, 5, 7, …
The next diagram would need 9 tiles.

Example

Write the next two terms in each sequence.

a 4, 5, 7, 10, 14, 19, …   **b** 0.6, 0.7, 0.8, 0.9, …

a 4, 5, 7, 10, 14, 19, … The terms increase by 1, then 2, then 3. so the next two terms are 19 + 6 = 25 and 25 + 7 = 32.

b 0.6, 0.7, 0.8, 0.9, … The terms increase by 0.1. The next two terms are 0.9 + 0.1 = 1.0 and 1.0 + 0.1 = 1.1.

You may be tempted to continue 0.7, 0.8, 0.9, 0.10, … but 0.10 is less than 0.9!

Example

Name each of these sequences.

a 1, 3, 5, 7, 9, …   **b** 25, 36, 49, 64, …

a These are the odd numbers.   **b** These are the square numbers, starting at 5 × 5 (or 5^2).

Example

How many tiles would be in the tenth diagram in this pattern?

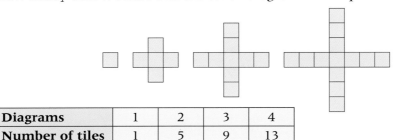

Diagrams	1	2	3	4
Number of tiles	1	5	9	13

A table is a good way of organising results so that you can see what is happening.

The number of tiles increases by 4 each time. If you continue this, you get 1, 5, 9, 13, 17, 21, 25, 29, 33, **37**.
The tenth diagram would have 37 tiles.

1 Copy each sequence and add the next two terms.

   a 4, 9, 14, 19, 24, ___, ___   **b** 100, 93, 86, 79, 72, ___, ___

   c 1, 2, 4, 7, 11, ___, ___   **d** 9, 99, 999, 9999, 99 999, ___, ___

   e 1, 1, 2, 3, 5, 8, ___, ___   **f** 54, 27, 13.5, 6.75, ___, ___

2 Copy and fill in the missing numbers in each sequence.

   a 4, ___, 10, ___, 16, ...   **b** 4, ___, ___, 32, 64, ...

   c 95, ___, ___, ___, 87, ...   **d** 1, ___, 27, 64, ___, ...

3 Write the first five terms of each of these well-known number patterns.

   a Multiples of 3   **b** Powers of 2

   c Prime numbers   **d** Square numbers over 100

4 The triangular numbers form a sequence.

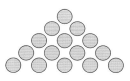

   a Copy the table and use the diagram to complete it.

Diagrams	1	2	3	4	5
Number of dots	1	3			

   b Hence, write the first 10 triangular numbers.

   c Why do you think the square numbers (1, 4, 9, 16, 25, ...) got their name?

   d How did the cube numbers get their name? Write the first five cube numbers.

5 Find the tenth term in each of these number patterns.

   a (1×2), (2×3), (3×4), (4×5), ...

   b $\frac{1}{2}$, $\frac{2}{3}$, $\frac{3}{4}$, $\frac{4}{5}$, ...

   c (5×2), (5×4), (5×8), (5×16), ...

   d (1×1), (4×8), (9×27), (16×64), ...

6 Look at this number pattern.

   a Write the next two lines in the pattern.

   b What is $66\ 666\ 666\ 667^2$?

   c What is $\sqrt{4\ 444\ 444\ 444\ 888\ 888\ 889}$?

$$7^2 = 49$$
$$67^2 = 4489$$
$$667^2 = 444\ 889$$

This spread will show you how to:

- Generate sequences from a general term
- Describe the behaviour of a sequence, using the relevant vocabulary

Keywords
Converge
Diverge
Limit
*n*th term
Oscillate

- You can generate a sequence from a general term.

If the general term $T_n = n^2 + 3$,
to find T_1 you substitute $n = 1$: $T_1 = 1^2 + 3 = 4$
to find T_5 you substitute $n = 5$: $T_5 = 5^2 + 3 = 28$

T_1 is the 1st term,
T_2 is the 2nd term,
T_n is the *n*th term.

- You can generate the terms of a sequence to observe its behaviour. For example:
  - 5, 8, 11, 14, 17, ... each term is larger than the one before. The sequence **diverges**.

  - 2, −4, 6, −8, 10, −12, ... the terms alternate between positive and negative. The sequence **oscillates**. The difference between terms is increasing, so the sequence diverges.

  - 1, $\frac{1}{2}$, $\frac{1}{3}$, $\frac{1}{4}$, $\frac{1}{5}$, ... each term is smaller than the one before, but will never be equal to zero. The sequence **converges** towards a **limit** of zero.

Example

For each of these sequences
i find the first five terms
ii describe its behaviour.

a $T_n = 5n^2$ **b** $T_n = (-2)^n$

a i $T_n = 5n^2$
  $T_1 = 5 \times 1^2 = 5$
  $T_2 = 5 \times 2^2 = 20$
  $T_3 = 5 \times 3^2 = 45$
  $T_4 = 5 \times 4^2 = 80$
  $T_5 = 5 \times 5^2 = 125$
  First five terms are 5, 20, 45, 80, 125
ii The sequence is diverging.

b i $T_n = (-2)^n$
  $T_1 = (-2)^1 = -2$
  $T_2 = (-2)^2 = 4$
  $T_3 = (-2)^3 = -8$
  $T_4 = (-2)^4 = 16$
  $T_5 = (-2)^5 = -32$
  First five terms are −2, 4, −8, 16, −32
ii The sequence is diverging and oscillating.

BIDMAS: calculate indices before multiplying by 5.

$(-2)^2 = (-2) \times (-2)$
$= +4$
$(-2)^3 = (-2) \times (-2)$
$\times (-2) = -8$ etc.
Positive terms are getting larger, negative terms are getting smaller.

1 Generate the first five terms of each of these sequences.

a $T_n = 7n - 2$ **b** $T_n = n^2 - 4$ **c** $T_n = n^3$

d $T_n = 3n^2 + 5$ **e** $T_n = n^2 + 2n + 1$ **f** $T_n = 10 - 3n$

g $T_n = (-n)^2$ **h** $T_n = (n+1)(n+2)$

2 Generate the first five terms of each of these sequences.
Hence, name the sequences.
For example, $T_n = 2n$ ... so you get 2, 4, 6, 8, 10, ... the even numbers.

a $T_n = 2n - 1$ **b** $T_n = 7n$ **c** $T_n = 2^n$

d $T_n = 10^n$ **e** $T_n = \dfrac{n(n+1)}{2}$ **f** $T_n = (11 - n)^2$

g $T_n = n$ **h** $T_n = T_{(n-1)} + T_{(n-2)}$ where $T_1 = 1$ and $T_2 = 1$

3 Choose as many of the following words as necessary to describe
these sequences. You will need to generate some terms of the
sequence first in order to observe how it is behaving.

> Convergent   Divergent   Oscillating   Has a limit

a $T_n = \dfrac{2n + 1}{2n}$ **b** $T_n = 3^n$ **c** $T_n = 2^{-n}$

d $T_n = \dfrac{1}{n + 1}$ **e** $T_n = \sin(90n)°$ **f** $T_n = n!$

In question 3f, $n!$ is
'n factorial.
You should find
this function on
your calculator.
Try to work out
what it does.

4 Match these sequences with the sketch graphs, where n
(term number) is displayed on the x-axis and T_n (term) is
displayed on the y-axis.

a $T_n = \dfrac{1}{n}$ **b** $T_n = 5^n$ **c** $T_n = 10 - n$ **d** $T_n = (n+1)(n-2)$

Try substituting
some values for n,
and see what
happens to T_n.

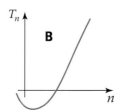

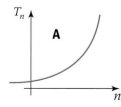

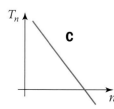

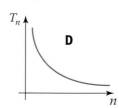

The *n*th term of a linear sequence

This spread will show you how to:

- Find the *n*th term of a linear sequence
- Justify the general term of a linear sequence by looking at the context from which it arises

Keywords
Arithmetic sequence
Linear sequence
*n*th term

- In an **arithmetic** or **linear sequence** the difference between successive terms is constant.

- You can find the *n*th term of a linear sequence by comparing the sequence to the multiples of its constant difference.

For example, for the sequence, 3, 10, 17, 24, 31, ...
Constant difference = 7, so compare to the multiples of 7.

Term number (*n*)	1	2	3	4	5
7*n*	7	14	21	28	35
Term	3	10	17	24	31

Each term is 4 less than 7*n*.
The *n*th term is $7n - 4$.

The terms go up or down by the same amount each time.
12, 21, 30, 39, 48, ... goes up in 9s.
20, 18, 16, 14, 12, ... goes down in 2s.

Multiples of 7 = 7*n*

- Sometimes you can find a rule from the pattern that generates the sequence, for example

To find the number of dots (*D*) in a square of side *L* you add the number of dots along the top and along the bottom ($L + L$) to the number of dots in the sides ($(L - 2) + (L - 2)$).
$D = 2L + 2(L - 2)$.

Thinking big will help you see the pattern. Number of dots in the 100th square would be: 100 across top + 100 across bottom + 98 down left + 98 down right = $2 \times 100 + 2 \times (100 - 2)$.

Example

Find the *n*th term of the sequence: 100, 95, 90, 85, 80, ...

Term number (*n*)	1	2	3	4	5
−5*n*	−5	−10	−15	−20	−25
Term	100	95	90	85	80

$T_n = 105 - 5n$

Constant difference is −5, so compare to the multiples of −5.

Each term is 105 more than the multiple of −5.

Example

Find the rule for this pattern.

Term 1: 3 matches
Term 2: 3 + 2 matches
Term 3: 3 + 2 × 2 matches
Term 4: 3 + 3 × 2 matches
Term 100: 3 matches + 99 × 2 matches
Term *n*: 3 + 2(*n* − 1) matches

Number of matches = $M = 3 + 2(n - 1) = 3 + 2n - 2 = 2n + 1$
so $M = 2n + 1$.

1 Find the *n*th term formula for each of these linear sequences.

    a 13, 16, 19, 22, 25, ...   **b** 2, 5, 8, 11, 14, ...

    c 25, 30, 35, 40, 45, ...   **d** 4, 5, 6, 7, 8, ...

    e 1, 1.2, 1.4, 1.6, 1.8, ...   **f** 50, 47, 44, 41, 38, ...

    g Counting down in 4s from 10

2 Explain the rule for each pattern.

    a

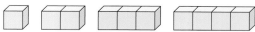

$G = 4L + 2$
G = number of green faces visible
L = length of strip

    b

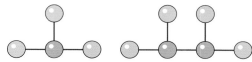

$A = H(H + 1)$
A = area
H = height of rectangle

    c

$B = R + 2$
B = number of blue beads
R = number of red beads

$L = 2R + 1$
L = number of links
R = number of red beads

3 By considering the sequence of patterns, write a formula to connect the quantities given.

    a

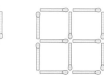

    Relate the number of edges E to the number of hexagons H.

    b

    Relate the number of matches M with the length of the square L.

> Account for the rows first, then the columns.

4 Imagine a $3 \times 3 \times 3$ cube made from unit cubes stuck together. The outside is painted black. If this is then dismantled, how many cubes have 0, 1, 2, 3, ... faces painted black? Repeat for an $n \times n \times n$ cube and an $m \times n \times p$ cuboid.

Solving quadratic equations

This spread will show you how to:
- Solve quadratic equations by factorisation

- **Quadratic** equations contain a squared term as the highest power, for example $2x^2 + 7x - 9 = 0$.

- Quadratic equations can have 0, 1 or 2 solutions.

For example	$x^2 = 100$	$x^2 = 0$	$x^2 = -25$
Solution(s)	$x = 10$ or -10	$x = 0$	Impossible
Number of solutions	2	1	0

- Many quadratic equations can be solved by:
  - rearranging so that one side equals zero
  - **factorising**.

Example

Solve $x^2 = 6x$.

$x^2 - 6x = 0$
$x(x - 6) = 0$
Either $x = 0$ or $(x - 6) = 0$
Either $x = 0$ or $x = 6$

Spot the x^2 term ... the equation is quadratic.

Rearrange so one side = zero.

Factorise.

Example

Solve **a** $x^2 + 5x + 6 = 0$ **b** $4x^2 = 81$.

a $x^2 + 5x + 6 = 0$
$(x + 3)(x + 2) = 0$
Either $x + 3 = 0$ so $x = -3$
or $x + 2 = 0$ so $x = -2$

b $4x^2 = 81$
$4x^2 - 81 = 0$
$(2x - 9)(2x + 9) = 0$
Either $2x - 9 = 0$ so $2x = 9$ so $x = 4\frac{1}{2}$
or $2x + 9 = 0$ so $2x = -9$ so $x = -4\frac{1}{2}$

If two expressions **multiply** to give zero, one of them must be zero.

This equation has two solutions.

Factorise – DOTS.

Example

Solve $5x^2 + 14x - 3 = 0$.

$5x^2 + 14x - 3 = 0$
$(5x - 1)(x + 3) = 0$

Either $5x - 1 = 0$ or $x + 3 = 0$
$x = \frac{1}{5}$ $x = -3$

1 First factorise these quadratic equations, by using the common factor, then solve them.

    a $x^2 - 3x = 0$ **b** $x^2 + 8x = 0$ **c** $2x^2 - 9x = 0$ **d** $3x^2 - 9x = 0$

    e $x^2 = 5x$ **f** $x^2 = 7x$ **g** $12x = x^2$ **h** $4x = 2x^2$

    i $6x - x^2 = 0$ **j** $9y - 3y^2 = 0$ **k** $0 = 7w - w^2$

2 Solve these quadratic equations by factorising into double brackets.

    a $x^2 + 7x + 12 = 0$ **b** $x^2 + 8x + 12 = 0$ **c** $x^2 + 10x + 25 = 0$

    d $x^2 + 2x - 15 = 0$ **e** $x^2 + 5x - 14 = 0$ **f** $x^2 - 4x - 5 = 0$

    g $x^2 - 5x + 6 = 0$ **h** $x^2 - 12x + 36 = 0$ **i** $2x^2 + 7x + 3 = 0$

    j $3x^2 + 7x + 2 = 0$ **k** $2x^2 + 5x + 2 = 0$ **l** $6y^2 + 7y + 2 = 0$

    m $x^2 = 8x - 12$ **n** $2x^2 + 7x = 15$ **o** $0 = 5x - 6 - x^2$

    p $x(x + 10) = -21$

3 Solve these quadratic equations by factorising using the difference of two squares.

    a $x^2 - 16 = 0$ **b** $x^2 - 64 = 0$ **c** $y^2 - 25 = 0$ **d** $9x^2 - 4 = 0$

    e $4y^2 - 1 = 0$ **f** $x^2 = 169$ **g** $4x^2 = 25$ **h** $36 = 9y^2$

4 Solve these quadratic equations.

    a $3x^2 - x = 0$ **b** $x^2 - 2x - 15 = 0$

    c $3x^2 - 11x + 6 = 0$ **d** $9y^2 - 16 = 0$

    e $25 = 16x^2$ **f** $x^2 = x$

    g $20x^2 = 7x + 3$ **h** $8x = 12 + x^2$

5 Solve these equations.

    a $5 = x + 6x(x + 1)$ **b** $(x + 1)^2 = 2x(x - 2) + 10$

    c $10x = 1 + \dfrac{3}{x}$ **d** $\dfrac{2}{x - 2} + \dfrac{4}{x + 1} = 0$

    e $x^4 - 13x^2 + 36 = 0$

6 A rectangle has a length that is 7 cm more than its width, w. The area of the rectangle is 60 cm^2.

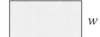 w

    a Write an algebraic expression for the area of the rectangle.

    b Show that $w^2 + 7w - 60 = 0$.

    c Find the dimensions of the rectangle.

The quadratic formula

This spread will show you how to:

● Solve quadratic equations using the quadratic formula

Keywords

Coefficient
Constant
Quadratic

Some **quadratic** equations do not factorise.

● You can solve *all* quadratic equations of the form $ax^2 + bx + c = 0$ using the quadratic equation formula:

$$x = \frac{-b \pm \sqrt{b^2 - 4ac}}{2a}$$

In the formula
– a is the **coefficient** of the x^2-term
– b is the coefficient of the x-term
– c is the **constant**.

$x^2 - 7x + 11 = 0$
does not factorise.
You cannot find
two numbers that
multiply to give 11
and add to give −7.

The constant is the
number on its own.

Example

Solve the quadratic equation $3x^2 - 5x = 1$.

$$3x^2 - 5x = 1$$
$$3x^2 - 5x - 1 = 0$$

In the formula: $a = 3$, $b = -5$ and $c = -1$

$$x = \frac{-b \pm \sqrt{b^2 - 4ac}}{2a}$$

$$x = \frac{5 \pm \sqrt{(-5)^2 - 4 \times 3 \times (-1)}}{2 \times 3}$$

$$x = \frac{5 \pm \sqrt{25 + 12}}{6}$$

Either $x = \dfrac{5 + \sqrt{37}}{6} = 1.847\ 127\ 088\ \ldots = 1.85$ (to 3 sf)

Or $\quad x = \dfrac{5 - \sqrt{37}}{6} = -0.180\ 460\ 421\ \ldots = -0.180$ (to 3 sf)

First rearrange into
the form
$ax^2 + bx + c = 0$.

Write the values of
a, b and c.

b is −5, so −b is
+5.

Example

Given that the side of a rectangle measures $(x + 4)$ and its area satisfies the equation $x^2 + 3x - 14 = 0$, find the length to 2 dp.

$$x^2 + 3x - 14 = 0$$
$a = 1$, $b = 3$ and $c = -14$, so

$$x = \frac{-3 \pm \sqrt{9 - 4 \times 1 \times (-14)}}{2 \times 1}$$

Hence, $x = \dfrac{-3 + \sqrt{65}}{2} = 2.531 \ldots$ or $x = \dfrac{-3 - \sqrt{65}}{2} = -5.531 \ldots$

Since the length is $x + 4$, $x = 2.531$ gives a length of 6.53 (to 2 dp).

The formula gives
two answers, but a
negative value of x
does not make
sense in the
context of the
question.

1 Solve these quadratic equations using the quadratic formula.
Where necessary, give your answers to 3 significant figures.

a $3x^2 + 10x + 6 = 0$ **b** $5x^2 - 6x + 1 = 0$

c $2x^2 - 7x^2 - 15 = 0$ **d** $x^2 + 4x + 1 = 0$

e $2x^2 + 6x - 1 = 0$ **f** $6y^2 - 11y - 5 = 0$

g $3x^2 + 3 = 10x$ **h** $6x + 2x^2 - 1 = 0$

i $20 - 7x - 3x^2 = 0$

2 Solve these equations using the quadratic formula.
Give your answers to 3 sf.

a $x(x + 4) = 9$

b $2x(x + 1) - x(x + 4) = 11$

c $(3x)^2 = 8x + 3$

d $y + 2 = \dfrac{14}{y}$

e $\dfrac{3}{x + 1} + \dfrac{4}{2x - 1} = 2$

> Rearrange the equation in the form
> $ax^2 + bx + c = 0$.

3 Here is a rectangle.

$x - 2$

$x + 7$

a Write an expression for the area of this rectangle.

b Given that the area of the rectangle is 20 cm², show that
$$x^2 + 5x - 34 = 0.$$

c Solve the equation to find the dimensions of the rectangle to 2 dp.

4 **a** Solve both $2x^2 + 11x + 12 = 0$ and $2x^2 + 11x + 10 = 0$, using the quadratic equation formula.

b Now try and solve the equations by factorisation.

c Compare your answers to parts **a** and **b**. Which part of the formula helps you to decide whether a quadratic will factorise or not?

d Use your suggestion to decide if these equations will factorise.

   i $3x^2 + 19x - 12$

   ii $12x^2 + 14x - 3$

> Factorisation leads to exact answers, where the formula gives rounded answers.

5 Use the quadratic formula to solve $2x^2 + 7x + 9 = 0$.
What do you notice?
Why does this happen?

Exam review

Key objectives

- Generate common integer sequences (including sequences of odd or even integers, squared integers, powers of 2, powers of 10, triangular numbers)
- Solve quadratic equations by factorisation, completing the square and using the quadratic formula

1 a Factorise $3x^2 + 41x - 94$. (2)

  b Hence solve $3x^2 + 41x - 94 = 0$. (2)

2 a Write down an expression, in terms of n, for the nth multiple of 5. (1)

  b Hence or otherwise

    i prove that the sum of two consecutive multiples of 5 is always an odd number

    ii prove that the product of two consecutive multiples of 5 is always an even number. (5)

(Edexcel Ltd., 2004)

D2

This unit will show you how to

- Draw and interpret cumulative frequency diagrams for grouped data
- Estimate the averages and range for grouped data using a cumulative frequency diagram

Before you start ...

You should be able to answer these questions.

1 Work out:

    a $\frac{1}{2}$ of 124   **b** 50% of 120

    c $\frac{1}{2}$ of $(36 + 1)$ **d** 25% of 60

    e $\frac{1}{4}$ of $(27 + 1)$ **f** 25% of 120

    g $\frac{3}{4}$ of 144   **h** 75% of 200

2 Here is a graph showing the cost per day to hire a power tool.

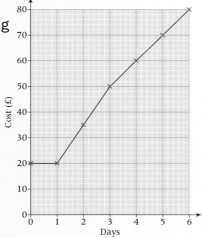

How much does it cost to hire the power tool for   **a** 3 days   **b** 5 days?

c Mike has £40. What is the maximum number of days he can hire the power tool?

Review

Unit N3

Unit N3

Cumulative frequency diagrams

This spread will show you how to:

Draw and interpret cumulative frequency diagrams for grouped data

Keywords
Cumulative
  frequency
Upper bound

You can represent grouped data on a **cumulative frequency** diagram by:

- calculating the cumulative frequencies and recording them in a cumulative frequency table
- plotting the cumulative frequency against the **upper bound** for each class
- joining the points with a smooth curve.

> The cumulative frequencies are the running totals.

Example

The table shows the heights of 120 men.

Height, h cm	$160 \leqslant h < 165$	$165 \leqslant h < 170$	$170 \leqslant h < 175$	$175 \leqslant h < 180$	$180 \leqslant h < 185$
Frequency	9	26	47	33	5

a Draw a cumulative frequency table for these data.
b Draw a cumulative frequency diagram for these data.
c Estimate the number of men over 168 cm tall.

a

Height, h cm	< 165	< 170	< 175	< 180	< 185
Cumulative frequency	9	35	82	115	120

$9 + 26 = 35$

> Upper bound of each class.

> Add frequencies to get cumulative frequency (CF).

b

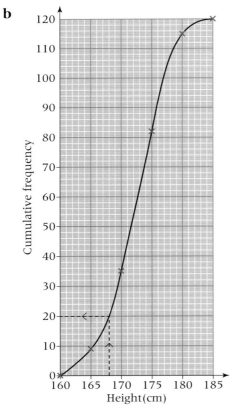

> Plot:
> (upper bound, CF).

> Lower bound of 1st class is 160, so plot (160, 0).

> Read up from 168 on the horizontal axis to the curve. Read across to the vertical axis to find the number of men *less than* 168 cm.

c From the graph, $120 - 20$ men are over 168 cm, that is, 100 men.

You will need the cumulative frequency tables and diagrams you draw in this exercise for Exercises D2.3 and D2.4.

1 The heights of 100 women are given in the table.

Height, h cm	$150 \leqslant h < 155$	$155 \leqslant h < 160$	$160 \leqslant h < 165$	$165 \leqslant h < 170$	$170 \leqslant h < 175$
Frequency	9	27	45	16	3

  a Draw up a cumulative frequency table for these data.

  b Draw a cumulative frequency diagram for these data.

2 The table gives information about the ages of staff in a large store.

Age, A	$20 \leqslant A < 30$	$30 \leqslant A < 40$	$40 \leqslant A < 50$	$50 \leqslant A < 60$	$60 \leqslant A < 70$
Frequency	20	36	51	27	11

  a Draw a cumulative frequency table for these data.

  b Draw a cumulative frequency diagram for these data.

3 The table gives information about journey times to work.

Time, t minutes	$0 \leqslant t < 10$	$10 \leqslant t < 20$	$20 \leqslant t < 30$	$30 \leqslant t < 40$	$40 \leqslant t < 50$	$50 \leqslant t < 60$
Frequency	6	18	29	35	21	11

Draw a cumulative frequency diagram to represent this data.

4 The table gives information about the birth weights of babies.

Weight, w grams	$2500 \leqslant w < 3000$	$3000 \leqslant w < 3500$	$3500 \leqslant w < 4000$	$4000 \leqslant w < 4500$	$4500 \leqslant w < 5000$
Frequency	7	23	34	25	11

Draw a cumulative frequency diagram for these data.

5 The table gives information about height of maize plants in a crop trial.

Height, h cm	$40 \leqslant h < 60$	$60 \leqslant h < 80$	$80 \leqslant h < 100$	$100 \leqslant h < 120$	$120 \leqslant h < 140$	$140 \leqslant h < 160$
Frequency	3	19	29	40	21	8

  a Draw a cumulative frequency diagram for these data.

  b Estimate how many maize plants were taller than 130 cm.

6 The table gives information about test results of a group of students.

Test result, t %	$30 \leqslant t < 40$	$40 \leqslant t < 50$	$50 \leqslant t < 60$	$60 \leqslant t < 70$	$70 \leqslant t < 80$	$80 \leqslant t < 90$
Frequency	5	16	25	33	18	3

  a Draw a cumulative frequency diagram for these data.

  b Marks over 65% are awarded grade A. Estimate how many students were awarded grade A.

This spread will show you how to:

● Draw and interpret cumulative frequency diagrams for grouped data

● Estimate the averages and range for grouped data using a cumulative frequency diagram

Keywords
Cumulative frequency
Interquartile range
Median
Quartiles

For grouped data:

● You can estimate the mean

● You can state which class contains the median.

● You can estimate the **median** and **quartiles** from a **cumulative frequency** diagram.

This gives a more accurate estimate of the median.

Example

The weights of 100 rats are recorded in the table:

Weight, w grams	$100 \leqslant w < 120$	$120 \leqslant w < 140$	$140 \leqslant w < 160$	$160 \leqslant w < 180$	$180 \leqslant w < 200$	$200 \leqslant w < 220$
Frequency	8	19	23	27	18	5

a Draw a cumulative frequency diagram to represent the data.
b Use your diagram to estimate
  i the median
  ii the **interquartile range**.
c Use your diagram to estimate the number of rats that weigh between 165 g and 190 g.

a

Weight, w, grams	< 120	< 140	< 160	< 180	< 200	< 220
Cumulative frequency	8	27	50	77	95	100

For large data sets, use $\frac{1}{2}n$ (not $\frac{1}{2}(n+1)$) for the median.

b From the graph
  i Median = 160 g
  ii IQR = 177 − 138
       = 39 g
c From the graph:
56 rats weigh < 165 g
88 rats weigh < 190 g
So, 88 − 56 = 32 rats weigh between 165 g and 190 g.

For 100 pieces of data:
● median is the 50th
● LQ is the 25th
● UQ is the 75th

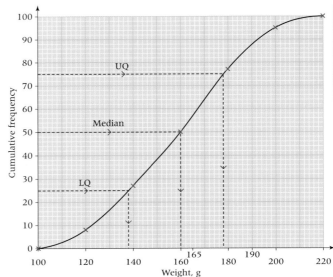

You will need some of your results from this exercise in Exercise D2.4.

1 Use your table and graph from Exercise D2.2 question **1** to estimate
  a the median   **b** the interquartile range
  c the number of women taller than 163 cm.

2 Use your table and graph from Exercise D2.2 question **2** to estimate
  a the median   **b** the interquartile range
  c the number of staff under 25.

3 Use your table and graph from Exercise D2.2 question **3** to estimate
  a the median   **b** the interquartile range
  c the number of journeys that take between 15 and 35 minutes.

4 Use your table and graph from Exercise D2.2 question **4** to estimate
  a the median   **b** the interquartile range
  c the number of babies that weighed less than 2600 g or more than 3300 g.

5 Use your table and graph from Exercise D2.2 question **5** to estimate
  a the median   **b** the interquartile range
  c the number of maize plants between 75 and 110 cm tall.

6 The cumulative frequency graph shows the IQ scores of a sample of 100 adults.
Estimate from the graph
  a the median
  b the interquartile range
  c the number in the sample with an IQ score
    i less than 85
    ii greater than 115.

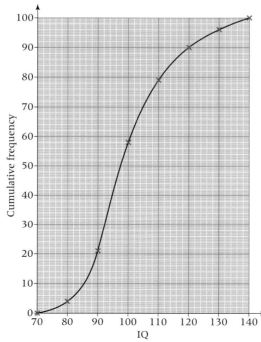

Key objectives

- Draw and produce cumulative frequency tables and diagrams for grouped continuous data
- Find the median, quartiles and interquartile range for large data sets

1 Here are the results, to the nearest minute, of the times of runners in a 10 kilometre race.

60　54　69　55　60　72　52　53　61　59
61　52　55　58　49　61　58　70　52　50

  a Work out the

    i median   **ii** interquartile range   **iii** range.   (3)

2 The cumulative frequency graph gives information about the examination marks of a group of students.

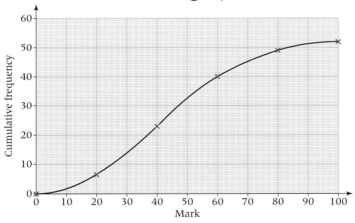

  a How many students were in the group?   (1)

  b Use the graph to estimate the inter-quartile range.   (2)

    The pass mark for the examination was 56.

  c Use the graph to estimate the number of students who passed the examination.   (2)

(Edexcel Ltd., (Spec.))

This unit will show you how to

- Understand that the form $y = mx + c$ represents a straight line and that m is the gradient and c is the value of the y-intercept
- Understand that parallel lines have the same gradient
- Find the equation of a straight-line by considering its gradient and axes intercepts
- Calculate the gradient of a line segment
- Find the y-intercept of a straight-line graph when the gradient and a point on the line are known
- Find the equation of a line when given two points on the line
- Find the midpoint of a line segment with known end points
- Solve simple inequalities in one or two variables

Before you start ...

Review

You should be able to answer these questions.

1 a On axes labelled −10 to 10, plot and join these points.

    (2, 5)   (−6, 5)   (−6, −2)   (2, −2)

  b Find the area of the resulting shape.

CD – A4

2 On axes labelled from −8 to +8, plot these lines.

  a $y = 6$ **b** $x = -2$ **c** $y = 2x + 1$ **d** $x + y = 8$

CD – A4

3 Decide if these shapes have any parallel sides, indicating how many pairs you can find.

  a Rectangle   **b** Regular hexagon
  c Rhombus   **d** Right-angled triangle
  e Isosceles trapezium

CD – A4

4 Insert the symbol <, > or = into the ☐ to make them true.

  a 10% of 360 ☐ 15% of 250

  b $(-2)^2$ ☐ $(-2)^3$ **c** $-\frac{3}{8}$ ☐ $-\frac{1}{3}$

  d $(-2) \times 15$ ☐ $60 \div (-2)$

Unit N3

Straight line graphs

This spread will show you how to:

- Recognise and understand the form of equations corresponding to horizontal, vertical and diagonal line graphs

Keywords

Diagonal
Horizontal
Intersect
Vertical

A straight line can be **diagonal**, **vertical** or **horizontal**.

The x-coordinate of every point on this vertical line is 2.

The y-coordinate can have any value.

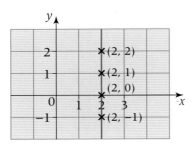

The equation of the line is x = 2

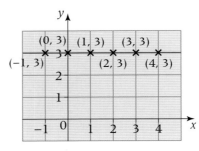

The equation of the line is y = 3

The y-coordinate of every point on this horizontal line is 3.

The x-coordinate can have any value.

- Horizontal lines have equations of the form x = c.
- Vertical lines have equations of the form y = c.

c stands for a number.

Example

Give three points that would lie on each of these lines
a y = 5　　　**b** x = −2.

a y = 5
   Since the y-coordinate is 5, possible points are (1, **5**), (2, **5**) and (17, **5**)
b x = −2
   Since the x-coordinate is −2, possible points are (−**2**, 1), (−**2**, 2) and (−**2**, 11)

Example

Where do the graphs x = 4 and y = −1 intersect?

When lines **intersect** they cross.

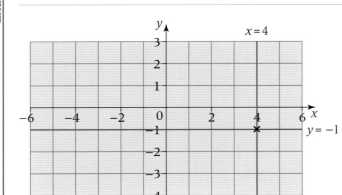

All points on the line x = 4 have x-coordinate 4.

All points on the line y = −1 have y-coordinate −1.

The lines intersect at (4, −1).

1 a Copy and complete the table, deciding if each equation is that of a horizontal, vertical or diagonal line, or none of these.

$x = 9$ $y = 2x - 1$ $x = -0.5$ $y = 7$ $y = x^2 + x$

Horizontal	Vertical	Diagonal	None of these

b Add an equation of your own to each column in the table.

2 Match each line with its equation.

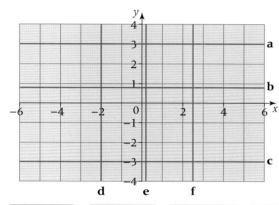

$y = -3$ $x = -2$ $y = 3$ $x = 2.5$ $x = \frac{1}{4}$ $y = \frac{3}{4}$

3 On one set of axes labelled from −6 to +6, plot these graphs.

a $x = 5$ **b** $y = 2$ **c** $x = 1.6$ **d** $y = -3$ **e** $y = 1$ **f** $x = -1\frac{1}{4}$

4 Where do these graphs intersect? Plot the graphs if you need to.

a $x = 5$ and $y = 2$ **b** $x = 4$ and $y = -3$

c $x = -2$ and $y = 9$ **d** $y = -4$ and $x = -2$

5 a Give the equations of four lines which, when plotted, form the sides of a rectangle.

b Repeat part **a** for a square.

c Repeat part **a** for an isosceles right-angled triangle.

6 a Which point with integer coordinates fits these clues?

Above $y = -1$, below $y = 3$, below $y = 2x + 1$, above $y = 2 - x$ and left of $x = 2$.

b Write your own clues to describe the point (3, 4).

This spread will show you how to:

● Understand that the form $y = mx + c$ represents a straight line and that m is the gradient and c is the value of the y-intercept

● The equation of a straight line is of the form $y = mx + c$, where m is the **gradient** and c is the **y-intercept**.

The y-intercept is the y-value where the graph cuts the y-axis.

● Two lines that have the same gradient are **parallel**.

A line with a positive gradient slopes upwards.

A line with a negative gradient slopes downwards.

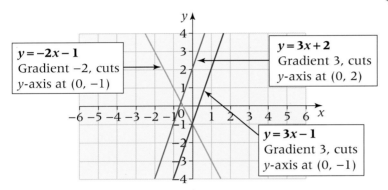

$y = -2x - 1$
Gradient -2, cuts y-axis at $(0, -1)$

$y = 3x + 2$
Gradient 3, cuts y-axis at $(0, 2)$

$y = 3x - 1$
Gradient 3, cuts y-axis at $(0, -1)$

● The coordinates of any point on a line satisfy the equation of the line.

Example

Describe the line $2y = 8 - 3x$.

$2y = 8 - 3x$
$y = 4 - \frac{3}{2}x$
$y = -\frac{3}{2}x + 4$

The line has gradient $-\frac{3}{2}$ and cuts the y-axis at $(0, 4)$.

First rearrange the equation into the form $y = mx + c$.

Example

The straight line L_1 has equation $y = 3x + 4$.
The straight line L_2 is parallel to L_1.
The straight line L_2 passes through $(3, 1)$.
Find an equation of the straight line L_2.

L_2 is parallel to L_1, so L_2 has gradient 3.
L_2: $y = 3x + c$
At $(3, 1)$: $1 = 3 \times 3 + c$
            $-8 = c$
The equation of L_2 is $y = 3x - 8$.

Substitute $x = 3$ and $y = 1$ into the equation and solve to find c.

1 Describe each of these lines in reference to gradient and y-intercept.

   a $y = 3x - 2$ **b** $y = 7 + \frac{1}{2}x$

   c $3y = 9x - 6$ **d** $2y - 4x = 5$

2 Find the equations of these lines.

   a Gradient 6, passes through $(0, 2)$

   b Gradient -2, passes through $(0, 5)$

   c Gradient -1, passes through $(0, \frac{1}{2})$

   d Gradient -3, passes through $(0, -4)$

3 For each of these lines, give the equation of a line parallel to it.

   a $y = 2x - 1$ **b** $y = -5x + 2$ **c** $y = -\frac{1}{4}x + 2$

   d $y = 7 - 4x$ **e** $y = 6 + \frac{3}{4}x$ **f** $2y = 9x - 1$

4 Here are the equations of several lines.

   $\boxed{y = 3x - 2}$ $\boxed{y = 4 + 3x}$ $\boxed{y = x + 3}$

       $\boxed{y = 5}$ $\boxed{2y - 6x = -3}$ $\boxed{y = 3 - x}$

   a Which three lines are parallel to one another?

   b Which two lines cut the y-axis at the same point?

   c Which a line has a zero gradient?

   d Which a line passes through $(2, 4)$?

   e Which a pair of lines are reflections of one another in the y-axis?

5 Find the equations of these lines.

   a A line parallel to $y = -4x + 3$ and passing through $(-1, 2)$

   b A line parallel to $2y - 3x = 4$ and passing through $(6, 7)$.

6 True or false?

   a $y = 2x - 1$ and $y = 2 + 2x$ are parallel.

   b $y = 3x - 4$ and $y = 6 - x$ both pass through $(2, 4)$.

   c $x = 6$ and $y = 2x - 8$ never meet.

   d $y = x + 1$ and $y = x^2$ meet once.

7 Find the coordinates of the points where the graphs of these lines meet.

   a $y = 3x + 2$ and $2y + 3x = 13$

   b $y = 2x - 2$ and $x^2 + y^2 = 25$

This spread will show you how to:

- Find the equation of a straight line by considering its gradient and axes intercepts

Keywords
Gradient
y-intercept

- The **gradient** of a line segment is calculated as

$$\frac{\text{change in the } y\text{-direction}}{\text{change in the } x\text{-direction}}.$$

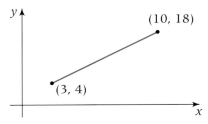

$$\text{Gradient} = \frac{\text{rise}}{\text{run}}$$

From (3, 4) to (10, 18):
Change in y-direction $= +14$
Change in x-direction $= +7$
Gradient $= \frac{14}{7} = 2$

- The gradient m of the line joining (x_1, y_1) to (x_2, y_2) is

$$m = \frac{y_2 - y_1}{x_2 - x_1}$$

- Once you know the gradient, you can use one of the points to find the value of the **y-intercept**, c.

Line joining (3, 4) to (10, 18):
$y = 2x + c$
Substitute $x = 3$, $y = 4$:
$4 = 6 + c \rightarrow c = -2$
Equation: $y = 2x - 2$

Example

Find **a** the gradient **b** the equation
of the line passing through (3, 5) and (9, 35).

a $m = \frac{35 - 5}{9 - 3} = \frac{30}{6} = 5$

b $y = 5x + c$
At the point (3, 5), $5 = 3 \times 5 + c$
$-10 = c$

Equation is $y = 5x - 10$.

$m = \frac{y_2 - y_1}{x_2 - x_1}.$
Learn this gradient formula.

Example

The line L is parallel to the graph of $y = 2x - 1$.
Find the equation of the line labelled L.

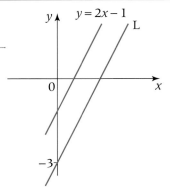

L is parallel to $y = 2x - 1$, so L has gradient 2.
For L: $y = 2x + c$
From the diagram, the y-intercept is -3.
Hence: $y = 2x - 3$.

1 Find the gradients of the line segments joining these pairs of points.

    a (4, 9) to (8, 25)   **b** (5, 6) to (10, 16)   **c** (2, 1) to (5, 2)

2 Sketch each of these graphs, labelling the point where the graph crosses the y-axis.

    a $y = 7x - 2$ **b** $y = 4 - 2x$ **c** $2y - x = 8$ **d** $2x + 3y = 4$

3 Find the equation of each line on this graph.

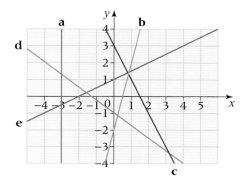

4 Find the equations of these line segments.

    a A line joining (0, 5) to (3, 17)

    b A line joining (2, 8) to (12, 13)

    c

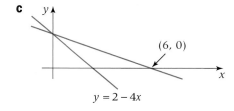

(6, 0)

$y = 2 - 4x$

5 **a** The gradient of the line segment joining (2, p) to (6, $4p$) is 4. Find the value of p.

    b The gradient of the line segment joining (m^2, m) to (5, 8) is 2. Use algebra to find two possible values of m.

6 Use the information given in the diagram to find the equation of the line on each graph.

    a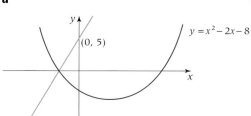

(0, 5)

$y = x^2 - 2x - 8$

    b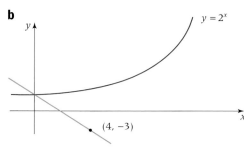

$y = 2^x$

(4, −3)

Regions

This spread will show you how to:
- Solve simple inequalities in one variable

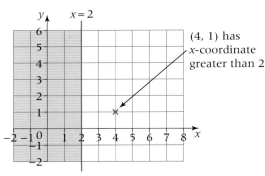 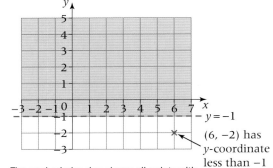

Keywords
Inequality
Region

- An **inequality** is a mathematical statement using one of these signs:
  < less than   ≤ less than or equal to
  > greater than   ≥ greater than or equal to

$5 > 3$ and $3 < 5$ are inequalities.

- You can represent inequalities as **regions** on a graph.

(4, 1) has x-coordinate greater than 2

(6, −2) has y-coordinate less than −1

The *unshaded* region shows all points with x-coordinate greater than 2, or the inequality $x \geq 2$.

The line $x = 2$ is solid, to show that points on this line are included in the region required.

The *unshaded* region shows all points with y-coordinate less than −1 or the inequality $y < -1$.

The line $y = -1$ is dashed, to show that points on this line are *not* included in the region required.

Example

On a pair of axes, construct an *unshaded* region to represent the inequalities $x > 1$ and $y \geq 2$.

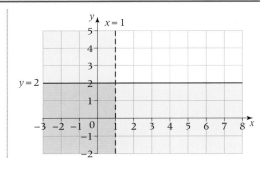

First draw the lines $x = 1$ and $y = 2$.

Use a solid line for ≤ or ≥.

Use a dashed line for < or >.

Example

Write the inequality represented by the *shaded* region in this diagram.

$-1 < x \leq 4$

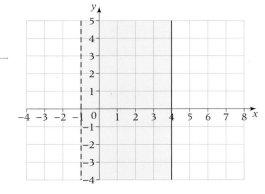

Examiner's tip
Some questions ask you to shade the region required. Some ask you to leave it unshaded. Read the question carefully.

1 Draw suitable diagrams to show these inequalities. You should leave the required region *unshaded* and label it **R**.

  a $x \geqslant 4$ **b** $y \leqslant 3$ **c** $x > -2$

  d $y < -3$ **e** $-1 \leqslant x \leqslant 6$ **f** $1 < y < 8$

  g $x > 8$ or $x \leqslant -2$ **h** $y < 2$ or $y > 9$ **i** $2x \geqslant 5$

  j $8 > y$ **k** $-2.5 < x < 7$ and $0 \leqslant y \leqslant 5\frac{1}{4}$

2 For each diagram, write the inequalities shown by the *shaded* region.

  a
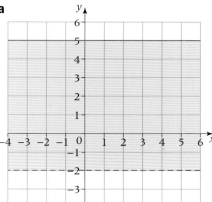

  b
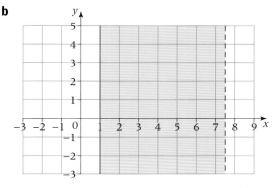

3 List all the points with integer coordinates that satisfy both the inequalities
$-2 < x \leqslant 1$ and $2 > y \geqslant 0$.

4 Draw diagrams and write inequalities that when shaded would give

  a a shaded rectangle

  b an unshaded square

  c a shaded rectangle with length double its width.

5 Here are two graphs of quadratic equations, with horizontal lines drawn as shown.
Write the inequalities represented by the *shaded* regions.

  a
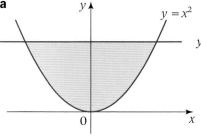

  b
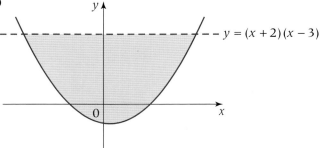

This spread will show you how to:

● Solve simple inequalities in one or two variables

Keywords
Inequality
Point test
Region

● You can represent **inequalities** in two variables on a graph.

For example, to draw $y > 2x - 1$, start by drawing the line $y = 2x - 1$ (dashed for $>$).

● You can use a **point test** to decide which side of the line is the required **region**.

Choose one point and see if it fits the inequality. If it does, this is the region you need.

For $y > 2x - 1$ choose $(0, 0)$:

Is $0 > 2 \times 0 - 1$?

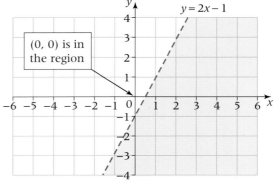

$(0, 0)$ is in the region

Yes ... so the point $(0, 0)$ is in the region required, shown unshaded on the graph.

Example

Show the region of points with coordinates that satisfy the four inequalities $y > 0$, $x > 0$, $2x < 5$ and $3y + 2x < 9$.

Draw dashed lines at $y = 0$ (x-axis), $x = 0$ (y-axis) and $x = \frac{5}{2}$.

$2x = 5 \rightarrow x = \frac{5}{2}$

Plot the graph of $3y + 2x = 9$.

x	−3	0	3
y	5	3	1

Find the y-values for three values of x and plot the points. Draw the line dashed.

Choose $(1, 1)$ for the point test for all the inequalities.

$y > 0$: Is $1 > 0$? Yes.
$x > 0$: Is $1 > 0$? Yes.

$2x < 5$: is $2 \times 1 < 5$? Yes.
$3y + 2x < 9$: Is $3 \times 1 + 2 \times 1 < 9$? Yes.

Choose a point that does not lie on any of the graphs.

$(1, 1)$ satisfies all the inequalities, so it is in the required region.

Required region left unshaded.

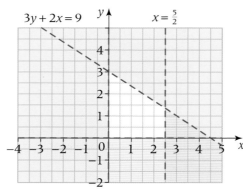

1 Shade the regions satisfied by these inequalities.

a $y \leqslant x + 5$ **b** $y > 2x + 1$ **c** $y \geqslant 1 - 3x$ **d** $y < \frac{1}{4}x - 2$

e $2y \geqslant 3x + 4$ **f** $3y < 9x - 7$ **g** $x + y < 9$ **h** $2x + 4y > 7$

2 What inequalities are shown by the shaded region in each of these diagrams?

a
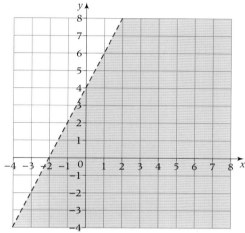

b
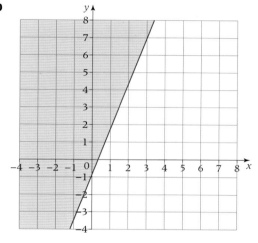

3 On one diagram, show the region satisfied by the three inequalities.
List all the integer coordinates that satisfy all three inequalities.

a

| $x \geqslant -1$ | $y < 2$ | $y \geqslant 3x - 1$ |

b

| $2x + 3y < 10$ | $x > 2$ | $2y + 1 > 0$ |

4 Draw the region satisfied by all three inequalities

$y \geqslant x^2 - 4,\ y < 2,\ y + 1 < x.$

5 300 students are going on a school trip and 16 adults will accompany them. The head teacher needs to hire coaches to take them on the trip. She can hire small coaches that seat 20 people or large coaches that seat 48 people. There must be at least 2 adults on each coach to supervise the students.
If x is the number of small coaches hired and y is the number of large coaches, explain why these inequalities model the situation.

$x \geqslant 0,\ y \geqslant 0,\ 5x + 12y \geqslant 79,\ x + y \leqslant 8$

6 Draw diagrams and write inequalities that when shaded would give

a a shaded, right-angled triangle with right angle at the origin

b an unshaded trapezium.

Exam review

Key objectives

- Understand that the form $y = mx + c$ represents a straight line and that m is the gradient of the lines and c is the value of y-intercept
- Explore the gradients of parallel lines
- Find the equation of a given straight line in the form $y = mx + c$
- Solve several linear inequalities in two variables and find the solution set

1 The equation of the straight line, L, is $y = 3x - 2$.
The line, M, is parallel to the line L and passes through the point (1, 1).
Find the equation of the line M, expressing your answer in the form $y = mx + c$. (3)

2 The line with equation $6y + 5x = 15$ is drawn on the grid.

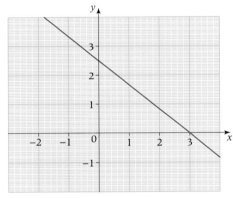

a Rearrange the equation $6y + 5x = 15$ to make y the subject. (2)

b The point $(-21, k)$ lies on the line.
Find the value of k. (2)

c i Copy the grid and shade the region of points whose coordinates satisfy the four inequalities:
$$y > 0, \quad x > 0, \quad 2x < 3, \quad 6y + 5x < 15.$$
Label this region R.
P is a point in the region R. The coordinates of P are both integers.

ii Write down the coordinates of P. (3)

(Edexcel Ltd., 2004)

This unit will show you how to

- Understand the probability scale
- Understand mutually exclusive events and their related probabilities
- Know that if two events, A and B, are mutually exclusive, then the probability of A or B occurring is P(A) + P(B)
- Understand theoretical and experimental probability and how estimates can be used in probability
- Calculate the expected frequency and relative frequency of events
- Understand that more trials give a more reliable estimate of probability
- Recognise independent events
- Know that if two events, A and B, are independent, then the probability of A and B occurring is P(A) × P(B)
- List possible outcomes systematically

Before you start ...

You should be able to answer these questions.

Review

1 Work out each of these.

a $1 - \frac{2}{5}$ **b** $1 - \frac{4}{7}$ **c** $1 - \frac{3}{8}$

Unit N3

2 Work out each of these.

a $\frac{2}{3} + \frac{1}{6}$ **b** $\frac{1}{5} + \frac{1}{4}$ **c** $\frac{1}{3} + \frac{5}{8}$

Unit N3

3 Work out each of these.

a $\frac{1}{4} \times \frac{2}{3}$ **b** $\frac{1}{5} \times \frac{3}{4}$ **c** $\frac{1}{3} \times \frac{2}{5}$

Unit N3

4 Change these fractions to decimals.

a $\frac{7}{10}$ **b** $\frac{3}{4}$ **c** $\frac{3}{8}$

d $\frac{2}{5}$ **e** $\frac{1}{3}$

Unit N3

This spread will show you how to:

- Understand the probability scale
- Understand mutually exclusive events and their related probabilities
- Know that if A and B are mutually exclusive, then the probability of A or B occurring is P(A) + P(B)

Probability is a measure of the likelihood that an **event** will happen.

- An event is one or more of the possible **outcomes**.
- The probability of an event lies between 0 and 1.
- Probability of an event happening $= \dfrac{\text{number of favouable outcomes}}{\text{total number of all possible outcomes}}$

You can write probability as a decimal, fraction or percentage.

- Events are **mutually exclusive** if they cannot happen at the same time.
- For two mutually exclusive events A and B:
  P(A or B) = P(A) + P(B)
- For an event A,
  P(not A) = 1 − P(A)   or   P(A) = 1 − P(not A)

P(A) and P(not A) are mutually exclusive events.

Example

A spinner has 10 equal sections:
3 green, 2 blue, 4 red and 1 white.
What is the probability that the spinner lands on

a green   **b** green or blue
c blue or white   **d** blue or red or white
e not green?

a $P(\text{green}) = \frac{3}{10}$

b P(green or blue) = P(green) + P(blue)

$= \frac{3}{10} + \frac{2}{10}$

$= \frac{5}{10} = \frac{1}{2}$

c P(blue or white) = P(blue) + P(white)

$= \frac{2}{10} + \frac{1}{10}$

$= \frac{3}{10}$

d P(blue or red or white) = P(blue) + P(red) + P(white)

$= \frac{2}{10} + \frac{4}{10} + \frac{1}{10}$

$= \frac{7}{10}$

e P(not green) = 1 − P(green)

$= 1 - \frac{3}{10}$

$= \frac{7}{10}$

1 A bag contains 6 green, 10 blue and 9 red cubes.
8 of the cubes are large, 17 are small.
One cube is chosen at random.
What is the probability that the cube is

    a red **b** not red **c** large **d** small **e** blue **f** not green?

2 Four teams are left in a hockey tournament.
Their probabilities of winning are

Colts	Falcons	Jesters	Hawks
0.17	0.38	x	$2x$

    a Compare the Hawks' chance of winning with
    the Jesters' chance of winning.

    b Work out the value of x.

3 The probabilities that a player will win, draw or lose at snap are

Win	Draw	Lose
x	$2x$	x

Work out the value of x.

4 A nine-sided spinner has sections numbered 1, 2, 3, 4, 5, 6, 7, 8, 9.
What is the probability that it lands on

    a 6   **b** not the number 6   **c** an even number

    d a multiple of 5   **e** a multiple of 3   **f** not a multiple of 3

    g a factor of 24   **h** not a factor of 24   **i** a prime number?

5 The table shows how many TVs are owned by some families.

Number of TVs	0	1	2	3	more than 3
Number of families	8	12	5	1	4

    a How many families were surveyed?

    b What is the probability that a family chosen at random from the
    sample will own

        i exactly 1 TV   **ii** no TV   **iii** either 1 or 2 TVs

        iv 2 or more TVs   **v** 3 or more TVs   **vi** 4 TVs?

6 The table shows staff absences over a two-month period.

Number of days off	0	1	2	3	more than 3
Number of staff	5	1	9	11	6

Find the probability that a member of staff chosen at random had

    a 1 day off   **b** 1 or 2 days off

    c 2 or more days off   **d** 2 or fewer days off.

Theoretical and experimental probability

This spread will show you how to:

- Understand theoretical and experimental probability and how estimates can be used in probability
- Calculate the expected frequency of events

Keywords
Expected
   frequency
Experimental
   probability
Theoretical
   probability

If all the outcomes are equally likely, you can calculate the **theoretical probability**.

- **Theoretical probability** = $\dfrac{\text{number of favourable outcomes}}{\text{total number of outcomes}}$

You use theoretical probability for fair activities, for example rolling a fair dice.

You can use data from a survey or experiment to estimate **experimental probability**.

- **Experimental probability** = $\dfrac{\text{number of successful trials}}{\text{total number of trials}}$

For unfair or biased activities, or where you cannot predict the outcome, you use experimental probability.

A fair spinner has 12 sides:
5 yellow, 2 red, 4 blue and 1 white.
Work out the probability that the spinner lands on

a yellow   **b** red or blue   **c** not white.

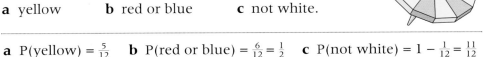

a P(yellow) = $\frac{5}{12}$ **b** P(red or blue) = $\frac{6}{12} = \frac{1}{2}$ **c** P(not white) = $1 - \frac{1}{12} = \frac{11}{12}$

P(red or blue) = P(red) + P(blue)

You can use probability to calculate the **expected frequency** of an event.

- **Expected frequency = number of trials × probability**

Finn carries out a survey on left-handedness. He asks 50 students at his school. Six of them are left-handed.

a Estimate the probability that a student at the school is left-handed.
b Estimate how many of the 1493 students at the school are left-handed.

a P(left-handed) = $\frac{6}{50}$
b Expected frequency = $1493 \times \frac{6}{50} = 179.16$
     Estimate: 179 students are left-handed.

'Number of trials' = 'number of students' in this context.

A biased coin is thrown 150 times. It lands on heads 105 times.

a Estimate the probability of the coin landing on heads next time.
b What number of heads would you expect in 400 throws?

a P(Head) = $\frac{105}{150} = 0.7$
b Estimated number of heads in 400 throws = $400 \times 0.7 = 280$

1 The probability that a biased dice will land on a six is 0.24.
Estimate the number of times the dice will land on a six if it is rolled 500 times.

2 A biased coin is thrown 140 times. It lands on tails 28 times.
Estimate the probability that this coin will land on tails on the next throw.

3 A tetrahedral dice is rolled 100 times. The table shows the outcomes.

Score	1	2	3	4
Frequency	16	28	33	23

a The dice is rolled once more. Estimate the probability that the dice will land on

i 4   **ii** 3   **iii** 1 or 2   **iv** not 2.

b If the dice is rolled another 200 times, how many times would you expect the dice to land on

i 2   **ii** 3 or 4?

4 There are 240 adults in the village of Hollowdown.

a In a survey of 30 adults from Hollowdown, 19 owned a bicycle.
How many adults in the village would you expect to own bicycle?

b In a survey of 40 adults, 21 owned a dog.
How many adults in the village would you expect to own a dog?

c In a survey of 48 adults, 43 felt the speed limit through the village was too high.
Estimate the number of adults in the village who think the speed limit is too high.

5 On a holiday, people chose one activity.

	Canoeing	Abseiling	Potholing	Total
Male	13		5	22
Female		9		
Total	20			50

a Copy and complete the table.

b What is the probability that a person chosen at random chose

i canoeing   **ii** not canoeing

iii canoeing or potholing   **iv** abseiling or potholing?

c There are 300 people in total on the holiday.
How many of the total group would you expect

i to choose canoeing   **ii** to be female?

Relative frequency and best estimate

This spread will show you how to:

- Calculate the expected frequency and relative frequency of events
- Understand that more trials give a more reliable estimate of probability

Experimental probability is also called **relative frequency**.

Relative frequency is the proportion of successful trials in an experiment.

- Relative frequency $= \dfrac{\text{number of successful trials}}{\text{number of trials}}$

The more trials you carry out, the more reliable the relative frequency will be as an estimate of probability.

Example

A double glazing representative telephones people to make appointments to demonstrate his product. One week he makes 50 calls each morning and creates a table to record the number of appointments.

Day	Mon	Tues	Wed	Thurs	Fri
Number of appointments	4	6	9	7	5

a Work out the relative frequency of the number of appointments for each day.
b Write down the best estimate of the relative frequency of setting up an appointment.

a
Day	Mon	Tues	Wed	Thurs	Fri
Number of appointments	4	6	9	7	5
Relative frequency	$4 \div 50 =$ 0.08	$6 \div 50 =$ 0.12	$9 \div 50 =$ 0.18	$7 \div 50 =$ 0.14	$5 \div 50 =$ 0.1

b Total number of calls made $= 250$
Best estimate of relative frequency $= \dfrac{4 + 6 + 9 + 7 + 5}{250} = \dfrac{31}{250} = 0.124$

For the best estimate of relative frequency, pool all the results.

You can compare theoretical probability with relative frequency.

Example

Kaseem rolled a dice 50 times and in 14 of those he scored a 2.

a What is the relative frequency of rolling a 2?
b Is the dice biased towards 2? Explain your answer.

a Relative frequency of rolling a 2 $= \dfrac{14}{50} = \dfrac{14}{50} = 0.28$
b Theoretical probability of rolling a 2 $= \dfrac{1}{6} = 0.166 \ldots$
In 50 rolls expected number of 2s $= 50 \times 0.166 \ldots = 8.333$
The relative frequency of rolling 2 is 0.28, which is much higher than the theoretical probability, 0.17.
The actual number of 2s is 14, which is much higher than the expected number, 8.
The dice does appear to be biased towards 2.

1 In a statistical experiment, a coin is flipped 360 times and lands on tails 124 times.
Is the coin fair? Explain your answer.

2 Surjeet picks a card from a deck of playing cards and records whether it is black or red.
She returns the card to the pack and shuffles it, before picking the next card.
In total she picks a card 420 times and records 213 black.
Is the deck of cards complete? Explain your reasons.

> In a complete deck, there are equal numbers of black and red cards.

3 The outcomes when a dice is rolled 200 times are

Score	1	2	3	4	5	6
Frequency	32	33	34	32	35	34

Is the dice fair? Explain your answer.

4 A circular spinner is divided into four sections: red, blue, yellow and green.
The table shows the outcomes from 100 spins.

Colour	Red	Blue	Yellow	Green
Frequency	19	18	47	16

Do you think the sections of the spinner are equally sized?
Explain your answer.

5 A group of students carry out an experiment to investigate the probability that a dropped drawing pin lands point up.
Each student drops a drawing pin 10 times and records the number of times it lands point up.

Point up   Point down

Student	1	2	3	4	5	6	7	8	9	10
Point up	2	3	4	1	1	2	3	2	1	3
Relative frequency										

a Copy the table and complete the relative frequency row.

b Write the best estimate of the drawing pin landing point up.

6 A four-sided dice is rolled 36 times. These are the results:

2	1	4	2	2	3	2	1	3	4	4	1
3	1	3	2	3	2	1	1	3	1	2	3
2	4	4	1	1	4	1	4	4	1	3	4

a Find the relative frequency of the dice landing on 1, 2, 3 and 4.

b Is the dice biased? Explain your answer.

143

Independent events

This spread will show you how to:

- Recognise independent events
- Know that if A and B are independent events, the probability of both A and B occurring is P(A and B) = P(A) × P(B)

- Two or more events are **independent** if when one **event** occurs it has no effect on the other event(s) occurring.

For example, when you roll two coins, the result for one coin does not affect the result for the other.

- For two independent events A and B
  P(A and B) = P(A) × P(B)

This is called the 'multiplication rule' or the 'AND rule'.

Example

A spinner has 9 equal sides:
2 green, 3 blue, 1 red and 3 white.
A fair coin is thrown and the spinner is spun.

a What is the probability of
  i head and green   **ii** tail and white
  iii tail and red   **iv** head and red?

b Comment on your answers to **iii** and **iv**.

a i P(head and green) = P(H) × P(green)
    $= \frac{1}{2} \times \frac{2}{9} = \frac{1}{9}$
ii P(tail and white) = P(T) × P(white)
    $= \frac{1}{2} \times \frac{3}{9} = \frac{1}{6}$
iii P(tail and red) = P(T) × P(red)
    $= \frac{1}{2} \times \frac{1}{9} = \frac{1}{18}$
iv P(head and red) = P(H) × P(red)
    $= \frac{1}{2} \times \frac{1}{9} = \frac{1}{18}$

b The answers to **iii** and **iv** are the same, because the coin is fair, so P(H) = P(T).

Example

Find the probability of rolling three consecutive sixes on a fair dice.

$P(6) = \frac{1}{6}$
P(6 and 6 and 6) = P(6) × P(6) × P(6)
    $= \frac{1}{6} \times \frac{1}{6} \times \frac{1}{6}$
    $= \frac{1}{216}$

Rolling a dice three times will give independent results

1 a Copy and complete the table to list all the outcomes, when a fair coin is thrown and a fair dice is rolled. One has been done for you.

		Dice					
		1	2	3	4	5	6
Coin	Head						
	Tail				Tail and 4		

b Find the probability of

i head and 3   **ii** tail and 6.

Comment on your answers.

2 A red and a blue dice are rolled.

a Draw a table to show all the possible outcomes.

b Find the probability of

i   1 on the red dice and 1 on the blue dice

ii 2 on the red dice and 5 on the blue dice

iii 3 on the red dice and an even number on the blue dice

iv 5 or greater on the red dice and 5 on the blue dice.

3 A fair dice is rolled twice.
Find the probability that the dice shows an odd number on the first roll and a number less than 3 on the second roll.

4 A spinner has ten equal sides, 4 show squares, 3 pentagons, 2 hexagons and 1 circle. The spinner is spun twice.
Find the probability of

a square on the first and second spins

b square on first and circle on second spin

c pentagon on first and hexagon on second spin

d hexagon on first and circle on second spin

e circle on first and pentagon on second spin.

5 A 10 pence and a 2 pence coin are spun.

a Draw a table to show all the possible outcomes.

b Find the probability that the 10p shows tails and the 2p shows heads.

6 The probability that Richard will catch a bus to work is 0.24.
The probability that Kerry will catch a bus to work is 0.15.
The probability that Richard and Kerry will catch the bus to work is 0.036.
Are the events 'Richard catches a bus to work' and 'Kerry catches a bus to work' independent? Explain your reasons.

This spread will show you how to:

● Know that if A and B are independent events, the probability of both A and B occurring is P(A and B) = P(A) × P(B)
● List possible outcomes systematically

Keywords
Event
Outcome

To calculate the probability of two **events** both occurring, you need to identify all the possible **outcomes** systematically.

Example

In a game of backgammon you can win or lose or draw.
The probability that Joshua will win any game of backgammon against Reuben is 0.4.

Joshua and Reuben play two games of backgammon.
Work out the probability that Joshua wins at least one game.

The only outcome where Joshua *does not* win at least one game is 'lose, lose'.

P(win at least one game) = 1 − P(lose, lose)
$$= 1 - 0.6 \times 0.6 = 1 - 0.36 = 0.64$$

The events of the two games are independent.
P(lose, lose) = P(lose) × P(lose)

Example

Two boxes of chocolates contain caramel, nut and cream centres.
Box X has 4 caramel, 7 nut and 9 cream centres.
Box Y has 8 caramel, 5 nut and 2 cream centres.

Jodie chooses one chocolate at random from box X and one from box Y.
Work out the probability that she picks two chocolates of different types.

X   Y

P(same type) = P(caramel, caramel) + P(nut, nut) + P(cream, cream)
$$= \left(\frac{4}{20} \times \frac{8}{15}\right) + \left(\frac{7}{20} \times \frac{5}{15}\right) + \left(\frac{9}{20} \times \frac{2}{15}\right)$$
$$= \frac{32}{300} + \frac{35}{300} + \frac{18}{300}$$
$$= \frac{85}{300} = \frac{17}{60}$$
P(different types) $= 1 - \frac{17}{60} = \frac{43}{60}$

The events 'from box X' and 'from box Y' are independent.

Example

The table shows the number of boys and the number of girls in Years 12 and 13 at a school.
Two students are to be chosen at random.

	Year 12	Year 13
Boys	48	96
Girls	72	54

a One student is chosen from Year 12 and one from Year 13.
   Calculate the probability that both students will be girls.
b One student is chosen from all the girls and one from all the boys.
   Calculate the probability that both students are in Year 13.

a P(girl Y12 and girl Y13) $= \frac{72}{120} \times \frac{54}{150} = \frac{27}{125}$
b P(Y13 girl and Y13 boy) $= \frac{54}{126} \times \frac{96}{144} = \frac{2}{7}$

The denominator for each fraction depends on the group from which the student is chosen.

1 In a game of chess you can win, lose or draw.
The probability that Edmund will win any game of
chess against Kevin is 0.9.
Kevin and Edmund play two games of chess.
Work out the probability that Edmund will win at
least one game.

2 James has two bags of marbles.
Bag A contains 4 red, 3 green and 5 blue marbles.
Bag B contains 7 red, 6 green and 2 blue marbles.
James chooses one marble at random from each bag.
Find the probability that the two marbles he chooses are
different colours.

3 Penny has two packets of photos.
Packet X has 7 pictures of her family, 11 of friends and 6 of her cat.
Packet Y has 9 pictures of her family, 12 of friends and 15 of her cat.
Penny chooses one picture at random from each packet.
Work out the probability that the pictures she chooses are of
different subjects.

4 The table shows the number of male and female staff in two
branches of a company.

	London	Manchester
Male	76	74
Female	84	66

Two people are to be chosen at random.

a One person is chosen from London and one from Manchester.
Calculate the probability that both are women.

b One woman and one man are chosen.
Calculate the probability that both are from Manchester.

5 The table shows how many boys and girls in Class 10Z wear glasses.

	Wears glasses	Does not wear glasses
Boys	6	10
Girls	4	13

Two students are to be chosen at random from class 10Z.
a One student is chosen who wears glasses and one student is chosen
who does not wear glasses. Calculate the probability that both
students will be boys.

b One student is chosen from all the girls and one student is chosen
from all the boys.
Calculate the probability that neither student wears glasses.

Exam review

Key objectives

- Identify different mutually exclusive outcomes and know that the sum of the probabilities of all these outcomes is 1
- Know when to add or multiply two probabilities: if A and B are mutually exclusive, then the probability of A or B occurring is $P(A) + P(B)$, whereas if A and B are independent events, the probability of A and B occurring is $P(A) \times P(B)$
- List all outcomes for single events, and for two successive events, in a systematic way
- Understand and use estimates or measures of probability from theoretical models, or from relative frequency

1 Laura rolls a dice 100 times and records the results in a table.

Number	1	2	3	4	5	6
Frequency	20	12	26	9	13	20

Do you think the dice is fair? (2)
Explain your answer.

2 The probability that Betty will be late for school tomorrow is 0.05. The probability that Colin will be late for school tomorrow is 0.06.

The probability that both Betty and Colin will be late for school tomorrow is 0.011.

Fred says that the events 'Betty will be late tomorrow' and 'Colin will be late tomorrow' are independent.

Justify whether Fred is correct or not. (2)

(Edexcel Ltd., 2003)

This unit will show you how to

- Calculate a proportion of an amount, writing it as a fraction of the total
- Understand ratio, direct proportion and inverse proportion
- Solve problems involving proportion, direct proportion and inverse proportion
- Calculate proportional and percentage changes
- Solve problems involving proportional change and interest
- Use ratio notation, including reduction to its simplest form and its various links to fraction notation
- Divide a quantity in a given ratio
- Solve problems involving ratio

Before you start ...

You should be able to answer these questions.

Review

1 Calculate each of these.

    a $\frac{2}{3}$ of 120   **b** 0.35×60

    c 28% of 70   **d** $\frac{3}{5}$ of 280

Unit N3

2 If three identical books weigh 288 g altogether, how much would seven of the same books weigh?

Unit N3

3 A car completes a particular journey in 4 hours, at an average speed of 30 miles per hour.
How long would it take to complete the same journey at an average speed of 40 miles per hour?

Unit N3

4 Divide these amounts in the ratios given.

    a 160 in the ratio 3 : 2

    b 24 in the ratio 3 : 2

    c 25 in the ratio 1 : 4

    d 36 in the ratio 5 : 4

Unit N3

This spread will show you how to:

- Calculate a proportion of an amount, writing it as a fraction of the total
- Solve problems involving proportion

Keywords

Part
Proportion
Total

- To calculate a **proportion**, write the **part** as a fraction of the **total**.

  In a cooking class with 14 girls and 16 boys, the proportion of girls is $\frac{14}{30} = \frac{7}{15} = 0.467 = 46.7\%$.

- If you know what proportion of the total a part is, you can find the actual size of the part by multiplying the proportion by the total.

  If a 450 g pie contains 12% chicken, the actual weight of chicken is $450 \text{ g} \times \frac{12}{100} = 54 \text{ g}$.

Example

a Find: **i** $\frac{3}{5}$ of 205 **ii** 31% of 490 **iii** 135% of 27 **iv** $\frac{15}{16}$ of 95

b What proportion is **i** 5 of 35 **ii** 7 of 31 **iii** 18 of 24?
   Give your answers as fractions and percentages.

a i Estimate: $\frac{3}{5}$ of 205 ≈ $\frac{3}{5}$ of 200 = 120

$\frac{3}{1\cancel{5}} \times \cancel{205}^{41} = \frac{3}{1} \times 41 = 123$

ii Estimate: 31% of 490 ≈ 30% of 500 = 150
By long multiplication, 31 × 49 = 1519
So 31% of 490 = 151.9

iii Estimate: 135% of 27 ≈ 100% + 30% of 30 = 30 + 9 = 39
By long multiplication, 27 × 135 = 3645
So 135% of 27 = 36.45

iv Estimate: $\frac{15}{16}$ of 95 ≈ 90

$\frac{15}{16} \times 95 = \frac{15 \times 95}{16} = 89.0625$ (by calculator)

b i $\frac{1\cancel{5}}{\cancel{35}7} = \frac{1}{7}$ **ii** $\frac{7}{31}$ **iii** $\frac{3\cancel{18}}{\cancel{24}4} = \frac{3}{4}$

$\frac{1}{7} = 14.3\%$ $\frac{7}{31} = 22.6\%$ $\frac{3}{4} = 75\%$

5 is $\frac{1}{7}$ or 14.3% of 35   7 is $\frac{7}{31}$ or 22.6% of 31   18 is $\frac{3}{4}$ or 75% of 24

You can use these techniques to solve proportion problems.

Example

Tom has £2800. He gives $\frac{1}{5}$ to his son and $\frac{1}{4}$ to his daughter.

How much does Tom keep himself? You must show all of your working.

$\frac{1}{5}$ of £2800 = 2 × £280 = £560

$\frac{1}{4}$ of £2800 = £700

Tom keeps £2800 − (£560 + £700) = £1540

1 Work out these proportions, giving your answers as
   i fractions in their lowest terms   ii percentages.
   a 7 out of every 20   **b** 8 parts in a hundred
   c 6 out of 20   **d** 75 in every 1000
   e 18 parts out of 80   **f** 9 parts in every 60

2 A 250 ml glass of fruit drink contains 30 ml of pure orange juice.
   What proportion of the drink is pure juice? Give your answer as
   a a fraction in its lowest terms
   b a percentage.

3 A soft drink comes in two varieties. The 'Regular' variety contains
   8.4% sugar by weight, and the 'Lite' version contains 2.5% sugar by
   weight. Find the weight of sugar in these amounts of soft drink.
   a 200 g of Regular   **b** 350 g of Lite   **c** 360 g of Lite
   d 800 g of Regular   **e** 750 g of Lite   **f** 580 g of Regular

4 Calculate each of these.
   a 90% of 50 kg   **b** 105% of 80 g   **c** 80% of 60 cm | Do these mentally.
   d 95% of 300 m   **e** 55% of 29 cc   **f** 65% of 400 mm

5 Use an appropriate method to work out these percentages.
   Show your method each time.
   a 50% of 270 kg   **b** 27.9% of 115 m   **c** 37.5% of £280
   d 25% of 90 cm^3 **e** 19% of 2685 g   **f** 27.5% of £60.00

6 Work out these.
   a $\frac{4}{5}$ of £800   **b** $\frac{3}{7}$ of €420   **c** $\frac{3}{8}$ of 640 kg
   d $\frac{2}{5}$ of 760 cc **e** $\frac{2}{3}$ of 2460 kg **f** $\frac{5}{6}$ of 28 m

7 Jo wins £3600 in a competition. She gives $\frac{1}{3}$ to her mother,
   and $\frac{1}{4}$ to her sister.
   a How much does Jo keep? Show your working.
   b What proportion of the prize money does she give away?
   Give your answer as
   i a fraction
   ii a percentage.

This spread will show you how to:

- Understand direct proportion
- Solve problems involving direct proportion

- Quantities which are allowed to change are called **variables**.

Amounts of red and white paint in a mixture of pink paint are variables.

- Two variables are in **direct proportion** if the **ratio** between them stays the same as the actual values vary.

One litre of a shade of pink paint is made by mixing 200 ml of red paint and 800 ml of white paint. The ratio of red paint to white paint is 1 : 4. To make the same shade of pink paint, this ratio should stay the same even if the quantities change.

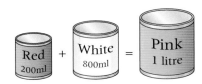

- When you multiply (or divide) one of the variables by a certain number, you have to multiply or divide the other variable by the same number.

250 ml of red paint has to be mixed with 4 × 250 = 1000 ml of white paint.

- You write 'y is proportional to x' as $y \propto x$. This can also be written as $y = kx$, where k is the **constant** of proportionality.

You can always find the amount of white paint by multiplying the amount of red paint by a fixed number, the constant of proportionality, in this case 4.

Example

If 35 metres of steel cable weigh 87.5 kilograms, how much do 25 metres of the same cable weigh?

If w = length of the cable (metres)
   w = weight (kilograms),

then $w = kx$, where the constant of proportionality k represents the weight of one metre of cable.

$87.5 = k \times 35 \Rightarrow k = 87.5 \div 35 = 2.5$

so $w = 2.5x$

Substitute $x = 2.5$ into the formula

$w = 2.5 \times 25$
$= 62.5$ kg

1 A 750 ml can of paint covers 5 m². What area will a 2.5 litre can of the same paint cover? Show your working.

2 A 4 metre length piece of pipe weighs 28.7 kg. How much does a 3.6 m length of the same pipe weigh?

3 The table shows corresponding values of the variables w, x, y and z.

w	3	6	9	15
x	8	14	20	32
y	5	10	15	25
z	4	7	10	16

Which of these statements could be true?

a $w \propto x$ **b** $z \propto x$ **c** $z \propto w$ **d** $y \propto w$

e $w = ky$, where k is a constant

4 Using the values from the table in question **3**, plot graphs of

a w against x **b** w against y **c** x against y **d** x against z.

Use your results to describe the key features of a graph showing the relationship between two variables that are in direct proportion.

5 The weight, w, of a piece of wooden shelving is directly proportional to its length, l.

a Write the statement 'w is directly proportional to l' using algebra.

b Rewrite your answer to part **a** as a formula including a constant of proportionality, k.

c Given that 2.5 m of the shelving weighs 6.2 kg, find the value of the constant k.

d Use your previous answers to calculate the weight of a 2.9 m length of the shelving.

6 450 ml of olive oil costs €2.45. Calculate the cost of 750 ml of the same oil.

7 Anna and Betty buy gravel from a garden centre. Anna buys 4.7 kg of gravel for £12.98. Betty buys 6.2 kg of the same gravel. How much does she pay? Show your working.

8 A store sells the same type of ribbon in two different packs.

Regular   **Super**

5 metres   12 metres

Cost £13.25   Cost £31.50

Which of these two packs gives the best value for money? You must show all your working.

This spread will show you how to:

- Understand inverse proportion
- Solve problems involving inverse proportion

Keywords

Constant
Inverse
  proportion
Product

When two variables are in direct proportion, as the value of one of the variables increases, the other one increases; as the value of one of the variables decreases, the other one decreases.

- When variables are in **inverse proportion**, one of the variables increases as the other one decreases, and vice-versa.
- 'y is inversely proportional to x' can be written as $y \propto \frac{1}{x}$, or $y = \frac{k}{x}$, where k is the **constant** of proportionality.
- If two variables are in inverse proportion, the **product** of their values will stay the same.

For example, the number of bricklayers building a wall and the time taken are in inverse proportion.

5 bricklayers × 4 hours = 20	5 people take 4 hours
10 bricklayers × 2 hours = 20	10 people take 2 hours
1 bricklayer × 20 hours = 20	1 person takes 20 hours

The amount of labour needed is often called 'man-hours'.

Example

A pot of paint, which will cover an area of 12 m², is used to paint a garage floor.
a Show that the width of the floor that can be painted is inversely proportional to its length.
b Find the width of the floor when the length is
  i 1 m   **ii** 3 m   **iii** 6.5 m

a Use l for the length of the floor and w for the width
$$lw = 12 \implies w = \frac{12}{l}$$
This equation shows that w is inversely proportional to l.
b Substitute the given values into the formula.
  i $w = \frac{12}{1} = 12$ m   **ii** $w = \frac{12}{3} = 4$ m   **iii** $w = \frac{12}{6.5} = 1.85$ m

Example

A trolley travels down a 10 m long test track at a constant speed.
a Show that the time required for the journey is inversely proportional to the speed of the trolley.
b Find the speed required to make the trolley complete the journey in exactly 4 seconds.

a Use v to represent the speed (in metres per second), and t to represent the time (in seconds). The relationship 'Speed × Time = Distance' gives:
$$vt = 10 \implies t = \frac{10}{v}$$
This equation shows that t is inversely proportional to v.
b The equation from part **a** can be rearranged to give $v = \frac{10}{t}$.

Substituting $t = 4$ gives $v = \frac{10}{4} = 2.5$ metres per second.

1 You are told that the variable y is directly proportional to the variable x. Explain what will happen to the value of y when the value of x is

   a doubled   **b** halved   **c** multiplied by 6

   d divided by 10   **e** multiplied by a factor of 0.7.

2 You are told that the variable w is inversely proportional to the variable z. Explain what will happen to the value of w when the value of z is

   a doubled   **b** halved   **c** multiplied by 6

   d divided by 10   **e** multiplied by a factor of 0.7.

3 You are told that y is inversely proportional to x, and that when $x = 4$, $y = 4$. Find the value of y when x is equal to

   a 8   **b** 2   **c** 40   **d** 1   **e** 100.

4 Given that $y \propto \frac{1}{w}$, and that $y = 10$ when $w = 50$, write an equation connecting y and w.

5 You are told that $y = \frac{k}{x}$, and that $y = 20$ when $x = 40$. Find the value of the constant k.

6 The number of hours, t, required to dig a hole is inversely proportional to n, the number of men digging the hole. It would take 2 men 5 hours to dig the hole.

   a Write a formula to give t in terms of n.

   b Find the time required to dig the hole when the number of men is

      i 1   **ii** 4   **iii** 8   **iv** 5   **v** 7

   c Do you think that the assumption $t \propto \frac{1}{n}$ is realistic? Explain your answer.

7 The electric current, I amps, flowing through a component is inversely proportional to the resistance, R ohms, of the component. When $R = 240$, $I = 1$. Find the value of I when R is equal to

   a 120   **b** 60   **c** 40   **d** 30

   e 24   **f** 360   **g** 480   **h** 100.

8 Use your answers from question **7** to plot a graph of R against I. Comment on the main features of your graph.

9 The variables u and v are in inverse proportion to one another. When $u = 6$, $v = 8$. Find the value of u when $v = 12$.

10 Prove that

   a if x is directly proportional to y, then y is directly proportional to x

   b if y is inversely proportional to x, then x is inversely proportional to y.

Repeated proportional change

This spread will show you how to:
- Calculate proportional and percentage changes
- Solve problems involving proportional change and interest

Keywords
Compound interest
Depreciation
p.a. (per annum)
Proportional change
Simple interest

- To work out a **proportional change**, first work out the size of the change, and then add to, or subtract from, the original amount.

  If fuel charges go up by $\frac{1}{4}$ and original charge = £48,
  increase = $\frac{1}{4}$ × £48 = £12 and new charge = £48 + £12 = £60.

- Percentage changes can be found using a decimal multiplier.

  To increase £20 by 15%, multiply £20 by 1.15.
  To reduce 18 kg by 12.5%, multiply 18 kg by 0.875.

100% − 12.5% = 87.5%

Example

Find the new total when

a £340 is increased by a third
b 210 cm is decreased by 45%
c VAT at 17.5% is added to a price of £297.

a £340 ÷ 3 = £113.33
  New total £340 + £113.33 = £453.33
b 210 cm × 0.55 = 115.5 cm   To decrease an amount by 45%, multiply by 0.55.
c £297 × 1.175 = £348.98   To increase an amount by 17.5%, multiply by 1.175.

- Repeated percentage changes can be found using powers of a decimal multiplier. A common application of this is in **compound interest**.

Example

Find the final amount when an initial sum of £1000 is invested for 12 years at

a 5% p.a. simple interest   b 5% p.a. compound interest.

p.a. (per annum) means 'per year'.

a In **simple interest**, the original amount invested (the principal) earns interest, but the interest is not added to the principal.
  The principal earned £50 per year for 12 years = £600.
  Final amount = £1000 + £600 = £1600.
b In **compound interest**, the interest payments are added to the principal, and will themselves earn interest. The sum in the account after 12 years will be £1000 × 1.05^{12} = £1795.86.

Some items, like cars, decrease in value as they age. This is called **depreciation**.

Example

Sue has had her car for 3 years. Each year, the car has depreciated in value by 15%. Now the car is worth £5220.
How much was the car worth when Sue bought it?

Original cost = C
$C × 0.85^3$ = £5220
so C = £5220 ÷ 0.85^3 = £8500 (to the nearest pound)

1 Copy and complete the table to show the result of some proportional changes.

Original quantity	Proportional change	Result
235 mm	Decrease by $\frac{1}{5}$	
38 litres	Increase by $\frac{1}{8}$	
295 cm³	Decrease by $\frac{2}{9}$	
£96.55	Increase by $\frac{5}{6}$	

2 Calculate these percentage changes.

a £375 decreased by 12%   b £290 decreased by 15%

c £885 decreased by 9%   d £439 increased by 22%

3 Find the simple interest earned on

a £860 at 5.3% p.a. for 4 years

b £350 000 at 4.75% for 18 years.

4 Find the compound interest earned on

a £720 at 3.8% p.a. for 5 years

b £410 000 at 5.15% for 12 years.

5 Charlene invests £950 for 5 years in an account that pays 4.5% per year simple interest.

a How much interest will Charlene earn over 5 years?

b Charlene's Dad tells her that it would be better to invest the money in an account that paid 3.5% compound interest. Is he right? Explain your answer.

c Would your answer for part **b** have been different if the money was invested for 10 years instead of 5? Explain your reasoning.

6 £6500 is invested at 3.9% compound interest per annum. How many years will it take for the investment to exceed £8000?

7 During the 21 years that Pat owned a house, it increased in value at an average rate of 3.5% per year. When she sold the house, it was worth £245 000. Pat's sister says, 'Your house must have doubled in value since you bought it.' Is she right? Show your working.

8 A special savings account earns 7% per annum compound interest.

a Liz invests £2400 in the special account. How much will she have in her account after 2 years?

b Paul invests some money in the same account. After earning interest for 1 year, he has £1669.20 in his account. How much money did Paul invest?

Ratio problems

This spread will show you how to:

- Use ratio notation, including reduction to its simplest form and its various links to fraction notation
- Divide a quantity in a given ratio
- Solve problems involving ratio

Keywords

Ratio
Simplify

- **Ratios** are used to compare quantities.

If there are 18 girls and 14 boys in a class, the ratio of girls to boys = 18 : 14.
This simplifies to 9 : 7.

- **To divide a quantity in a given ratio**
  - first find the total number of parts
  - find the size of one part
  - multiply to find each share.

Simplify a ratio by cancelling common factors, for example 8 : 12 simplifies to 2 : 3.

Example

Share £400 between Tom and Ed in the ratio 3 : 2.

Total number of parts = 3 + 2 = 5

Size of one part = £400 ÷ 5 = £80
$$3 \times £80 = £240$$
$$2 \times £80 = £160$$

Tom gets £240, Ed gets £160.

Example

Divide

a 460 cm in the ratio 3 : 2

b £2500 in the ratio 3 : 2 : 5.

a　　3 + 2 = 5
　　460 ÷ 5 = 92
　　　2 × 92 = 184
　　　3 × 92 = 276
　　The two parts are 276 cm and 184 cm.

b　　3 + 2 + 5 = 10
　　2500 ÷ 10 = 250
　　　250 × 3 = 750
　　　250 × 2 = 500
　　　250 × 5 = 1250
　　The three parts are £750, £500 and £1250.

Example

Andy and Brenda share a bingo prize of £65 in the ratio of their ages. Andy is 38, and Brenda is 42. How much does each get?

Ratio of ages = 38 : 42 = 19 : 21
　　19 + 21 = 40

Andy receives (£65 ÷ 40) × 19 = £30.88
Brenda receives (£65 ÷ 40) × 21 = £34.12.

One part is
£65 ÷ 40.

1 Simplify these ratios.

   a $8:4$ **b** $15:3$ **c** $6:4$

   d $2:4:6$ **e** $45:18$ **f** $14:28:84$

2 Divide each of these amounts in the ratio given. Show your working.

   a £55 in the ratio $3:2$ **b** 120 cm in the ratio $5:3$

   c 96 seats in the ratio $4:3:1$ **d** 42 tickets in the ratio $9:3:2$

   e 144 books in the ratio $8:3:1$ **f** 160 hours in the ratio $8:5:3$

3 Copy and complete the table to show how the quantities can be divided in the ratios given. Show your working.

Quantity	Ratio	First share	Second share
£140	$4:3$		
85 cm	$3:2$		
18 hours	$5:4$		
49 cc	$1:6$		
45 minutes	$7:8$		
€720	$19:5$		

4 Mrs Jackson wins £800 in a competition. She decides to keep half of the money, and share the rest between her two children, Amber (who is 8 years old) and Benny (who is 10) in the ratio of their ages. Work out how much each child receives. Show your working.

5 Prize money of £9600 is shared between Peggy, Grant and Mehmet in the ratio $8:9:7$ respectively. How much do they each receive?

6 Steven, Will and Phil divide prize money of £12 000 between them in the ratio $10:6:3$. Find the amount that each person receives, giving your answers to the nearest penny.

7 Copy and complete the table to show how each quantity can be divided in the ratio given. Give your answers to a suitable degree of accuracy.

Quantity	Ratio	Share 1	Share 2	Share 3
200 cm^2	$12:4:1$			
38 cm	$3:2:1$			
450 m	$6:3:4$			
720 mm	$2:5:4$			
$95	$8:3:5$			

8 John and Janine entered a quiz. John came first with 32 points, and Janine came second with 27 points. They shared prize money of £45 in the ratio of their scores. How much did they each receive?

Compound measures

This spread will show you how to:

- Round answers to an appropriate degree of accuracy
- Solve problems involving compound measures, including speed and density

Keywords
Compound
Density
Speed

Compound measures involve a combination of measurements and units.

- The **density** of a material is its mass divided by its volume.
- **Speed** is the distance travelled divided by the time taken.

- A formula triangle can be a useful way of remembering the relationships between the different parts of a compound measure.
  For example:
  speed = distance ÷ time
  distance = speed × time
  time = distance ÷ speed

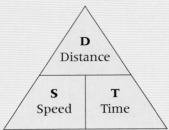

The units for density can be grams per cubic centimetre (g/cm³), or kilograms per cubic metre (kg/m³).

Speed can be measured in miles per hour (mph), kilometres per hour (kph) or metres per second (m/s).

Example

a A cube of side 3.5 cm has a mass of 600 g. Find the density of the cube in g/cm³, correct to 3 significant figures.

b A car travels 240 miles in 3 hours 45 minutes. Find the average speed of the car in miles per hour.

c Lubricating oil has a density of 0.58 g/cm³. Find
   i   the mass of 2.5 litres of this oil
   ii the volume of 10 grams of the oil.

Volume of a cube = length³.

---

a Volume = 3.5^3 cm³ = 42.875 cm³
   Density = 600 g ÷ 42.875 cm³ = 13.994 ... g/cm³ = 14.0 g/cm³ to 3 sf

$$Density = \frac{mass}{volume}$$

b Average speed = $\dfrac{\text{total distance}}{\text{total time}}$

   $= \dfrac{240}{3.75}$

   = 64 mph

Put the time in hours.

c i   Mass = density × volume = 0.58 g/cm³ × 2500 cm³ = 1450 g
   ii Volume = mass ÷ density = 10g ÷ 0.58g/cm³ = 17.2 cm³

1 litre ≡ 1000 cm³

Example

A metal cuboid is 95 cm long and has a length of 2 cm, width 4 cm and height 6 cm.

a Find the density of the cuboid.

b If the mass of the cuboid is 7.8 kg, find its density in g/cm³.

---

a Volume of cuboid = 2 × 4 × 6 = 48 cm³

Volume of a cuboid = length × width × height.

b Density of cuboid = $\dfrac{\text{mass}}{\text{volume}} = \dfrac{7800}{48}$ = 163 g/cm³ to 3 sf

Change the mass into grams.

1 Rod cycles 18 miles in 2 hours. Find his average speed, in miles per hour (mph).

2 If 4 metres of fabric costs £8.40, find the price of the fabric in pounds per metre.

3 A car travels 24 miles in 45 minutes. Find the average speed of the car in miles per hour (mph).

4 A train leaves Euston at 8:57 a.m. and arrives at Preston at 11:37 a.m. If the distance is 238 miles find the average speed of the train.

5 Copy and complete the table to show speeds, distances and times for five different journeys.

Speed (kph)	Distance (km)	Time
105		5 hours
48	106	
	84	2 hours 15 minutes
86		2 hours 30 minutes
	65	1 hour 45 minutes

6 A cube of side 2 cm weighs 40 grams.

   a Find the density of the material from which the cube is made, giving your answer in g/cm³.

   b A cube of side length 2.6 cm is made from the same material. Find the mass of this cube, in grams.

> Volume of cube = length³.

7 A box has a length and width of 22.50 mm, and a height of 3.15 mm. It has a mass of 9.50 g.

   a Find the density of the metal from which the box is made, giving your answer in g/cm³.

   b How many boxes can be made from 1 kg of the material?

> Volume of cuboid = length × width × height.

8 Emulsion paint has a density of 1.95 kg/litre. Find

   a the mass of 4.85 litres of the paint.

   b the number of litres of the paint that would have a mass of 12 kg.

9 A steel cable weighs 2450 kg.
   The cable has a uniform circular cross-section of radius 0.85 cm.
   The steel from which the cable is made has a density of 7950 kg/m³.
   Find the length of the cable.

> Volume of a cylinder = $\pi r^2 \times$ length.

0.85 cm ─ l ─

Key objectives

- Represent proportional changes using a multiplier raised to a power
- Calculate an unknown quantity from quantities that vary in direct or inverse proportion
- Use ratio notation, including reduction to its simplest form and its various links to fraction notation

1 Louise invests £x in a bank account.

The account pays 4% per year compound interest.

After two years the total in her account is £270.40.

What was Louise's original investment? (3)

2 Rosa prepares the ingredients for pizzas.

She uses cheese, topping and dough in the ratio 2 : 3 : 5.

Rosa uses 70 grams of dough.

Work out the number of grams of cheese and the number of grams of topping Rosa uses. (3)

(Edexcel Ltd., 2004)

This unit will show you how to

- Identify properties preserved under reflection, rotation and translation
- Understand congruence
- Describe reflections, using mirror lines
- Understand that rotations are specified by a centre and an (anticlockwise) angle
- Understand and use vector notation
- Describe translations by giving a distance and direction (or vector)
- Transform 2-D shapes by translation, rotation and reflection and combinations of these transformations
- Recognise and construct enlargements, understanding the effect on lengths and angles
- Understand the significance of the scale factor of an enlargement, including fractional scale factors
- Apply the intersecting chord theorem

Before you start ...

You should be able to answer these questions.

Review

1 Write the coordinates of these points.

CD – S3

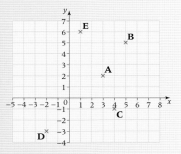

2 Write the equations of these lines.

Unit A4

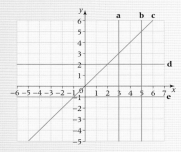

3 Draw these lines on a grid.

Unit A4

  a $x = 4$ **b** $x = -2$ **c** $y = 3$

  d $y = 0$ **e** $x = 0$

Congruence

This spread will show you how to:

Keywords
Congruent
Hypotenuse

- Understand congruence in the context of transforming 2-D shapes
- Recognise the properties of congruent triangles

- In **congruent** shapes:
  - corresponding lengths are equal
  - corresponding angles are equal.

Congruent shapes are rotations, reflections or translations of each other.

Rotation, reflection and translation preserve lengths and angles.

- You can prove that two triangles are congruent by showing they satisfy one of four sets of conditions.

SSS: three sides are the same

SAS: two sides and the included angle are the same

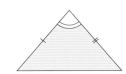

ASA: two angles and the included side are the same

RHS: Right-angled triangles with **hypotenuse** and one other side the same.

Example

Explain why the following pairs of triangles are congruent.

a

7 cm
3 cm
7 cm
3cm

b

6 cm
32°

74°
6 cm

---

a The triangles are both right-angled.
The hypotenuses are both 7 cm.
One other side is the same, 3 cm.
The triangles are congruent by RHS.

b The triangles are isosceles.
The angles in the triangles are:

32°
x   x

74° 74°
y

ASA (32°, 6 cm, 74°) so the triangles are congruent.

Base angles in an isosceles triangle are equal.
$2x + 32° = 180°$
$x = \frac{148}{2} = 74°$
$y + 2 \times 74° = 180°$
$y = 32°$

You could also use: SAS (6 cm, 32°, 6 cm)

1 Explain whether or not these pairs of triangles are congruent.

a
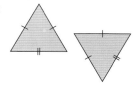

b
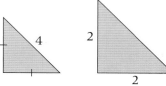

c
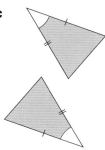

2 ABCD is a rectangle.

Prove that triangles ABD and CDB are congruent.

3 Explain why triangles PQR and WXT are congruent.

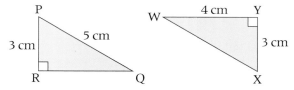

4 KLMN is a kite.

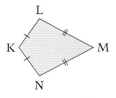

a Explain why triangles KLN and MLN are not congruent.

b Explain why triangles KLM and KNM are congruent.

5 EFGH is a parallelogram.

a Find two different pairs of congruent triangles.

b Explain how you know they are congruent.

Reflection

This spread will show you how to:

- Identify properties preserved under reflection
- Understand congruence
- Describe reflections, using mirror lines

Keywords
Congruent
Mirror line
Perpendicular
Reflection

Just as you can see your reflection in a mirror, you can reflect a shape in a mirror line.

Corresponding points on the object and image are the same distance from the **mirror line**.

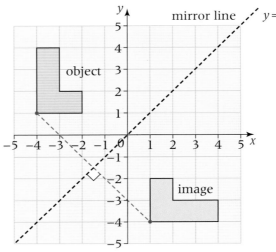

Corresponding angles and lengths are the same in the image and the object.

The object and the image are **congruent**.

Reflection flips the shape over.

The line joining a point and its image is **perpendicular** to the mirror line.

Example

Reflect the triangle with vertices at (1, 2), (2, 7) and (4, 7) in the line $x = 1$.

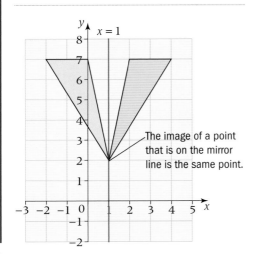

The image of a point that is on the mirror line is the same point.

Example

Reflect the pink triangle in the line $y = 3$.

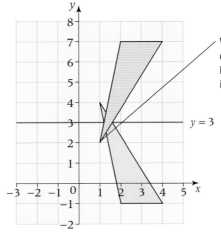

When an object crosses the mirror line, so does its image.

1 Copy this diagram.

  a Reflect the triangle T in the line $x = 2$.
  Label the image U.

  b Reflect the triangle T in the line $y = 1$.
  Label the image V.

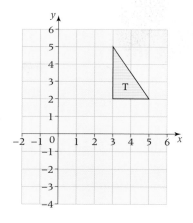

2 Copy this diagram.

  a Reflect the kite K in the line $x = -1$.
  Label the image L.

  b Reflect the kite K in the line $y = 1$.
  Label the image M.

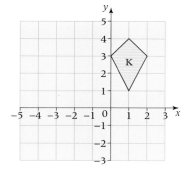

3 Copy this diagram and extend the y-axis to -8.

  a Reflect the quadrilateral Q in the x-axis.
  Label the image R.

  b Reflect the quadrilateral Q in the line $y = x$.
  Label the image S.

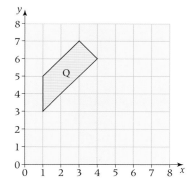

4 Copy this diagram.

  a Reflect triangle A in the y-axis.
  Label the image B.

  b Reflect triangle B in the y-axis.
  Label the image C.

  c What do you notice? Does this always happen?
  Check with some other reflections.

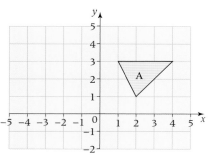

Rotation

This spread will show you how to:

- Identify properties preserved under rotation
- Understand that rotations are specified by a centre and an (anticlockwise) angle

Keywords
Angle
Centre of
   rotation
Congruent
Rotation

You can rotate a shape by turning it about a fixed point – the **centre of rotation**.

This object is rotated 90° anticlockwise. Every point on the shape moves through the same **angle**.
The centre of rotation is (1, 0).

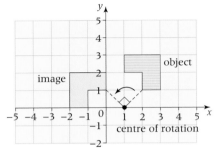

The object and the image are **congruent**.

- In a rotation, corresponding angles and lengths are the same in the image and the object.

Corresponding points on the object and image are the same distance from the centre of rotation.

Example

Rotate the triangle through −90° about centre (1, 2).

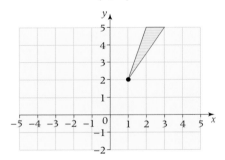

Rotate each vertex through −90° Use tracing paper to help.

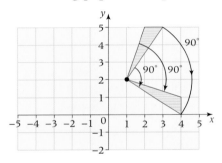

You measure angles anti-clockwise, so −90° means 90° clockwise.

A vertex at the centre of rotation does not move.

Rotation through 180° is a half turn, the same clockwise or anticlockwise.

Example

Rotate the pink triangle through 180° about centre O.

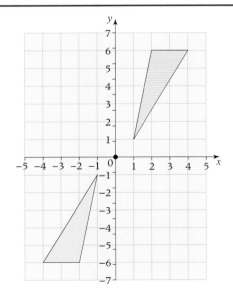

O is the origin, (0, 0).

1 Copy this diagram.

  a Rotate the triangle T through 180° about (0, 0). Label the image U.

  b Rotate the triangle T through 180° about (2, 2). Label the image V.

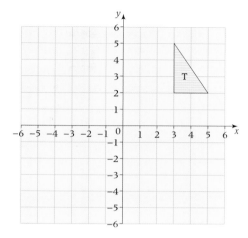

2 Copy this diagram.

  a Rotate the kite K through 180° about (1, 1). Label the image L.

  b Rotate the kite K through −90° about (2, 0). Label the image M.

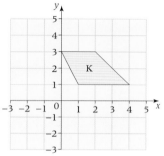

3 Copy this diagram.

  a Rotate the triangle A through 90° clockwise about (1, 0). Label the image B.

  b Rotate the triangle A through 90° anticlockwise about (0, 1). Label the image C.

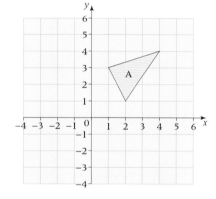

4 Copy this diagram.

  a Rotate shape P through −90° about (2, 1). Label the image Q.

  b Rotate image Q through −90° about (2, 1). Label the image R.

  c Rotate shape P through 180° about (2, 1). What do you notice? Explain why this happens.

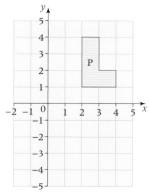

Translation

This spread will show you how to:

- Identify properties preserved under translation
- Understand and use vector notation
- Describe translations by giving a distance and direction (or vector)

Keywords

Translation
Vector

A **translation** is a sliding movement.
All points on the shape slide the same distance in the same direction.

- In a translation the object and the image are congruent.

You can describe a translation using a **vector**.

- Vector $\begin{pmatrix} a \\ b \end{pmatrix}$ means moving a units in the

  x-direction and b units in the y-direction.

  The L-shape is translated by the vector $\begin{pmatrix} 4 \\ 3 \end{pmatrix}$.

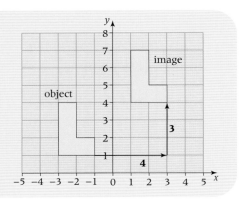

Example

Translate triangle A by the vector $\begin{pmatrix} -3 \\ 4 \end{pmatrix}$.

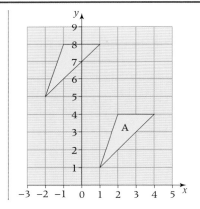

$\begin{pmatrix} -3 \\ 4 \end{pmatrix}$ means move

−3 in the
x-direction, so 3
squares to the left.

Move 4 in the
y-direction, so 4
squares up.

Example

Translate triangle T by the vector $\begin{pmatrix} 0 \\ -3 \end{pmatrix}$.

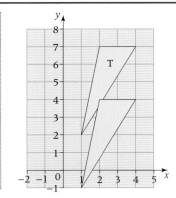

$\begin{pmatrix} 0 \\ -3 \end{pmatrix}$ means move 0

in the x-direction, so
the shape only moves
in the y-direction.

Move −3 in the
y-direction, so 3
squares down.

1 Copy this diagram.

a Translate the triangle T by the vector $\begin{pmatrix} 3 \\ 4 \end{pmatrix}$.

Label the image U.

b Translate the triangle T by the vector $\begin{pmatrix} 5 \\ -2 \end{pmatrix}$.

Label the image V.

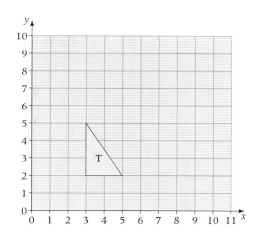

2 Copy this diagram.

a Translate the kite K by the vector $\begin{pmatrix} -4 \\ 3 \end{pmatrix}$.

Label the image L.

b Translate the kite K by the vector $\begin{pmatrix} 2 \\ -4 \end{pmatrix}$.

Label the image M.

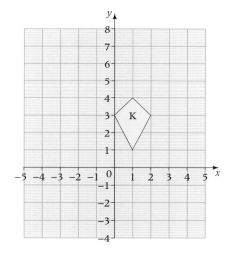

3 Copy this diagram.

a Translate shape A by the vector $\begin{pmatrix} 6 \\ -4 \end{pmatrix}$.

Label the image B.

b Translate the image B by the vector $\begin{pmatrix} -6 \\ 4 \end{pmatrix}$.

What do you notice? Does this always happen? If so, why?

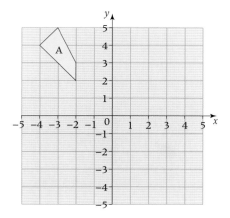

Describing transformations

This spread will show you how to:

- Describe reflections, using mirror lines
- Understand that rotations are specified by a centre and an (anticlockwise) angle
- Describe translations by giving a distance and direction (or vector)

Keywords

Maps
Reflection
Rotation
Transformation
Translation

- To describe a **reflection**, you give the equation of the line.
- To describe a **rotation**, you give the centre and the angle of rotation.
- To describe a **translation**, you give the distance and direction or you specify the vector.

Reflections, rotations and translations are all **transformations**.

Example

Describe the transformation that maps shape A on to

a shape B   **b** shape C   **c** shape D.

Maps means changes.

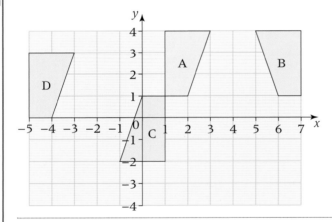

a In a reflection the mirror line bisects the line joining corresponding points on the object and image.

Shape B is a reflection of shape A in the line $x = 4$.

b The vertex (1, 1) does not move during the rotation, so it must be the centre of rotation.

Shape C is a rotation of shape A through $180°$.

c Shape D is a translation of shape A by the vector $\begin{pmatrix} -5 \\ -1 \end{pmatrix}$.

1 Describe fully the
transformation that maps

   a shape A onto shape B

   b shape A onto shape C

   c shape A onto shape D

   d shape B onto shape D

   e shape C onto shape D

   f shape D onto shape C.

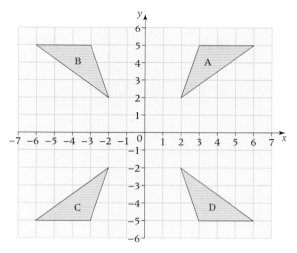

2 Describe fully the
transformation that maps

   a shape J onto shape K

   b shape L onto shape K

   c shape M onto shape K

   d shape L onto shape M

   e shape J onto shape M

   f shape M onto shape J.

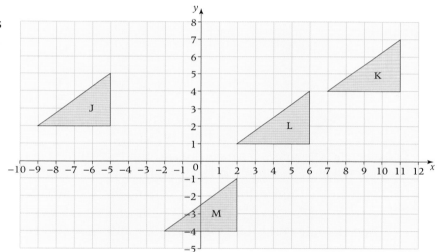

3 Describe fully the
transformation that maps

   a shape W onto shape X

   b shape W onto shape Y

   c shape W onto shape Z

   d shape Z onto shape W

   e shape X onto shape Y

   f shape Z onto shape X.

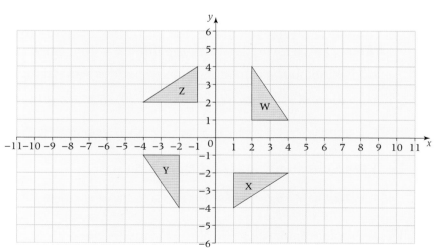

Combining transformations

This spread will show you how to:

● Transform 2-D shapes by translation, rotation and reflection and combinations of these transformations

Keywords
Reflection
Rotation
Transformations
Translation

You can combine **transformations** by doing one after the other.
You can describe a combination of transformation as a single transformation.

Example

In this diagram, triangle A undergoes three pairs of transformations.

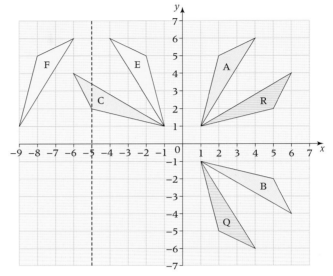

1. Triangle A is rotated 90° clockwise about (0, 0) to triangle B.
Then triangle B is rotated 180° about (0, 0) to triangle C.

What single transformation maps triangle A onto triangle C?

2. Triangle A is reflected in the line y = 0 (the x-axis) to triangle Q.
Then triangle Q is rotated through 90° anticlockwise about (0, 0) to triangle R.

What single transformation maps triangle A onto triangle R.

3. Triangle A is reflected in the line x = 0 (the y-axis) to triangle E.
Then triangle E is reflected in the line x = −5 to triangle F.

What single transformation maps triangle A onto triangle F?

1. A rotation of 90° anticlockwise about (0, 0) maps A onto C.

● A combination of rotations that have the same centre is equivalent to a single **rotation**.

2. A reflection in the line y = x maps A onto R.

● A combination of a reflection and a rotation is equivalent to a single **reflection**.

3. A translation by the vector $\begin{pmatrix} -10 \\ 0 \end{pmatrix}$ maps A onto F.

● A combination of reflections is a **translation** when the mirror lines are parallel.

1 Copy this diagram.

  a Rotate triangle A 90° anticlockwise about centre (0, 0). Label the image B.

  b Rotate triangle B 90° anticlockwise about centre (0, 0). Label the image C.

  c Describe fully the single transformation that takes triangle A to triangle C.

  d Rotate triangle B 90° clockwise about centre (2, 1). Label the image D.

  e Describe fully the single transformation that takes triangle A to triangle D.

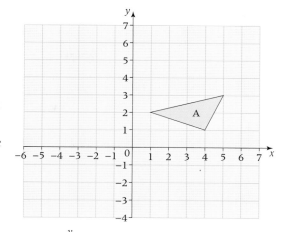

2 Copy this diagram.

  a Reflect triangle E in the y-axis. Label the image F.

  b Reflect triangle F in the x-axis. Label the image G.

  c Describe fully the single transformation that takes triangle G to triangle E.

  d Reflect triangle F in the line $x = 2$. Label the image H.

  e Describe fully the single transformation that takes triangle E to triangle H.

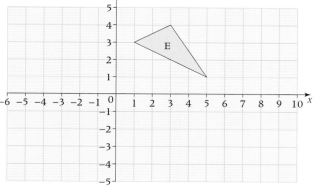

3 Copy this diagram.

  a Reflect triangle J in the x-axis. Label it K.

  b Rotate triangle K 180° about centre (0, 0). Label it L.

  c Describe fully the single transformation that takes triangle L to triangle J.

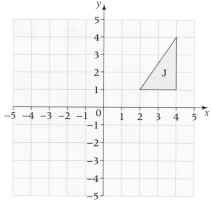

4 Copy this diagram.

  a Translate trapezium R by the vector $\begin{pmatrix} 5 \\ -3 \end{pmatrix}$. Label the image S.

  b Translate trapezium S by the vector $\begin{pmatrix} -1 \\ 2 \end{pmatrix}$. Label the image T.

  c Describe fully the single transformation that takes trapezium T to trapezium R.

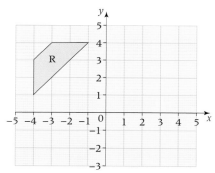

Enlargement

This spread will show you how to:

- Enlarge objects, given a centre of enlargement and scale factor, including fractional scale factors
- Recognise that enlargements preserve angle but not length
- Understand how enlargement affects perimeter

Keywords

Centre of
  enlargement
Enlargement
Scale factor
Similar

To enlarge a shape, you multiply all the lengths by the same scale factor.

- In an **enlargement**
  - corresponding angles are the same
  - corresponding lengths are in the same ratio.

You draw an enlargement from a centre.

In the diagram, △PQR is enlarged by **scale factor** 2 from the centre (0, 1).

The image, P'Q'R', is **similar** to the object, PQR.

All the distances are × 2 so CQ' = 2CQ, CP' = 2CP, CR' = 2CR

Lengths on the image are 2 × corresponding lengths on the object so P'R' = 2PR and so on.

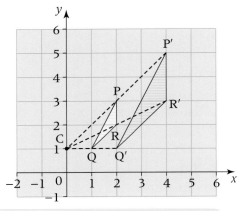

- Perimeter of image = scale factor × perimeter of object.

Example

a Enlarge the white triangle by scale factor 3, centre (2, 1).
b How much larger is the perimeter of the image than the perimeter of the object?

a

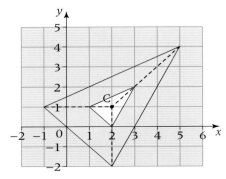

Lengths on the image are 3 × the corresponding lengths on the object.

The distance from C to a point on the image is 3 × the distance from C to the corresponding point on the object.

b Perimeter of image = 3 × perimeter of object.

1 a Copy this diagram, but extend both axes to 16.

 b Enlarge triangle T by scale factor 2, centre (0, 0). Label the image U.

 c Enlarge triangle T by scale factor 3, centre (0, 0). Label the image V.

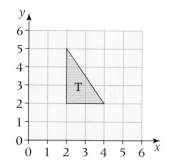

2 a Copy this diagram, but extend both axes from −7 to 7.

 b Enlarge kite K by scale factor 2, centre (1, 3). Label the image L.

 c Enlarge kite K by scale factor 4, centre (1, 3). Label the image M.

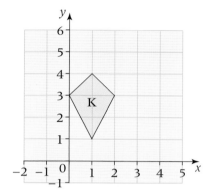

3 a Draw a grid with an *x*-axis from −6 to 6 and a *y*-axis from −3 to 15. Plot the points (1, 3) (1, 5) (3, 7) (4, 6). Join them to make quadrilateral Q.

 b Enlarge quadrilateral Q by scale factor 2, centre (4, 8). Label the image R.

 c Enlarge quadrilateral Q by scale factor 2, centre (0, 6). Label the image S.

 d How much larger is the perimeter of R than the perimeter of Q?

4 a Draw a grid with *x*- and *y*-axes from 0 to 13. Plot the points (1, 3) (4, 4) (2, 1). Join them to make triangle A.

 b Enlarge triangle A by scale factor 3, centre (0, 0). Label the image B.

 c Enlarge triangle A by scale factor 2, centre (2, 1). Label the image C.

 d How much larger is the perimeter of B than the perimeter of A?

Fractional scale factors

This spread will show you how to:

● Enlarge objects, given a centre of enlargement and scale factor, including fractional scale factors

Keywords

Centre of enlargement
Enlargement
Scale factor

A map or a scale drawing is an enlargement by a fractional scale factor.

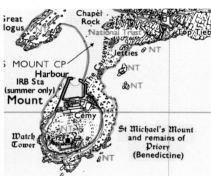

This map is an enlargement of St Michael's Mount by scale factor $\frac{1}{25\,000}$.

● Enlargement by a scale factor less than 1 produces a smaller image.

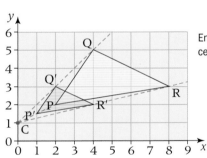

Enlargement by scale factor $\frac{1}{2}$, centre $(0, 1)$

All the distances are multiplied by $\frac{1}{2}$ so $CQ' = \frac{1}{2}CQ$, $CP' = \frac{1}{2}CP$, $CR' = \frac{1}{2}CR$.

Lengths on the image are half the corresponding lengths on the object so $P'R' = \frac{1}{2}PR$ and so on.

Enlargement by scale factor $\frac{1}{2}$ is the inverse of enlargement by scale factor 2.

Example

Enlarge triangle A by scale factor $\frac{1}{2}$, centre $(-4, 2)$.

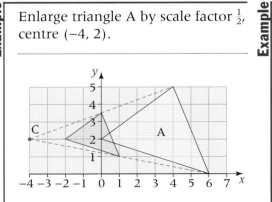

Example

Enlarge triangle B by scale factor $\frac{1}{3}$ centre $(-2, -2)$.

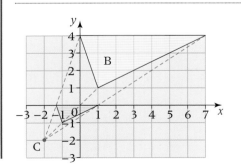

Lengths on the image are $\frac{1}{3}$ corresponding lengths on the object.

1 a Copy this diagram.

b Enlarge triangle T by scale factor $\frac{1}{2}$, centre (0, 0). Label the image U.

c Enlarge triangle T by scale factor $\frac{1}{2}$, centre (2, 2). Label the image V.

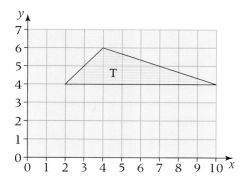

2 a Copy this diagram.

b Enlarge kite K by scale factor $\frac{1}{3}$, centre (0, 0). Label the image L.

c Enlarge kite K by scale factor $\frac{1}{3}$, centre (3, 3). Label the image M.

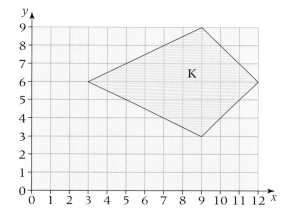

3 a Draw a grid with an x-axis from −3 to 5 and a y-axis from −2 to 10. Plot the points (1, 3) (1, 6) (3, 9) (4, 6). Join them to make quadrilateral Q.

b Enlarge quadrilateral Q by scale factor $\frac{1}{3}$, centre (4, 3). Label the image R.

c Enlarge quadrilateral Q by scale factor $\frac{1}{2}$, centre (−2, −2). Label the image S.

4 a Draw a grid with x-axis from −6 to 6 and y-axis from 0 to 11. Plot the points (−6, 6) (0, 6) (3, 3). Join them to make triangle A.

b Enlarge triangle A by scale factor $\frac{1}{3}$, centre (0, 0). Label the image B.

c Enlarge triangle A by scale factor $\frac{1}{2}$, centre (0, 10). Label the image C.

Describing an enlargement

This spread will show you how to:

- Describe an enlargement by giving the scale factor and centre of enlargement
- Understand, identify and use scale factors

Keywords

Centre of
 enlargement
Enlargement
Scale factor

To describe an **enlargement** you give the **scale factor** and the **centre of enlargement**.

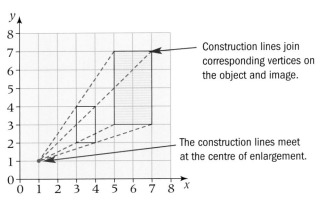

Construction lines join corresponding vertices on the object and image.

The construction lines meet at the centre of enlargement.

This is an enlargement of scale factor $\frac{1}{2}$, centre (1, 1).

The scale factor of an enlargement is the ratio of corresponding sides.

- **Scale factor** $= \dfrac{\text{length of image}}{\text{length of original}}$

You can write this as a ratio
length of image : length of object

Example

Describe fully the single transformation that maps triangle PQR onto P′Q′R′.

The construction lines meet at (−3, −2)

Scale factor $= \dfrac{P'R'}{PR} = \dfrac{6}{2} = 3$.

The transformation that maps PQR onto P′Q′R′ is an enlargement, centre (−3, −2), scale factor 3.

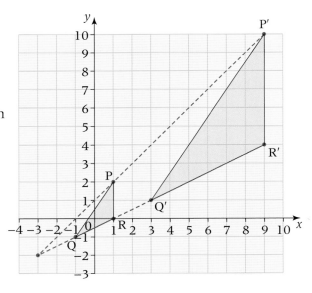

1 Describe fully the single transformation that maps

   a triangle P onto triangle Q

   b triangle Q onto triangle P.

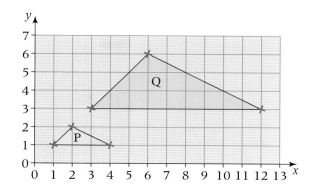

2 a Describe fully the single transformation that maps

    i rectangle R onto rectangle S

    ii rectangle S onto rectangle R.

   b The perimeter of rectangle R is 8 units. What is the perimeter of rectangle S?

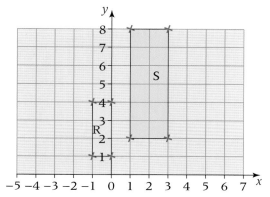

3 Describe fully the single transformation that maps

   a triangle W onto triangle X

   b triangle X onto triangle W.

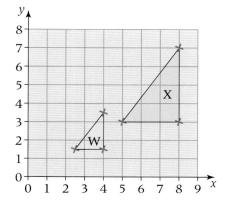

4 a Describe fully the single transformation that maps

    i triangle B onto triangle A

    ii triangle B onto triangle C

    iii triangle A onto triangle C.

   b How many times bigger is the perimeter of triangle C than triangle A?

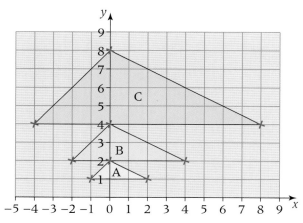

181

Similar shapes

This spread will show you how to:

- Understand similarity of 2-D shapes, using this to find missing lengths and angles

Keywords

Ratio
Similar

In an enlargement, the object and image are **similar**.

- In similar shapes:
  - corresponding angles are equal
  - corresponding sides are in the same **ratio**.

You can use ratio to find missing lengths in similar shapes.

Example

These two pentagons are similar.
Work out the side lengths a and b.

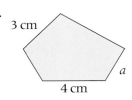
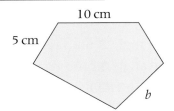

Side 4 cm corresponds with side 10 cm.

The ratio of the sides is 4 : 10 or $\frac{4}{10}$.
Side a cm corresponds to side 5 cm.
So $\frac{a}{5} = \frac{4}{10}$
$a = \frac{5 \times 4}{10} = 2$ cm

Side b cm corresponds to side 3 cm.
So $\frac{3}{b} = \frac{4}{10}$ Invert both sides of the equation.
$\frac{b}{3} = \frac{10}{4}$
$b = \frac{10 \times 3}{4} = 7.5$ cm

Sketch the pentagons the same way up, to help you identify corresponding sides and angles.

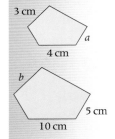

Example

Find the length of QS.

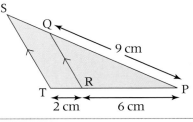

$\angle QPR = \angle SPT$
$\angle QRP = \angle STP$ (corresponding angles TS//RQ)
$\angle PQR = \angle PST$ (corresponding angles TS//RQ)
So triangle PQR is similar to triangle PST.

PT corresponds to PR.
$PT : PR = \frac{PT}{PR} = \frac{8}{6}$
So $\frac{PS}{PQ} = \frac{8}{6} = \frac{PS}{9}$
$PS = \frac{9 \times 8}{6} = 12$
$QS = PS - PQ = 12 - 9 = 3$ cm

PST is an enlargement of PQR, centre P.

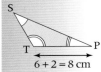

Write all the ratios in the correct order.
These are all smaller : larger.

The symbol // means 'parallel to'.

1 The two trapeziums are similar.
Find the lengths *a* and *b*.

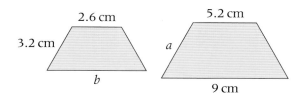

2 The two quadrilaterals are similar.
Find the lengths *c* and *d*.

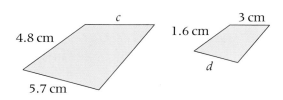

3 In the diagram PQ is parallel to TR.
Work out the lengths RT, QR and QS.

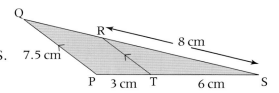

4 In the diagram WZ is parallel to XY.
  a Work out the lengths XY and VY.
  b Work out the perimeter of the
  trapezium WXYZ.

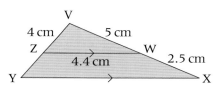

5 KN is parallel to LM.

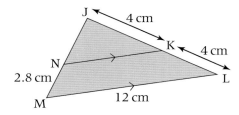

Work out the perimeters of triangle JKN and trapezium KLMN.

6 A and B are similar circles.

  a What is the ratio of their radii?
  b What is the ratio of their circumferences?
  c Explain why all circles are similar.

Similarity and circles

This spread will show you how to:

● Solve problems involving similarity and circles

Keywords
Similar
Chord

Figure 1 shows two chords that intersect inside a circle.
AED and CEB are similar triangles.

$$\frac{AE}{CE} = \frac{DE}{BE} \quad \text{or} \quad AE \times BE = CE \times DE$$

This result is called the intersecting chords theorem.

Figure 1

Figure 2 shows two chords that intersect outside the circle.
ABD and AEC are similar triangles

$$\frac{AD}{AC} = \frac{AB}{AE} \quad \text{or} \quad AD \times AE = AC \times AB$$

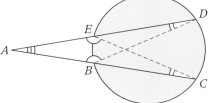

Figure 2

Figure 3 shows a tangent to a circle intersecting line BAC with B and A on the circle.
CBD and CDA are similar triangles

$$\frac{CB}{CD} = \frac{CD}{CA} \quad \text{or} \quad CA \times CB = CD^2$$

Figure 3

Example

$AB = 9$ cm; $AC = x$ cm; $CD = 5$ cm
CD is a tangent

Find x

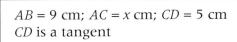

$$x(x + 9) = 5^2$$
$$x^2 + 9x - 25 = 0$$ rearrange to get a standard quadratic

$$x = \frac{-9 \pm \sqrt{9^2 - 4 \times 1 \times -25}}{2}$$ using the quadratic formula

$$x = 2.23$$ reject the negative solution

1 Find x

a

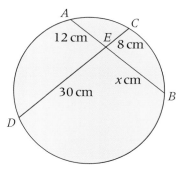

Diagram **NOT** accurately drawn

b

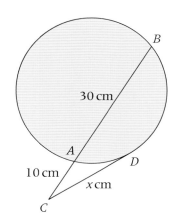

2 CD is a tangent. Find x

3 *CD* and *CE* are tangents. Find x and y

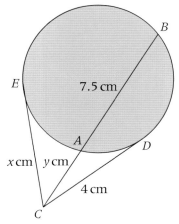

4 Find x

a

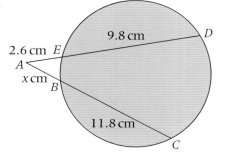

b

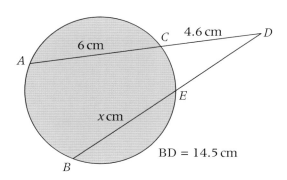

BD = 14.5 cm

Similar shapes – area and volume

This spread will show you how to:

● Understand and use the effect of enlargement on areas and volumes of 3-D shapes and solids

Keywords
Area
Length
Ratio
Similar
Volume

In **similar** shapes, corresponding sides are in the same **ratio**.
You can use the ratio to work out **areas** and **volumes** in similar shapes and solids.

● For similar shapes with length ratio $1 : n$
  – the ratio of the areas is $1 : n^2$ – the ratio of the volumes is $1 : n^3$

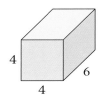

Area 6

Area 24

Length ratio is $1 : 2$
Area ratio is $6 : 24 = 1 : 4 = 1 : 2^2$

Volume 12

Volume 96

Length ratio is $1 : 2$
Volume ratio is $12 : 96 = 1 : 8 = 1 : 2^3$

Area = length × length (square units)
Volume = length × length × length (cubic units)

Example

Two similar Russian dolls are on display.

a The surface area of the smaller doll is 7.2 cm^2.
Work out the surface area of the larger doll.
b The volume of the larger doll is 145.8 cm^3.
Work out the volume of the smaller doll.

3.6 cm

1.2 cm

Ratio of lengths, smaller : larger = $1.2 : 3.6 = 1 : 3$

a Area ratio is $1 : 3^2 = 1 : 9$
Surface area of larger doll = $9 \times 7.2 = 64.8 \text{ cm}^2$
b Volume ratio is $1 : 3^3 = 1 : 27$
Volume of smaller doll = $145.8 \div 27 = 5.4 \text{ cm}^3$

Example

P and Q are similar shapes.
Calculate the surface area of shape Q.

Volume 260 cm³
Surface area 144 cm²

Volume 877.5 cm³
Surface area = ?

Volume ratio

$$\frac{Q}{P} = \frac{877.5}{260} = \frac{27}{8}$$

Length ratio

$$\frac{Q}{P} = \frac{\sqrt[3]{27}}{\sqrt[3]{8}} = \frac{3}{2}$$

Area ratio

$$\frac{Q}{P} = \frac{3^2}{2^2} = \frac{9}{4}$$

Surface area of Q = $144 \times \frac{9}{4} = 324 \text{ cm}^2$

Work out the length, area and volume ratios first.

Rearrange the area ratio equation and substitute in the area of P.

186

1 P and Q are two similar cuboids.

   a The surface area of cuboid P is 17.2 cm^2.
Work out the surface area of cuboid Q.

   b The volume of cuboid P is 12.4 cm^3.
Work out the volume of cuboid Q.

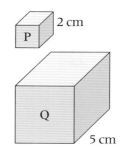

2 J and K are two similar boxes.
The volume of J is 702 cm^3.
The volume of K is 208 cm^3.
The surface area of J is 549 cm^2.
Calculate the surface area of K.

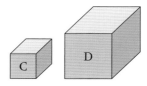

3 X and Y are two similar solids.
The total surface area of X is 150 cm^2.
The total surface area of Y is 216 cm^2.
The volume of Y is 216 cm^3.
Calculate the volume of X.

4 C and D are two cubes.

   a Explain why any two cubes must be similar.

   b Explain why any two cuboids are not necessarily similar.

   The ratio of the side lengths of cube C and cube D is 1 : 7.

   c Write down the ratio of **i** their surface areas **ii** their volumes.

5 Two model cars made to different scales are mathematically similar.
The overall widths of the cars are 3.2 cm and 4.8 cm respectively.

   a What is the ratio of the radii of the cars' wheels?

   The cars are packed in mathematically similar boxes so that they just fit inside the box.

   b The surface area of the larger box is 76.5 cm^2.
Work out the surface area of the smaller box.

   c The volume of the smaller box is 24 cm^3.
Work out the volume of the larger box.

6 At the local pizzeria Gavin is offered two deals for £9.99.

> Deal A: One large round pizza with radius 18 cm
> Deal B: Two smaller round pizzas each with radius 9 cm

Which deal gives the most pizza?

Exam review

Key objectives

- Understand similarity of triangles and of other plane figures, and use this to make geometric inferences
- Understand and use SSS, SAS, ASA and RHS conditions to prove the congruence of triangles using formal arguments, and to verify standard ruler and compass constructions
- Recognise, visualise and construct enlargements of objects using positive and fractional scale factors
- Understand and use the effect of enlargement on areas and volumes of shapes and solids
- Recall and use the properties of intersecting chords

1 Copy the diagram and enlarge triangle *T* by a scale factor 3, through the point (3, 4).   (3)

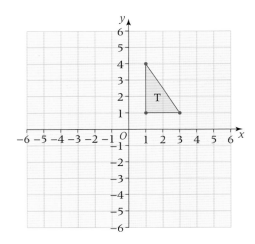

2 A wax statue of a spaceman is on display in a museum.
Wax models are to be sold in the museum shop.
The statue and the wax models are similar.
The height of the statue is 1.8 m.
The height of the model is 15 cm.
The area of the flag in the model is 10 cm².

  a Calculate the area of the statue's flag.   (3)

   The volume of the wax of the original statue is 172.8 litres.
  b Calculate the volume of wax used to make the model.   (2)
   Give your answer in ml.

(Edexcel Ltd., (Spec.))

This unit will show you how to

- Understand and use index notation and index laws, including integer, fractional and negative powers
- Rationalise a denominator, including when it involves a surd form
- Recognise the difference between rational and irrational numbers
- Use algebraic techniques to expand and simplify brackets containing surds
- Use surds and π in exact calculations, without a calculator
- Understand that the rules of algebra obey and generalise the rules of arithmetic
- Use the commutative, associative and distributive laws of addition, multiplication and factorisation

Before you start ...

You should be able to answer these questions.

Review

1 Find the results of these calculations, giving the answers in index form.

    a $2^3 \times 2^4$ **b** $3^5 \div 3^2$

    c $5^2 \times 5^3 \times 5^2$ **d** $6^6 \div 6^4$

    e $7^8 \div (7^2 \times 7^3)$ **f** $(4^6 \times 4^2) \div (4^3 \times 4)$

Unit N1

2 Write the value of each.

    a 4^0 **b** 6^0 **c** 5^1 **d** 2^{-1}

Unit N1

3 Simplify each expression.

    a $4\pi + 3\pi$ **b** $\pi(3^2 + 4)$

    c $4\pi(5^2 - 4^2)$ **d** $\sqrt{(5^2 - 3^2)}$

    e $\sqrt{2} + 2\sqrt{2}$ **f** $5\sqrt{3} - \sqrt{3}(4^2 - 3 \times 4)$

Unit N1

This spread will show you how to:

● Understand and use index notation and index laws

Keywords

Base
Index
Index laws
Power

In 10^4, the base is 10 and the index is 4.

You can write **powers** of a number in **index** form.

$$10 \times 10 \times 10 \times 10 = 10^4 \text{ (10 to the power 4)}$$

In algebra, x^n means the **base** x raised to the power n.

The **index laws** tell you how to use powers.

● To multiply powers of the same base, add the indices.

$$x^m \times x^n = x^{m+n}$$
$$4^2 \times 4^3 = (4 \times 4) \times (4 \times 4 \times 4)$$
$$= 4^{2+3} = 4^5$$

● To divide powers of the same base, subtract the indices.

$$x^m \div x^n = x^{m-n}$$
$$7^5 \div 7^2 = \frac{7 \times 7 \times 7 \times 7 \times 7}{7 \times 7}$$
$$= 7^{5-2} = 7^3$$

● To raise a power, multiply the indices.

$$(x^m)^n = x^{m \times n}$$
$$(9^2)^3 = (9 \times 9) \times (9 \times 9) \times (9 \times 9)$$
$$= 9^{2 \times 3} = 9^6$$

● $x^0 = 1$, for all values of x except $x = 0$ $1 = x^n \div x^n = x^{n-n} = x^0$

You can only use the index laws with powers of the same base.

● $x^m \times x^n \times y^m = x^{m+n} \times y^m$

Example

Write the answers to these in index form.

a $5^5 \times 5^3$ **b** $8^2 \times 8$ **c** $4^7 \div 4^3$ **d** $7^5 \div 7$ **e** $(6^3)^4$

a $5^5 \times 5^3 = 5^{5+3} = 5^8$
b $8^2 \times 8 = 8^{2+1} = 8^3$
c $4^7 \div 4^3 = 4^{7-3} = 4^4$
d $7^5 \div 7 = 7^{5-1} = 7^4$
e $(6^3)^4 = 6^{3 \times 4} = 6^{12}$

You can write 8 as 8^1.

Example

For the calculation $2^3 \times 5^4$, Jack wrote:

$2 \times 5 = 10$, and $3 + 4 = 7$.
So, the answer is 10^7.

Is he correct?

No. The index laws only apply when both numbers have the same base. For this calculation you need to evaluate each term and then multiply them together:

$2^3 = 8$ and $5^4 = 625$ $2^3 \times 5^4 = 8 \times 625 = 5000$

1 Evaluate these.

a 4^3 **b** 5^4 **c** 2^6 **d** 3^3

e 9^2 **f** 7^4 **g** 12^0 **h** 8^0

2 Find the value of the letters in these equations.

a $u = 4^4$ **b** $32 = 2^v$ **c** $3^4 = w^2$ **d** $144 = 2^x \times y^2$

3 Evaluate these, giving your answers in index form.

a $3^2 \times 3^2$ **b** $5^2 \times 5^3$ **c** $6^3 \times 6^4$ **d** $7^2 \times 7^7$

e $2^8 \times 2^7$ **f** $4^4 \times 4^6$ **g** $10^4 \times 10^{10}$ **h** $9^8 \times 9^3 \times 9^2$

4 Copy and complete the multiplication table for powers of x.

$\times$	x	x^3		x^9
x^2		x^6		
	x^7			
x^3				
		x^8		

5 Give the answers to these as powers.

a $4^4 \div 4^2$ **b** $5^7 \div 5^4$ **c** $9^5 \div 9^2$ **d** $6^8 \div 6^4$

e $7^6 \div 7^2$ **f** $8^5 \div 8^3$ **g** $9^8 \div 9^4$ **h** $3^7 \times 3^2 \div 3^4$

6 Simplify these expressions, giving your answers in index form.

a $\dfrac{3^4}{3^2}$ **b** $\dfrac{4^3 \times 4^4}{4^2}$ **c** $\dfrac{7^8}{7^3 \times 7^2}$ **d** $\dfrac{6^2 \times 6^7}{6^3 \times 6^4}$

e $5^2 \times \dfrac{5^7}{5^3 \times 5^2}$ **f** $\dfrac{2^3 \times 2^4}{2^5 \times 2^2} \times 2^5$ **g** $\dfrac{6^9 \div 6^3}{6^3 \times 6^2}$ **h** $\dfrac{4^5 \div 4}{4^8 \div 4^7}$

7 Use the relationship $(x^m)^n = x^{mn}$ to simplify these expressions, giving your answers in index form.

a $(2^3)^2$ **b** $(4^2)^5$ **c** $(7^2)^2$ **d** $(5^5)^3$

e $(3^4)^4$ **f** $(6^2)^2$ **g** $(5^7)^3$ **h** $(10^4)^4$

8 Find the value of the letter in each of these equations.

a $u = (4^3)^2$ **b** $64 = (v^2)^3$ **c** $81 = (3^w)^2$ **d** $x = (3^2)^3 + (2^3)^2$

9 Simplify these expressions, giving your answers in index form.

a $\left(\dfrac{4^4}{4^2}\right)^2$ **b** $\dfrac{(3^3)^2 \times 3^5}{3^7}$ **c** $\dfrac{6^3 \times (6^2)^4}{(6^5 \div 6^3)^2}$ **d** $\left(\dfrac{5^8 \div 5^2}{5^4 \div 5^2}\right)^3$

e $2^8 \times \left(\dfrac{2^4 \div 2^2}{2^9 \div 2^8}\right)^2$ **f** $\left(\dfrac{7^{15} \div 7^2}{7^8 \times 7^2}\right)^4$ **g** $\dfrac{(3^3)^5}{3^4 \times 3^2} \div (3^2)^3$ **h** $9^2 \times \left(\dfrac{9^7 \div 9}{9^2 \times 9^3}\right)^3$

More index laws

This spread will show you how to:

- Understand and use index notation and index laws, including integer, fractional and negative powers

The **index** laws also apply for fractional and negative indices.

- Fractional indices represent **roots**.
  - $x^{\frac{1}{2}} = \sqrt{x}$ for all values of x
  - $x^{\frac{1}{n}} = \sqrt[n]{x}$ (the nth root of x)
  - $x^{\frac{m}{n}} = (x^{\frac{1}{n}})^m = (\sqrt[n]{x})^m$, or
    $x^{\frac{m}{n}} = (x^m)^{\frac{1}{n}} = \sqrt[n]{x^m}$

- Negative indices represent **reciprocals**.
  - $x^{-n} = \dfrac{1}{x^n}$

$5^{\frac{1}{2}} \times 5^{\frac{1}{2}} = 5^{\frac{1}{2}+\frac{1}{2}} = 5^1 = 5$,
so $5^{\frac{1}{2}} = \sqrt{5}$

$x^{\frac{1}{3}} = \sqrt[3]{x}$

$\dfrac{1}{9^5} = 1 \div 9^5$
$= 9^0 \div 9^5$
$= 9^{0-5}$
$= 9^{-5}$

The square root of a number can be positive or negative.

$\sqrt{4} = \pm 2$ means the answer is +2 or −2.
$+2 \times +2 = 4 \qquad -2 \times -2 = 4$

The same is true for the square root of **any** positive number.

Example

Find the value of

a 4^{-1} **b** $4^{\frac{1}{2}}$ **c** $4^{-\frac{1}{2}}$ **d** $9^{\frac{3}{2}}$ **e** $8^{\frac{2}{3}}$ **f** $4^{-\frac{5}{2}}$

a $4^{-1} = \dfrac{1}{4^1} = \dfrac{1}{4}$

b $4^{\frac{1}{2}} = \sqrt{4} = \pm 2$

c $4^{-\frac{1}{2}} = \dfrac{1}{4^{\frac{1}{2}}} = \dfrac{1}{\sqrt{4}} = \pm\dfrac{1}{2}$

d $9^{\frac{3}{2}} = (9^{\frac{1}{2}})^3 = (\sqrt{9})^3 = (\pm 3)^3 = \pm 27$

e $8^{\frac{2}{3}} = (8^{\frac{1}{3}})^2 = (\sqrt[3]{8})^2 = 2^2 = 4$

f $4^{-\frac{5}{2}} = \dfrac{1}{4^{\frac{5}{2}}} = \dfrac{1}{(4^{\frac{1}{2}})^5} = \dfrac{1}{(\sqrt{4})^5} = \dfrac{1}{(\pm 2)^5} = \pm\dfrac{1}{32}$

Examiner's tip
On the Edexcel Higher Tier papers, you are expected to give both ± answers where appropriate.

Example

To find the value of $16^{-\frac{1}{2}}$ Alex wrote:

$16^{\frac{1}{2}} = 4$, so the answer must be −4.

Is she correct?

No. $16^{\frac{1}{2}} = 4$ is correct, but it could also be −4.
$16^{-\frac{1}{2}} = \frac{1}{4}$ or $-\frac{1}{4}$. You could write $16^{-\frac{1}{2}} = \pm(\frac{1}{4})$.

Examiner's tip
Rather than relying on learning the rules for fractional and negative indices 'by heart', make sure you understand the underlying relationships:

- $x^{\frac{1}{2}} \times x^{\frac{1}{2}} = x \Rightarrow x^{\frac{1}{2}} = \sqrt{x}$
- $x^{-n} = x^0 \div x^n = 1 \div x^n = \frac{1}{x^n}$

1 Find the value of each of these.
   a $16^{\frac{1}{2}}$ **b** $9^{\frac{1}{2}}$ **c** $27^{\frac{1}{3}}$ **d** $0^{\frac{1}{2}}$ **e** $1000^{\frac{1}{3}}$

2 Find the value of the letter in each of these equations.
   a $\sqrt{5} = 5^a$ **b** $\sqrt[3]{600} = 600^b$
   c $\sqrt{100} = 1000^c$ **d** $2 = 16^d$

3 Evaluate these.
   a $25^{\frac{1}{2}}$ **b** $25^{-\frac{1}{2}}$ **c** 25^0 **d** $25^{\frac{3}{2}}$ **e** 25^{-1}

4 Evaluate these.
   a $4^{\frac{3}{2}}$ **b** $8^{\frac{2}{3}}$ **c** $9^{\frac{5}{2}}$ **d** $100^{-\frac{1}{2}}$ **e** $16^{-\frac{3}{2}}$
   f $1000^{\frac{2}{3}}$ **g** $400^{-\frac{1}{2}}$ **h** $169^{\frac{3}{2}}$ **i** $81^{-\frac{3}{2}}$ **j** $4^{-\frac{3}{2}}$

5 Write these as powers of 4.
   a $\frac{1}{4}$ **b** 16 **c** $\frac{1}{16}$ **d** 8 **e** 32

6 Write each of the numbers given in question **5** as a power of 16.

7 Write these as powers of 10.
   a 100 **b** $\frac{1}{10}$ **c** $\sqrt{10}$ **d** $(\sqrt{10})^3$ **e** $\dfrac{1}{(\sqrt{10})^5}$

8 Write these as powers of 5.
   a $\dfrac{1}{25}$ **b** $\dfrac{1}{\sqrt[3]{5}}$ **c** $\sqrt[3]{25}$ **d** $\dfrac{1}{\sqrt{125}}$ **e** $\dfrac{1}{\sqrt[3]{625}}$

9 **a** Copy and complete the table to show the value of 9^x for various values of x.

x	-1	$-\frac{1}{2}$	0	$\frac{1}{2}$	1
9^x					

   b Use your table to plot a graph to show the value of 9^x, for values of x in the range -1 to $+1$.

   c On the same set of axes, plot the graph of 4^x for the same values of x. Comment on any similarities and differences between the two graphs.

This spread will show you how to:

- Rationalise a denominator, including when it involves a surd form
- Recognise the difference between rational and irrational numbers

- A number is **irrational** if it cannot be written as an exact fraction.

$\sqrt{2}$ and π are irrational.
If n is not a square number, then $\sqrt{n}$ is irrational.

- When approximated as decimals, irrational numbers have a non-recurring sequence of decimal digits.

  $\sqrt{2} = 1.414\,213$ $\pi = 3.141\,592$ There are no repeating patterns in the decimal digits.

Say whether each of these numbers is rational or irrational.

a $\frac{3}{17}$ **b** $\sqrt{16}$ **c** $\sqrt{17}$ **d** $\sqrt[3]{8}$ **e** $\frac{2\pi + 1}{3}$ **f** $\sqrt[3]{7}$

---

a $\frac{3}{17}$ rational $\frac{3}{17}$ is an exact fraction

b $\sqrt{16} = 4$ rational   16 is a square number

c $\sqrt{17}$ irrational   17 is not a square number

d $\sqrt[3]{8} = 2$ rational   8 is a cube number

e $\frac{2\pi + 1}{3}$ irrational π is irrational

f $\sqrt[3]{7}$ irrational   7 is not a cube number

Any multiple of π is irrational.

If you have an expression with an irrational number in the **denominator**, you should **rationalise** it.

- To rationalise the denominator in a fraction, multiply numerator and denominator by the denominator.

For example:
$$\frac{2}{\sqrt{3}} = \frac{2}{\sqrt{3}} \times \frac{\sqrt{3}}{\sqrt{3}} = \frac{2\sqrt{3}}{3}$$

Rewrite each of these expressions without roots in the denominator.

a $\dfrac{5}{\sqrt{2}}$ **b** $\dfrac{7}{\sqrt{8}}$

---

a $\dfrac{5}{\sqrt{2}} = \dfrac{5}{\sqrt{2}} \times \dfrac{\sqrt{2}}{\sqrt{2}}$

$= \dfrac{5\sqrt{2}}{2}$

b $\dfrac{7}{\sqrt{8}} = \dfrac{7}{\sqrt{8}} \times \dfrac{\sqrt{8}}{\sqrt{8}} = \dfrac{7\sqrt{8}}{8}$

$= \dfrac{7\sqrt{8}}{8}$

1   Explain whether each of these numbers is rational or irrational.

   a $\frac{1}{3}$ **b** 2π **c** $\frac{3}{19}$ **d** $\sqrt{3}$ **e** $\sqrt{25}$ **f** $\sqrt[3]{10}$

2   Nina says, 'I worked out $\frac{4}{19}$ on my calculator, and there was no repeating pattern. This means that $\frac{4}{19}$ is irrational.' Explain why Nina is wrong.

3   Give an example of

   a an irrational number with a rational square

   b a number with a rational square root

   c a number with an irrational square, but a rational cube

   d a number with an irrational square, cube, square root and cube root.

4   There are two whole numbers less than 100 for which both the square root and the cube root are rational. One of the numbers is 1. Find the other one.

5   Karla says, 'I am thinking of a rational number. When I halve it, I get an irrational number.' Can Karla be correct? Explain your answer.

6   Jim says, 'I am thinking of an irrational number. When I add one to my number, the answer is rational.' Can Jim be correct? Explain your answer.

7   Javier says, 'If a whole number is not a perfect cube, then its cube root is irrational.' Is Javier correct? Explain your answer.

8   Rationalise the denominator of each of these fractions.

   a $\frac{1}{\sqrt{2}}$ **b** $\frac{1}{\sqrt{3}}$ **c** $\frac{1}{\sqrt{7}}$ **d** $\frac{1}{\sqrt{6}}$ **e** $\frac{1}{\sqrt{5}}$

9   Rewrite each of these fractions without roots in the denominator.

   a $\frac{2}{\sqrt{8}}$ **b** $\frac{2}{\sqrt{10}}$ **c** $\frac{3}{\sqrt{12}}$ **d** $\frac{5}{\sqrt{30}}$ **e** $\frac{8}{\sqrt{40}}$

10   Lisa and Simone write down some decimal numbers.

   Lisa writes: 0.121 121 112 111 121 111 12...
   Simone writes: 0.454 545 454 545 454 545 45...

   Simone says, 'My number is rational, because there is a repeating pattern in the digits.'
   Lisa says, 'My number also has a pattern in the digits, so it is rational as well.'
   Explain whether you agree with each of these statements.

Calculations with surds

This spread will show you how to:

- Use surds and π in exact calculations, without a calculator
- Use algebraic techniques to expand and simplify brackets containing surds

In calculations, you can use approximate decimals for **irrational** numbers such as $\sqrt{2}$ or π.

For more accurate results, you can carry out a calculation using **surds**.

$\pi \approx 3.142 \ldots$
$\sqrt{2} \approx 1.414 \ldots$

- The square root of a number is the product of the square roots of the number's factors.

A **surd** is a square root that is an irrational number, for example $\sqrt{2}$, $\sqrt{3}$, $\sqrt{7}$.

For example, $\sqrt{20} = \sqrt{4 \times 5} = \sqrt{4} \times \sqrt{5} = 2\sqrt{5}$.

Example

a Simplify $\sqrt{12} + \sqrt{48}$.
b Write $6\sqrt{2}$ in the form $\sqrt{n}$, where n is an integer.

a $\sqrt{12} + \sqrt{48} = \sqrt{4} \times \sqrt{3} + \sqrt{16} \times \sqrt{3}$
$= 2\sqrt{3} + 4\sqrt{3} = 6\sqrt{3}$
b $6\sqrt{2} = \sqrt{6^2 \times 2} = \sqrt{36 \times 2} = \sqrt{72}$

- You can use algebraic techniques to expand and simplify brackets containing surds.

$(1 + \sqrt{3})^2 = 1 + 2\sqrt{3} + (\sqrt{3})^2 = 4 + 2\sqrt{3}$

For a combination of rational and irrational numbers, write the rational number first. To remove the surd, multiply by the surd with the opposite sign.

$(3 + \sqrt{5})(3 - \sqrt{5}) = 9 - 3\sqrt{5} + 3\sqrt{5} - 5 = 9 - 5 = 4$

$(a - x)(a + x) = a^2 - x^2$

Example

a Expand the brackets $(1 + \sqrt{2})(1 - \sqrt{2})$
b Rewrite the fraction $\dfrac{1}{1 + \sqrt{3}}$ without surds in the denominator.

a $(1 + \sqrt{2})(1 - \sqrt{2}) = 1 - \sqrt{2} + \sqrt{2} - (\sqrt{2})^2 = 1 - 2 = -1$

b $\dfrac{1}{1 + \sqrt{3}} = \dfrac{1}{1 + \sqrt{3}} \times \dfrac{1 - \sqrt{3}}{1 - \sqrt{3}} = \dfrac{1 - \sqrt{3}}{1 - 3} = \dfrac{1 - \sqrt{3}}{-2}$

Multiply numerator and denominator by $(1 - \sqrt{3})$.

Example

Work out the height x of this triangle, giving your answer to 2 dp.

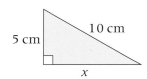

5 cm

10 cm

x

By Pythagoras's theorem:
$5^2 + x^2 = 10^2$
$x^2 = 10^2 - 5^2$
$= 100 - 25$
$x = \sqrt{75}$
$= \sqrt{25} \times \sqrt{3}$
$= 5\sqrt{3}$
$= 8.66$ cm (2 dp)

Simplify as much as possible, then use your calculator to approximate the answer at the end.

1 Write each of these as the square root of a single number.

   a $\sqrt{2} \times \sqrt{3}$ **b** $\sqrt{5} \times \sqrt{3}$ **c** $\sqrt{3} \times \sqrt{7} \times \sqrt{11}$

2 Write in the form $\sqrt{a}\sqrt{b}$, where a and b are prime numbers.

   a $\sqrt{14}$ **b** $\sqrt{33}$ **c** $\sqrt{21}$

3 Write in the form $a\sqrt{b}$, where b is a prime number.

   a $\sqrt{20}$ **b** $\sqrt{27}$ **c** $\sqrt{98}$

4 Write in the form $\sqrt{n}$, where n is an integer.

   a $4\sqrt{3}$ **b** $5\sqrt{2}$ **c** $4\sqrt{5}$

5 Without using a calculator, evaluate.

   a $\sqrt{12} \times \sqrt{3}$ **b** $\sqrt{18} \times \sqrt{8}$ **c** $\sqrt{33} \times \sqrt{132}$

6 Simplify these expressions.

   a $3\sqrt{5} + \sqrt{20}$ **b** $\sqrt{28} + 5\sqrt{7}$ **c** $7\sqrt{12} - 2\sqrt{27}$

7 **a** Use a calculator to evaluate your simplified expressions from question **6**, giving your answers to 3 decimal places.

   b Use a calculator to evaluate the non-simplified expressions from question **6**.

   c Compare your answers to parts **a** and **b**.

8 Expand the brackets and simplify.

   a $\sqrt{5}(1 + \sqrt{5})$ **b** $\sqrt{3}(2 - \sqrt{3})$ **c** $(1 + \sqrt{5})(2 + \sqrt{5})$

   d $(\sqrt{3} - 1)(2 + \sqrt{3})$ **e** $(4 - \sqrt{7})(6 - 2\sqrt{7})$ **f** $(5 + \sqrt{2})(5 - \sqrt{2})$

   g $(6 + 2\sqrt{5})(6 - 2\sqrt{5})$ **h** $(5 - 3\sqrt{7})(5 + 3\sqrt{7})$ **i** $(7 + 5\sqrt{3})(7 - 5\sqrt{3})$

9 Rewrite these fractions without surds in the denominators.

   a $\dfrac{1}{\sqrt{11}}$ **b** $\dfrac{1}{1 + \sqrt{2}}$ **c** $\dfrac{1}{1 - 2\sqrt{3}}$

   d $\dfrac{\sqrt{5}}{1 + \sqrt{5}}$ **e** $\dfrac{1 - \sqrt{2}}{1 + \sqrt{2}}$ **f** $\dfrac{4 + \sqrt{2}}{3 - 2\sqrt{2}}$

This spread will show you how to:

- Use the commutative, associative and distributive laws of addition, multiplication and factorisation
- Understand that the rules of algebra obey and generalise the rules of arithmetic

When you expand brackets you are using the **distributive** law.

- Multiplication is distributive over addition and subtraction.

$$a(b + c) = ab + ac \qquad 3(2 + 6) = 3 \times 2 + 3 \times 6 = 6 + 18 = 24$$
$$a(b - c) = ab - ac \qquad 4(3 - 1) = 4 \times 3 - 4 \times 1 = 12 - 4 = 8$$

- When you add or multiply two variables you can put the variables in any order. This is the **commutative** law.

For addition: $a + b = b + a$ $\qquad 10 + 5 = 5 + 10$

For multiplication: $a \times b = b \times a$ $\qquad 3 \times 6 = 6 \times 3$

Subtraction and division are *not* commutative.

- When you multiply together three variables you can use the **associative** law.

For addition: $\qquad\qquad (6 + 5) + 7 = 11 + 7 = 18$
$(a + b) + c = a + (b + c) \qquad 6 + (5 + 7) = 6 + 12 = 18$

For multiplication: $\qquad\quad (4 \times 2) \times 3 = 8 \times 3 = 24$
$(a \times b) \times c = a \times (b \times c) \qquad 4 \times (2 \times 3) = 4 \times 6 = 24$

It does not matter which pair you add (or multiply) first.

Subtraction and division are *not* associative.

Example

Prove that, for any two fractions $\dfrac{a}{b}$ and $\dfrac{c}{d}$,

$$\frac{a}{b} \div \frac{c}{d} = \frac{ad}{bc}.$$

$b \neq 0$ and $d \neq 0$ (you cannot have a fraction with zero denominator).

$$\frac{a}{b} \div \frac{c}{d} = \frac{ad}{b} \div c \qquad \text{Multiply both fractions by } d$$

$$\frac{ad}{b} \div c = ad \div bc = \frac{ad}{bc} \qquad \text{Multiply both fractions by } b$$

Multiplying both terms in a division by any non-zero number leaves the result unchanged.

Example

a Calculate $83^2 - 17^2$.
b Without using a calculator, evaluate 350^2.

Use DOTS:
$x^2 - y^2 \equiv (x + y)(x - y)$

a $83^2 - 17^2 = (83 + 17)(83 - 17)$
$\qquad\qquad\quad = 100 \times 66 = 6600$

b $x^2 - y^2 \equiv (x + y)(x - y)$
$\qquad x^2 \equiv (x + y)(x - y) + y^2$
$\quad 350^2 = (350 + 50)(350 - 50) + 50^2$
$\qquad\quad = 400 \times 300 + 2500$
$\qquad\quad = 120\,000 + 2500 = 122\,500$

$x = 350$
You could choose any number for y.
$y = 50$ gives 'easy' numbers in the brackets to multiply.

1 Charlie says, 'When I work out a multiplication like 7×17, I just do 7×10 and 7×7, and add the answers.' Which law is Charlie using? Explain your answer carefully.

2 Multiplication is distributive over subtraction. Show how this idea can be used to calculate 24×19 mentally.

$19 = 20 - 1$

3 Nancy says, 'I am working out $36 \div (2 + 3)$. I'll get the same result if I do $36 \div 2$ and $36 \div 3$, and then add the answers together.' Explain why Nancy is wrong.

4 Give numerical examples to show that
   a subtraction is *not* commutative   **b** division is *not* commutative
   c subtraction is *not* associative   **d** division is *not* associative.

5 Is the 'square root' operation distributive over addition? That is, is it true that $\sqrt{a + b} \equiv \sqrt{a} + \sqrt{b}$? Explain your answer carefully.

6 Evaluate these, without using a calculator.
   a $73^2 - 27^2$ **b** $6.4^2 - 3.6^2$ **c** $191^2 - 9^2$

   Follow the method used to evaluate $83^2 - 17^2$ in the example.

7 Without using a calculator, work out these.
   a 35^2 **b** 29^2 **c** 4.1^2

   Follow the method used to work out 350^2 in the example.

8 Given that $4.7^2 = 22.09$, show how you can find the value of 5.7^2 without further multiplication or squaring.

9 Given that $15.6^2 = 243.36$, show how you can use the identity $(x - 1)^2 = x^2 - 2x + 1$ to find the value of 14.6^2.

10 Petra has noticed a pattern in the squares of decimal numbers. She writes:

   > $1.5^2 = 2.25$, $2.5^2 = 6.25$, $3.5^2 = 12.25$, and so on.
   > To find any square like this, for example 7.5^2,
   > multiply the unit number by the number +1, that's
   > $7 \times 8 = 56$. Then just add 0.25, so $7.5^2 = 56.25$.

   Use algebra to prove that this pattern always works.

Exam review

Key objectives

- Use index laws to simplify and calculate the value of numerical expressions involving multiplication and division of fractional and negative powers
- Use surds and π in exact calculations
- Rationalise a denominator
- Understand that the transformation of algebraic entities obeys and generalises the well-defined rules of generalised arithmetic

1 **a** Expand and simplify the following, expressing your answers in their simplest from:

    i $(\sqrt{2} + \sqrt{3})(\sqrt{2} - \sqrt{3})$

    ii $(\sqrt{x} + 2)(\sqrt{x} - 1)$.   (3)

  b Rationalise the denominator of $\dfrac{2}{\sqrt{5}}$ (2)

  Give your answer in its simplest form.

2 **a** Find the value of:

    i 64^{0}

    ii $64^{\frac{1}{2}}$

    iii $64^{-\frac{2}{3}}$.   (4)

  b $3 \times \sqrt{27} = 3^{n}$. Find the value of n.   (2)

(Edexcel Ltd., 2005)

This unit will show you how to

- Know the meaning of and use the words 'equation', 'formula', 'identity' and 'expression'
- Use formulae from mathematics and other subjects
- Substitute numbers into a formula
- Generate a formula
- Change the subject of a formula, including cases where the subject appears twice
- Use inverse operations to rearrange a formula
- Understand the difference between a practical demonstration and a proof
- Use counter-examples to show that a statement is false
- Derive proofs using short chains of deductive reasoning

Before you start ...

You should be able to answer these questions.

Review

1 The cost, in pounds, C of a mobile phone bill is found using the formula
$$C = 0.5m + 12$$
where m is the number of minutes used.
Use the formula to find:
a The cost of using 39 minutes
b The number of minutes used if the cost is £39.

Unit A1

2 Solve these equations.
a $2x - 7 = 23$ **b** $10 - 2x = 15$
c $\frac{5}{x} - 2 = 15$ **d** $3x + 2 = 5 - 4x$

Unit A2

3 a Write the area formula for a circle.
b Use the formula to find the radius of a circle with an area of 50 cm^2.
c Write your own formula for the total area of this shape.
The curved end is a semi-circle.

Unit S1

Identities, formulae and equations

This spread will show you how to:

- Use the words 'equation', 'formula', 'identity' and 'expression'
- Use formulae from mathematics and other subjects
- Substitute numbers into a formula
- Generate a formula

Keywords
Equation
Expression
Formula(e)
Identity
Substitute
Variable

- $3x + 7 = 13$ is an **equation**. It has an equals sign.
  It is only true when $x = 2$.
- $3x + 7$ is an **expression**. It has two terms, $3x$ and 7.
  There is no equals sign.
- $3(x + 7) \equiv 3x + 21$ is an **identity**. It is true for all the values of x.
- $A = \frac{1}{2}bh$ is a formula. It expresses the relationship between the variables A, the area, b, the base, and h, the height of a triangle.

$\equiv$ means 'is identical to'

- You can **substitute** numbers into formulae to work out the value of an unknown.
- You can use given information to generate a formula to represent a situation.

Example

The formula $v^2 = u^2 + 2as$ connects velocity (v) with initial speed (u), acceleration (a) and distance (s).
Find the final velocity of a race car that is stationary at the start line and then accelerates at 6 m/s² for 300 metres.

$u = 0$, $a = 6$ and $s = 300$.

$v^2 = u^2 + 2as$
$\quad = 0^2 + 2 \times 6 \times 300$
$\quad = 3600$
$v = \sqrt{3600} = 60$ m/s

The car is stationary, so initial speed $(u) = 0$.

Substitute the values into the formula.

Example

A company sells three brands of mobile phones.
The company's profits on each are:
    Brand A – £12 per phone
    Brand B – £15 per phone
    Brand C – £23 per phone
Write a formula for P, the company's profit in £s,
for different numbers of phones sold.

If the company sold 7 phones of Brand A, its profit would be $£(12 \times 7)$.
Suppose the company sells a of Brand A, b of Brand B and c of Brand C, then:

$\quad P = 12a + 15b + 23c$

Use variables to represent unknowns – here, the numbers of each phone sold.

1 Copy the table, ticking the most appropriate column for each statement in the left-hand column. Hence, decide whether each statement is an identity, an equation or neither.

	Not true for any value of x	True for one value of x	True for two values of x	True for all values of x
a $x^2 = 7x - 12$				
b $2x(x^2 - 3) = 2x^3 - 6x$				
c $3x^2 = -27$				
d $2x + 3 = 9x - 6$				

2 Use this formula to find the value of the required variable.

a $A = \sqrt{s(s-a)(s-b)(s-c)}$, where A is the area of a triangle with sides a, b and c, and s is half of the perimeter.
Find the area of the triangle:

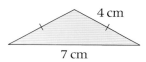

4 cm

7 cm

b $V = \frac{4}{3}\pi r^3$
Find the volume, V, of a sphere with diameter 6 mm.

c $T = 2\pi\sqrt{\dfrac{l}{g}}$

Find the length, l, of a pendulum that makes one complete swing in 8 seconds (T), given that gravity, g, is 9.8 m/s².

d $V = \dfrac{\pi^2(r_1 + r_2)(r_1 - r_2)^2}{4}$

Find the volume of a torus ('doughnut') with outer radius $r_1 = 10$ cm and inner radius $r_2 = 4$ cm.

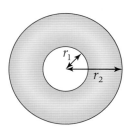

3 Show that the formula for each shaded area is

a $A = p^2(1 - \frac{\pi}{4})$

p

b $A = x^2 + 5x + 18$.

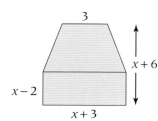

3

$x + 6$

$x - 2$

$x + 3$

Rearranging formulae

This spread will show you how to:

- Change the subject of a formula, including cases where a power of the subject appears

- The **subject** of a formula is the variable on its own, on one side of the equals sign.

For example, in the formula $A = \pi r^2$, A is the subject.

- You can **rearrange** a formula to make another variable the subject.

You use inverse operations to do this, as when rearranging equations.

Example

Make h the subject of $A = \frac{1}{2}(a + b)h$.

$A = \frac{1}{2}(a + b)h$

$2A = (a + b)h$ Multiply both sides by 2.

$\dfrac{2A}{a + b} = h$

h is multiplied by $(a + b)$. The inverse operation is 'divide by $(a + b)$'.
$\frac{2A}{a+b} = \frac{(a+b)h}{a+b}$

Example

Make x the subject of $V = p(w - ax^2y)$.

$V = p(w - ax^2y)$ Divide both sides by p

$\dfrac{V}{p} = w - ax^2y$ Add ax^2y to both sides

$\dfrac{V}{p} + ax^2y = w$ Subtract $\frac{V}{p}$ from both sides

$ax^2y = w - \dfrac{V}{p}$ Divide both sides by ay (or by a and then by y)

$x^2 = \dfrac{w - \frac{V}{p}}{ay}$ Take square roots of both sides

$x = \sqrt{\dfrac{w - \frac{V}{p}}{ay}}$ To simplify further, multiply each term in the fraction by p

Hence $x = \sqrt{\dfrac{pw - V}{apy}}$

Now the term including x is on its own.

Example

Make l the subject of $T = 2\pi\sqrt{\dfrac{l}{g}}$.

$T = 2\pi\sqrt{\dfrac{l}{g}}$ Divide both sides by 2π

$\dfrac{T}{2\pi} = \sqrt{\dfrac{l}{g}}$ Square both sides

$\dfrac{T^2}{4\pi^2} = \dfrac{l}{g}$ Multiply both sides by g

$\dfrac{T^2g}{4\pi^2} = l$

1 Make m the subject of each formula.

a $y = mx + c$

b $t = \dfrac{m - k}{w}$

c $m^2 + kt = p$

d $\sqrt[3]{m} - k = l$

2 Make x the subject of each formula.

a $y = \frac{1}{2}x + kw$

b $m = \frac{1}{4}(ax - t^2)$

c $2y = \sqrt{x}$

d $y = \sqrt{k - lx}$

e $k = \dfrac{t - a\sqrt{x}}{h}$

f $m = \dfrac{p}{x} - t$

g $w - \dfrac{p}{x} = c$

h $\dfrac{y}{ax} - b = j$

3 Clare and Isla have each tried to rearrange a formula to make x the subject. Each has made a mistake – where have they gone wrong?

Clare's attempt:

$C = p - ax$

$C - p = ax$

$\dfrac{C - p}{a} = x$

Isla's attempt:

$W = p - \dfrac{k}{x}$

$xW = p - k$

$x = \dfrac{p - k}{W}$

4 These boxes give the lines of the rearrangement of a formula but they have been scrambled up. Write them in the correct order to show the full rearrangement.

$\dfrac{c}{df(b - a)} = e$ $\quad$ $a + \dfrac{c}{def} = b$ $\quad$ $\dfrac{c}{def} = b - a$ $\quad$ $a = b - \dfrac{c}{def}$ $\quad$ $c = def(b - a)$

5 Show that each formula can be rearranged into the given form.

a $\dfrac{p(c - qt)}{m - x^2} = wr$ into $x = \sqrt{\dfrac{mwr + pqt - pc}{wr}}$

b $\dfrac{1}{a} + \dfrac{1}{b} = \dfrac{1}{c}$ into $b = \dfrac{ac}{a - c}$

c $\sqrt[4]{t - qx} = 2p$ into $x = \dfrac{t - 16p^4}{q}$

6 These three formulae are equivalent – true or false? Explain your answer.

$x = \dfrac{t}{p} - wx$ $\quad$ and $\quad$ $x = \dfrac{t - wpx}{p}$ $\quad$ and $\quad$ $\dfrac{w - px}{-p} = x$

7 Rearrange the lens formula $\dfrac{1}{f} = \dfrac{1}{u} + \dfrac{1}{v}$

a to make f the subject

b to make v the subject.

This spread will show you how to:

● Change the subject of a formula, including cases where the subject occurs twice

● You can **rearrange** formulae using inverse operations.

● When the **subject** appears twice, you can rearrange the formula by collecting like terms and then factorising.
For example, to make x the subject of $ax + b = cx + d$

$$ax + b = cx + d \longrightarrow ax - cx = d - b \longrightarrow x(a - c) = d - b \longrightarrow x = \frac{d - b}{a - c}$$

Collect x-terms   Factorise   Divide by
on one side $a - c$

Example

Rearrange these formulae to make x the subject.

a $k(p - x) = q(x + w)$ **b** $\dfrac{bx + c}{x} = 8$

a $k(p - x) = q(x + w)$ Expand first
     $kp - kx = qx + qw$ Collect x terms
     $kp - qw = qx + kx$ Factorise
     $kp - qw = x(q + k)$

     $\dfrac{kp - qw}{q + k} = x$

b $\dfrac{bx + c}{x} = 8$ Eliminate the fraction

     $bx + c = 8x$
         $c = 8x - bx$
         $c = x(8 - b)$

       $x = \dfrac{c}{8 - b}$

> Take the x terms to the side that avoids negatives: $qx + kx$ is easier to work with than $-kx - qx$

Example

Rearrange this formula to make R the subject.

$$\frac{2p + 10}{50} = \frac{R}{15 + R}$$

$(2p + 10)(15 + R) = 50R$

$30p + 150 + 2pR + 10R = 50R$

$30p + 150 + 2pR = 40R$

$30p + 150 = 40R - 2pR$

$30p + 150 = R(40 - 2p)$

$R = \dfrac{30p + 150}{40 - 2p} \quad \Rightarrow \quad R = \dfrac{15p + 75}{20 - p}$

1 Rearrange each formula to make x the subject.

    a $y = ax^2 + b$ **b** $t = \frac{1}{2}(x - c)$ **c** $k = c - bx$

    d $p = \dfrac{b}{x} - q$ **e** $\sqrt{axz - b} = 3t$ **f** $\dfrac{k - x^3}{t} = t$

2 Explain why you cannot make x the subject of $x^2 + 5x + 6 = 0$ using the method of collecting the x-terms on one side and then factorising.

3 Rearrange each formula to make w the subject.

    a $pw + t = qw - r$ **b** $a - cw = k - lw$ **c** $p(w - y) = q(t - w)$

    d $r - w = t(w - 1)$ **e** $w + g = \dfrac{w + c}{r}$ **f** $\dfrac{wx - t}{r} = 5 - w$

    g $\dfrac{w + t}{w - 5} = k$ **h** $\dfrac{w + p}{q - w} = \dfrac{3}{4}$ **i** $\sqrt{\dfrac{w - t}{w + q}} = 5$

4 **a** The diagram shows a cylinder with radius r and height h. Write a formula for the total volume of the cylinder.

    b Rearrange your formula to make r the subject.

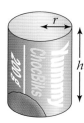

5 Repeat question **4** for a cone of height h and base radius r.

6 The quadratic equation formula can be used to solve quadratic equations.

$$x = \frac{-b \pm \sqrt{b^2 - 4ac}}{2a}$$

    a Use the formula to solve $3x^2 - 2x - 9 = 0$.

    b Rearrange the formula to make c the subject.

    c Rearrange the formula to make b the subject.

    d Why is it too difficult to make a the subject of this formula?

7 The double angle formula in trigonometry is:

    $\sin(A + B) = \sin A \cos B + \cos A \sin B$

Explain why you cannot rearrange this to give

    $\sin A = \dfrac{\cos A \sin B - \sin B}{1 - \cos B}$.

Hint:
Does $\sin(30° + 15°)$ equal $\sin 30° + \sin 15°$?

This spread will show you how to:

- Understand the difference between a practical demonstration and a proof
- Derive proofs using short chains of deductive reasoning
- Use counter-examples to show that a statement is false

Keywords
Counter-example
Demonstrate
Prove

- To **demonstrate** that a statement is true, you find examples that fit the statement.

- To **prove** that a statement is true you can use algebra to generalise it to all possible examples.

- To show that a statement is false, you can use a **counter-example**.

Some useful algebraic generalisations:

Even numbers	$2n$
Odd numbers	$2n + 1$
Consecutive even	$2n, 2n + 2, \ldots$
Consecutive odd	$2n + 1, 2n + 3, \ldots$

To disprove the statement 'All cube numbers are even', you could use the counter-example $125 (=5^3)$ is odd.

Example

'The square of an odd number is always odd.'

a Demonstrate this statement with an example.
b Use algebra to prove this statement.

a 3 is odd and $3^2 = 9$ is odd.
b Any odd number is one more than an even number, so you can represent any odd number as $2n + 1$, where n is an integer.
$(2n + 1)^2 = 4n^2 + 4n + 1$
$\qquad\qquad = 2(2n^2 + 2n) + 1$
For any integer n, $(2n^2 + 2n)$ is an integer, so
$(2n + 1)^2 = 2 \times integer + 1$, which is an odd number.

Find an expression for the square of an odd number.

Example

Prove that, for any three consecutive integers, the difference between the product of the first two and the product of the last two is always twice the middle number.

For example, for 6, 7, 8: $(8 \times 7) - (7 \times 6) = 56 - 42 = 14 = 2 \times 7$
Writing three consecutive integers as $n - 1$, n, $n + 1$:
$n(n + 1) - n(n - 1) = n^2 + n - n^2 + n$
$\qquad\qquad\qquad\qquad = 2n^2$
$\qquad\qquad\qquad\qquad = $ twice the middle number

It may help to demonstrate the statement to yourself using an example.

Exam Question

$y = x^2 + x + 11$ ... the value of y is prime when $x = 0, 1, 2$ or 3.
The following statement is *not* true:
'$y = x^2 + x + 11$ is *always* a prime number when x is an integer'.
Show that the statement is not true. *(Edexcel Ltd., 2005)*

For $x = 11$, $y = x^2 + x + 11 = 11^2 + 11 + 11 = 143$.
143 is not prime, since $143 = 11 \times 13$.
Therefore the statement is not true.

Try different values of x until you find a counter-example.

1 'For any three consecutive integers, the difference between the sum of the first two and the sum of the last two is always two.'

  a Demonstrate this statement with an example.

  b Letting the three consecutive integers be n, $n + 1$ and $n + 2$, write:

  i an expression for the sum of the first two

  ii an expression for the sum of the last two.

  c Using your expressions from part **b**, prove that the statement given is always true.

2 Find a counter-example to disprove each of these statements.

  a All square numbers are even.

  b The difference between two cube numbers is never odd.

  c Squaring a number will always give you a value greater than the number you started with.

  d $(a + b)^2 \neq a^2 + b^2$ for any values of a and b.

  e $\sin(A + B) \neq \sin A + \sin B$ for any values of A or B.

  f The value of $x^2 - 11x + 121$ is never a square number for any value of x.

$\neq$ means 'not equal to'.

3 Use algebraic generalisations to prove these statements.

  a For any three consecutive integers, the square of the middle integer is always one more than the product of the other two.

  b The product of two consecutive odd numbers is always one less than a multiple of four.

  c For any three consecutive integers, the difference between the square of the middle integer and the product of the other two is always equal to 1.

4 Prove that, when two ordinary dice are rolled, the sum of these four products is always equal to 49.

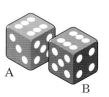

Investigate the relationship between numbers on opposite faces of a dice.

  • the product of the numbers on the top faces

  • the product of the numbers on the bottom faces

  • the product of the number on the top of dice A and the number on the bottom of dice B

  • the product of the number on the bottom of dice A and the number on the top of dice B.

5 Prove that
'The square of the mean of five consecutive integers differs from the mean of the squares of the same five consecutive integers by two.'

More proof

This spread will show you how to:

- Derive proofs using short chains of deductive reasoning

- You can **prove** results from any area of mathematics by generalising using algebra.

Example

Show that these triangles are similar *only* when x is equal to 3.

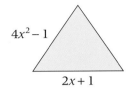

$4x^2 - 1$ $x^2 + 5x + 6$

$2x + 1$ $x + 3$

For the triangles to be similar, pairs of sides must be in the same ratio, that is

$$\frac{4x^2 - 1}{2x + 1} = \frac{x^2 + 5x + 6}{x + 3}$$

$$\frac{(2x - 1)(2x + 1)}{2x + 1} = \frac{(x + 2)(x + 3)}{x + 3}$$

$$2x - 1 = x + 2$$

$$x = 3$$

Consider how you would tackle a numerical example, for example

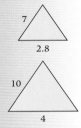

You would show that $\frac{10}{4} = \frac{7}{2.8}$

Exam Question

The distance of the point P from the point $(0, 2)$ is the same as the distance of the point P from the x-axis.

Show that $y = \frac{1}{4}x^2 + 1$.

[Distance of P from $(0, 2)$]$^2 = x^2 + (y - 2)^2$

Distance from P to x-axis $= y$.

Since two distances are equal

$$x^2 + (y - 2)^2 = y^2$$
$$x^2 + y^2 - 4y + 4 = y^2$$
$$x^2 - 4y + 4 = 0$$
$$x^2 + 4 = 4y$$

So, $\frac{1}{4}x^2 + 1 = y$ (as required)

(Edexcel Ltd., 2005)

Find the distance from P to $(0, 2)$ using Pythagoras:

1 Given that this triangle is right-angled, prove that $x^2 - 6x - 39 = 0$.

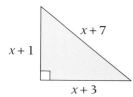

2 Given that n is a positive integer, prove that the perimeter of this triangle will always be even.

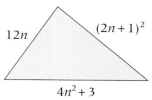

3 The formula $d = \sqrt{(x_2 - x_1)^2 + (y_2 - y_1)^2}$ can be used to find the distance between two points (x_1, y_1) and (x_2, y_2).

Show that the triangle joining A(2, 3) to B(4, 10) to C(7, 5) is scalene.

4 a Write a formula for the area of this trapezium.

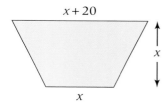

b Show that the area of the trapezium is 24 units^2 *only* when $x = 2$.

5 a A bucket contains m red balls and 6 blue balls.
Write an expression for the probability that when I pick a ball from the bucket it will be red.

b I pick a second ball without replacing the first one.
Write an expression for the probability that the second ball I pick is blue given that the first was red.

c Given that the probability that the two balls I picked in parts **a** and **b** were both red is $\frac{2}{11}$, show that $3m^2 - 11m = 20$.

d Hence, find the number of red balls in the bucket to start with.

6 Show that it is not possible for this triangle to be right-angled.

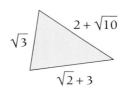

Exam review

Key objectives

- Use formulae from mathematics and other subjects.
- Change the subject of a formula including cases where the subject occurs twice.
- Generate a formula.
- Derive proofs using short chains of deductive reasoning.

1 The diagram shows the shape of a garden lawn.
The measurements are given in metres.

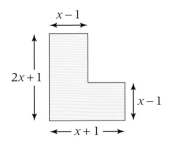

    a Write a formula for the perimeter, P, of the lawn.   (2)

    The area of the lawn is $52\,\text{m}^2$.

    b Show that: $2x^2 + x - 55 = 0$.   (3)

2 $P = \pi r + 2r + 2a$

   $P = 84$

   $r = 6.7$

    a Work out the value of a.
        Give your answer correct to 3 significant figures.   (3)

    b Make r the subject of the formula

        $P = \pi r + 2r + 2a$.   (3)

(Edexcel Ltd., 2005)

S4

This unit will show you how to

- Understand, recall and use Pythagoras's theorem in 2-D problems
- Given the coordinates of points A and B, calculate the length of the line segment AB, and its midpoint
- Understand, recall and use trigonometry in right-angled triangles
- Use trigonometry in right-angled triangles to solve problems
- Recall and use the properties of special types of quadrilaterals

Before you start ...

You should be able to answer these questions.

Review

1 Work out each of these.

a 7^2 **b** $4^2 + 6^2$ **c** $3^2 + 5^2$

d $8^2 - 4^2$ **e** $7^2 - 2^2$

Unit N5

2 Rearrange these equations to make x the subject.

a $y = \frac{x}{6}$ **b** $y = \frac{x}{5}$ **c** $y = \frac{x}{10}$

d $y = \frac{2}{x}$ **e** $y = \frac{5}{x}$ **f** $y = \frac{8}{x}$

Unit A5

3 a Solve $5x + 7 = 32$.

b Find y where $\frac{y}{7} = 11$.

Unit A2

4 a Solve $3x - 2.8 = 12$.

b Find y where $3y + 1 = 9$.

Unit A2

5 a Solve $\frac{x}{4} = 5.7$.

b Find y where $2 + 5y = 6$.

Unit A2

Quadrilaterals

This spread will show you how to:

- Recall the definitions of special types of quadrilaterals
- Classify quadrilaterals by their geometric properties

Keywords
Kite
Parallelogram
Quadrilateral
Rectangle
Rhombus
Square
Trapezium

- A **quadrilateral** is a shape with four straight sides.
- Angles in a quadrilateral add up to 360°.

Properties of quadrilaterals

	Square	Rhombus	Rectangle	Parallelogram	Trapezium	Kite
1 pair opposite sides parallel	✓	✓	✓	✓	✓	
2 pairs opposite sides parallel	✓	✓	✓	✓		
Opposite sides equal	✓	✓	✓	✓		
All sides equal	✓	✓				
All angles equal	✓		✓			
Opposite angles equal	✓	✓	✓	✓		
Diagonals equal	✓		✓			
Diagonals perpendicular	✓	✓				✓
Diagonals bisect each other	✓	✓	✓	✓		
Diagonals bisect the angles	✓	✓		✓		
2 pairs of adjacent sides equal	✓	✓				✓

Example

Work out the area of this rhombus.

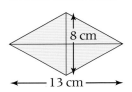

The diagonals bisect each other and are perpendicular.
Opposite angles are equal.
Diagonals bisect the angles.

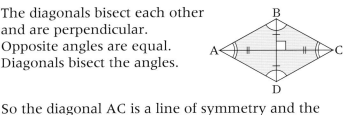

So the diagonal AC is a line of symmetry and the triangles ABC and ACD are congruent.

Each triangle has base 13 cm and height 4 cm.
Area of each triangle $= \frac{1}{2} \times 13 \times 4 = 26$ cm^2
Area of rhombus $= 2 \times 26 = 52$ cm^2

Recap properties of a **rhombus**.

A line of symmetry divides a shape into two congruent shapes.

Area of triangle $= \frac{1}{2}bh$

Example

Are these statements true or false?
Give reasons for your answer.

a All squares are rhombuses

b Parallelograms are rectangles

a True, all properties of a rhombus are also properties of a square.

b False, a **parallelogram** does not have all its angles equal, nor are its diagonals perpendicular.

A **square** is a special type of rhombus with all angles equal.

A **rectangle** is a special type of parallelogram with all angles equal.

1 The lengths of the diagonals of a rhombus are 5 cm and 9 cm.
Find the area of the rhombus.

2 A square has diagonals 10 cm long.

   a Sketch the square.

   b Find the area of the square.

3 Jenny draws these three kites each with diagonals 6 cm and 14 cm.

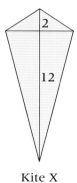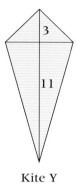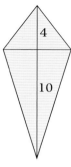

       Kite X   Kite Y   Kite Z

> **DID YOU KNOW?**
> The kite is a rare bird of prey that hovers in the wind, which influenced the naming of the toy kite. This in turn influenced the naming of the kite shape.

   a Find the areas of kites X, Y and Z.

   b Describe how to work out the area of any kite.

4 The lengths of the diagonals of a kite are 8 cm and 15 cm.
Find the area of the kite.

> Use your method from question **3**.

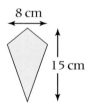

5 Here is a diagram of a parallelogram, P, and a rhombus, R.

   One diagonal been drawn inside each shape.
   Has either shape has been split into two congruent triangles?
   Give a reason for your answer.

6 Are these statements true or false? Give reasons for your answers.

   a All squares are rectangles.

   b All kites are rhombuses.

   c All rhombuses are rectangles.

Pythagoras's theorem and coordinates

This spread will show you how to:

- Understand, recall and use Pythagoras's theorem in 2-D problems
- Given the coordinates of points A and B, calculate the length of the line segment AB, and its midpoint

Keywords

Hypotenuse
Midpoint
Pythagoras's
  theorem
Right-angled
  triangle

The coordinates of the **midpoint** of two points are the means of their coordinates.

A(2, 3)   B(6, 5)   C(−2, 1)

A is the midpoint of the line joining C to B.

- The midpoint of (a, b) and (s, t) is $\left(\dfrac{a+s}{2}, \dfrac{b+t}{2}\right)$.

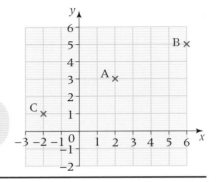

Example

Find the midpoint of   **a** AB   **b** AC.

a The midpoint of AB is $\left(\dfrac{2+6}{2}, \dfrac{3+5}{2}\right) = (4, 4)$

b The midpoint of AC is $\left(\dfrac{2+-2}{2}, \dfrac{3+1}{2}\right) = (0, 2)$

- **Pythagoras's theorem** for **right-angled triangles**:
  In a right-angled triangle, the square on the **hypotenuse** equals the sum of the squares on the other two sides.

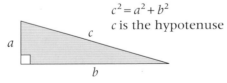

$$c^2 = a^2 + b^2$$
c is the hypotenuse

The hypotenuse is always opposite the right angle.

You can use Pythagoras's theorem to find the length of a line joining two points on a grid.

- sketch a diagram and label the right angle
- label the unknown side
- round your answer to a suitable degree of accuracy.

Unless the question tells you otherwise, round to 2 dp.

Example

Work out the length of the line joining A(2, 3) and C(−2, 1).

Draw a right-angled triangle and label the lengths of the two shorter sides.

$AC^2 = 2^2 + 4^2$
$AC^2 = 20$
  $AC = 4.47$ (2 dp)

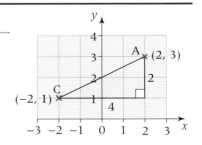

1 Find the midpoints of
   a A(6, 4) and B(−4, 5)   **b** C(−2, 5) and D(7, −1).

2 The sketch shows the points (3, 1) and (8, 4).
   Work out the length of the line joining
   these two points.

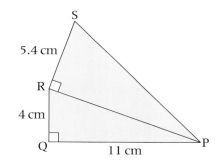

3 Use Pythagoras's theorem to work out the length of the line joining
   a (0, 6) and (1, −2)   **b** (−4, 3) and (5, −2).

   Sketch the line first.

4 Find the length of the diagonal of a square with side 10 cm.

5 **a** Find the length of the side of a square with diagonal 10 cm.
   b Write the area of a square with a diagonal 10 cm.

6 PQR and PRS are right-angled
   triangles.
   Work out the length PS.

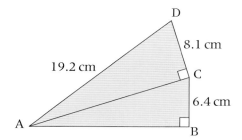

   Work out PR first.

7 ABC and ACD are right-angled
   triangles.
   Work out the length AB.

8 A cylinder has base diameter 7.5 cm and height
   18 cm.
   A thin rod just fits diagonally inside the cylinder.

   What is the longest length that the rod could be so
   that it fits exactly inside the cylinder?

Tangent ratio

This spread will show you how to:

● Understand, recall and use trigonometry in right-angled triangles

Keywords
Gradient
Right-angled
   triangle
Tangent

A road sign gives the **gradient** of a hill as a percentage.

Hills with the same percentage slope climb at the same angle.

The gradient is the ratio of the vertical climb to the horizontal distance.

You can draw a **right-angled triangle** to represent the hill.

The ratio $\dfrac{\text{vertical height}}{\text{horizontal distance}}$

is the same, however large or small you draw the triangle.

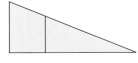

This ratio is called the **tangent** ratio.

● $\tan \theta = \dfrac{\text{opposite}}{\text{adjacent side}}$

You can use the tangent ratio in any right-angled triangle.

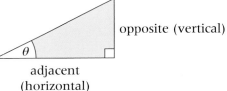

opposite (vertical)

θ

adjacent
(horizontal)

Adjacent means 'next to'.

θ is used to represent an angle.

Example

Find the missing sides in these triangles.

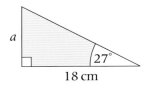

a

27°

18 cm

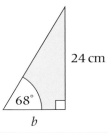

24 cm

68°

b

Remember to put your calculator into 'degree' mode.

$\tan 27° = \frac{a}{18}$
$18 \times \tan 27° = a$
$\qquad a = 9.17$ cm

$\tan 68° = \frac{24}{b}$

$b = \dfrac{24}{\tan 68°}$

$\quad = 9.70$ cm

You could use the other acute angle in this triangle:

$90° - 68° = 22°$

$\tan 22° = \dfrac{b°}{24°}$

$24 \times \tan 22° = b$
$\qquad\qquad b = 9.70$ cm

1 Find the missing side in each of these right-angled triangles.
Give your answers to 3 significant figures.

a

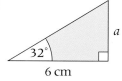

32°
6 cm
a

b

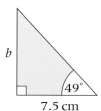

b
49°
7.5 cm

c

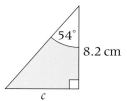

54°
8.2 cm
c

d

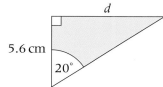

d
5.6 cm
20°

e

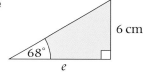

68°
e
6 cm

f

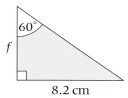

60°
f
8.2 cm

g

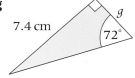

7.4 cm
g
72°

h

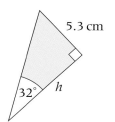

5.3 cm
32°
h

i

77°
15.8 cm
j

j

k
13 cm
70°

k

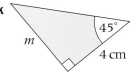

45°
m
4 cm

l

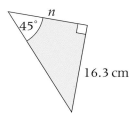

n
45°
16.3 cm

2 What type of triangles are question **1** parts **k** and **l**?
How could you find the missing sides *m* and *n* without using the
tangent ratio?

Sine and cosine ratios

This spread will show you how to:

● Understand, recall and use trigonometry in right-angled triangles

Keywords
Cosine
Hypotenuse
Right-angled
   triangle
Sine

The tangent ratio relates opposite and adjacent sides in a **right-angled triangle**.

Two other ratios you can use are **sine** and **cosine**.

$$\sin \theta = \frac{\text{opposite side}}{\text{hypotenuse}} \qquad \cos \theta = \frac{\text{adjacent side}}{\text{hypotenuse}}$$

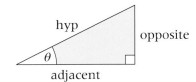

You can use the sine ratio and cosine ratio in any right-angled triangle.

Example

Find the missing sides in these triangles.

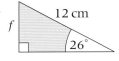

a

12 cm
f
26°

b

7 cm 18°
g

c

h
32 cm
70°

a $\sin 26° = \frac{f}{12}$
$f = 12 \times \sin 26°$
$= 5.26$ cm

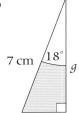

opp f 12 cm hyp
26°
adj

b $\cos 18° = \frac{g}{7}$
$g = 7 \times \cos 18$
$= 6.66$ cm

7 cm
hyp 18°
g adj

opp

Sketch the triangle and label the sides in relation to the angle given. Label the **hypotenuse** first.

c $\sin 70° = \frac{32}{h}$
$h \times \sin 70° = 32$

$$h = \frac{32}{\sin 70°}$$
$= 34.05$ cm

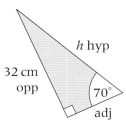

h hyp
32 cm
opp
70°
adj

You always need to divide by either $\sin \theta$ or $\cos \theta$ to find the hypotenuse.

Find the missing side in each of these right-angled triangles.
Give your answers to 3 significant figures.

1
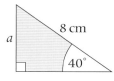
8 cm
a
40°

2
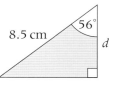
12 cm
b
60°

3
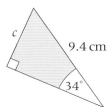
c
9.4 cm
34°

4
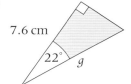
8.5 cm
56°
d

5
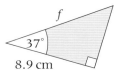
14.2 cm
48°
e

6
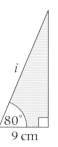
f
37°
8.9 cm

7
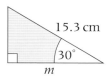
7.6 cm
22°
g

8
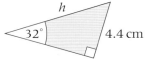
h
32°
4.4 cm

9
i
80°
9 cm

10
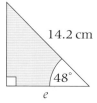
10.5 cm
75°
j

11
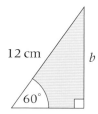
k
45°
11.7 cm

12
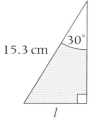
15.3 cm
30°
l

13
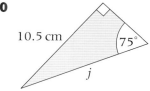
15.3 cm
30°
m

14
67°
n
6.4 cm

15
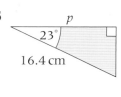
p
23°
16.4 cm

Finding angles in right-angled triangles

This spread will show you how to:

● Understand, recall and use trigonometry in right-angled triangles

Keywords
Cosine
Right-angled
  triangle
Sine
Tangent

You can use the **sine**, **cosine** and **tangent** ratios to find a missing angle in a **right-angled triangle**.

Choose the appropriate ratio for the angle you want to find.

● $\sin \theta = \dfrac{\text{opp}}{\text{hyp}}$ $\cos \theta = \dfrac{\text{adj}}{\text{hyp}}$ $\tan \theta = \dfrac{\text{opp}}{\text{adj}}$

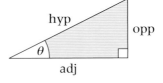

Use the inverse operations $\sin^{-1}$, $\cos^{-1}$ and $\tan^{-1}$ to find the angle.

You can also use the terms arcsin, arcos and arctan for $\sin^{-1}$, $\cos^{-1}$ and $\tan^{-1}$ respectively.

Example

Find the missing angles in these triangles.
Give your answers to 1 dp.

a
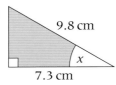
9.8 cm
7.3 cm
x

b

12.5 cm
5.6 cm
x

c
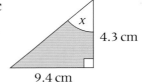
x
4.3 cm
9.4 cm

a opp and adj so use cos
$\cos x = \dfrac{7.3}{9.8}$
$x = \cos^{-1} \dfrac{7.3}{9.8}$
$= 41.9°$

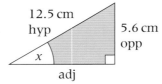
9.8 cm hyp
opp
7.3 cm
adj
x

Sketch the triangle and label the sides in relation to the sides given. Label the hypotenuse first.

b opp and hyp so use sin
$\sin x = \dfrac{5.6}{12.5}$
$x = \sin^{-1} \dfrac{5.6}{12.5}$
$= 26.6°$

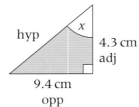
12.5 cm
hyp
5.6 cm
opp
x
adj

Sines and cosines are always proper fractions.

c opp and adj so use tan
$\tan x = \dfrac{9.4}{4.3}$
$x = \tan^{-1} \dfrac{9.4}{4.3}$
$= 65.4°$

hyp
x
4.3 cm
adj
9.4 cm
opp

Tan can be an improper fraction.

Find the missing angle in each of these right-angled triangles.
Give your answers to 3 significant figures.

1
11 cm
3 cm
a

2
4.7 cm
15 cm
b

3
13.4 cm
5.6 cm
c

4
7.6 cm
d
19.5 cm

5
7.5 cm
e
16.3 cm

6
4.3 cm
11.5 cm
f

7
8.8 cm
6 cm
g

8
6.5 cm
h
17.2 cm

9
7.6 cm
i
20.1 cm

10
j
12.4 cm
16 cm

11
5.8 cm
k
16.8 cm

12
9.6 cm
22 cm
l

13
2.4 cm
m
9 cm

14
11.4 cm
n
19.5 cm

15
23.6 cm
p
23.6 cm

16
q
10.8 cm
5.4 cm

17
24.6 cm
r
12.3 cm

Pythagoras's theorem and trigonometry

This spread will show you how to:

- Use trigonometry in right-angled triangles to solve problems
- Understand, recall and use Pythagoras's theorem in 2-D problems

Keywords
Pythagoras's
  theorem
Right-angled
  triangle

You often need to use **Pythagoras's theorem** and trigonometric ratios to solve problems involving **right-angled triangles**.

You should always sketch a diagram to represent the situation.
You may need to work out extra information to solve the problem.

Example

ABCD is a parallelogram.
AB = 8.2 cm, BC = 6.6 cm and angle ABC = 53°.
Work out the area of the parallelogram.

Sketch a diagram.
Area of a parallelogram = $b \times h$
You need to find the perpendicular height, h.

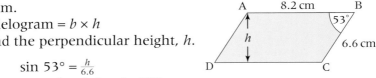

$$\sin 53° = \frac{h}{6.6}$$
$$h = 6.6 \times \sin 53°$$
$$h = 5.27 \ldots$$

$$\text{Area} = b \times h$$
$$= 8.2 \times 5.27 \ldots$$
$$= 43.2 \text{ cm}^2$$

The height h must be at right angles to the base b.

This is only an intermediate step. Do not round any values until the end of the calculation.

Example

A vertical flagpole FP has P at the top of the pole.
It is held by two wires, PX and PY, fixed on horizontal ground at X and Y.
Angle FXP = 36° and angle FYP = 49°.
FX is 20.5 m. Find the length of PY.

Sketch a diagram.

FP is vertical, so it is perpendicular to the horizontal ground.
It forms two right-angled triangles –
  PFX and PFY.
First find the height of PF.

$$\tan 36° = \frac{h}{20.5}$$
$$h = 20.5 \times \tan 36° = 14.894 \ldots$$

Use trigonometry to find the length PY.

$$\sin 49° = \frac{h}{\text{PY}}$$

$$\text{PY} = \frac{h}{\sin 49°} = \frac{14.894 \ldots}{\sin 49°} = 19.734 \ldots$$

So PY = 19.7 m

If you needed to find the length of PX you could use Pythagoras's theorem.

1 Find the missing lengths.

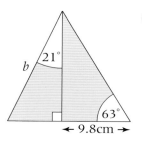

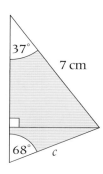

2 Find the missing angles.

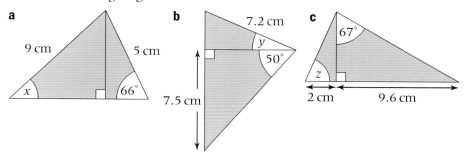

3 A parallelogram has sides of length 5 cm and 9 cm.
The smaller angles are both 45°.
Find the area of the parallelogram.

4 A rhombus has side length 8 cm and smaller angle 35°.
Find the area of the rhombus.

5 A chord AB, of length 12 cm, is drawn inside a circle with centre O
and radius 8.5 cm.
Find the angle AOB.

6 A chord PQ is drawn inside a circle with centre O and radius 6 cm,
such that angle POQ = 104°.
Find the length PQ.

7 An isosceles triangle has sides of length 7 cm, 7 cm and 9 cm.
Find the interior angles of the triangle.

8 Edina and Patsy are estimating the height of
the same tree.
Edina stands 20 m from the tree and
measures the angle of elevation of the
treetop as 52°.
Patsy stands 28 m from the tree and measures
the angle of elevation of the treetop as 44°.
Can they both be correct? Explain your reasoning.

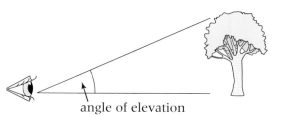

angle of elevation

Exam review

Key objectives

- Understand, recall and use Pythagoras's theorem in 2-D problems
- Understand, recall and use trigonometrical relationships in right-angled triangles, and use these to solve problems
- Recall and use the properties of special types of quadrilaterals

1 The diagram shows a rectangle.

$2x - 1$ **NOT** to scale

$3x - 1$

The length of a diagonal in the rectangle is $\sqrt{5}$ cm.
Calculate the area of the rectangle.   (4)

2 The diagram represents a vertical flagpole, AB.
The flagpole is supported by two ropes, BC and BD, fixed to the horizontal ground at C and at D.

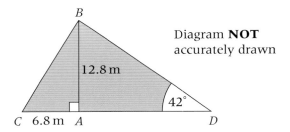

Diagram **NOT** accurately drawn

B

$12.8\,\text{m}$

$42°$

C $6.8\,\text{m}$ A D

AB = 12.8 m
AC = 6.8 m
Angle BDA = 42°

a Calculate the size of angle BCA.
Give your answer correct to 3 significant figures.   (3)

b Calculate the length of the rope BD.
Give your answer correct to 3 significant figures.   (3)

(Edexcel Ltd., 2003)

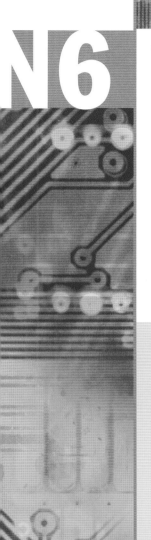

This unit will show you how to

- Check and estimate answers to problems
- Use π and surds in exact calculations
- Recognise limitations on the accuracy of data and measurements and calculate their upper and lower bounds
- Give answers to an appropriate degree of accuracy
- Use the correct order of operations in a calculation
- Use written calculation methods
- Use calculators effectively and efficiently for complex calculations, including trigonometrical functions
- Use calculators to calculate in standard form

Before you start ...

You should be able to answer these questions.

Review

1 Round these numbers to one significant figure (1 sf).

   a 38.5   **b** 16.08
   c 103.88   **d** 0.082
   e 0.38

Unit N2

2 Write a mental estimate for these calculations.

   a 18×53 **b** 3.77×89.5
   c $3870 \div 79$ **d** $642 \div 28.7$

Unit N2

3 Use a calculator to work out these calculations.
Write all of the digits on the calculator display, but do not write any intermediate answers.

   a $\dfrac{3.4}{56.2 - 18.9}$ **b** $\dfrac{5.9 - 8.2}{6.7 - 3.2}$
   c $\sqrt{5.3^2 - 4.6^2}$

Unit N2

4 Use a calculator to work out these calculations.
Give your final answers in standard form.

   a $(3.4 \times 10^4) \times (4.9 \times 10^2)$
   b $(7.9 \times 10^8) - (6.3 \times 10^7)$

Unit N2

This spread will show you how to:

● Check and estimate answers to problems
● Give answers to an appropriate degree of accuracy

Keywords

Approximation
Estimate
Significant
  figures
Standard form

● You can use **approximations** to one **significant figure** to make **estimates**.

You need to be careful when estimating powers.
For example, 1.3 is quite close to 1, but 1.3^7 is not close to 1^7.

Example

Estimate the value of these calculations.

a $\dfrac{563 + 1.58}{327 - 4.72}$ **b** $\dfrac{3.27 \times 4.49}{1.78^2}$ **c** $\dfrac{\sqrt{2485}}{1.4^3}$ **d** $\dfrac{2.45^3}{2.5 - 2.4}$

a Estimate: $600 \div 300 = 2$
b 1.78^2 is 'a bit more than 3', so it cancels with 3.27
  to give an estimate of 4.5.
c $2485 \approx 2500$ and $\sqrt{2500} = 50$
  $1.4^2 \approx 2$, so $1.4^3 \approx 1.4 \times 2 = 2.8 \approx 3$
  Estimate: $50 \div 3 \approx 17$
d $2.45 \approx 2.5$ and $25^2 = 625$, so $2.5^2 \approx 6$
  $2.45^3 \approx 6 \times 2.5 = 15$
  $2.5 - 2.4 = 0.1$, so estimate is: $15 \div 0.1 = 150$

In part **a**, ignore the relatively small amounts added and subtracted.

In part **b**, $\sqrt{3} = 1.73$ to 2 dp.

In part **d**, to 1 sf the denominator is $2 - 2 = 0$, which is not possible.

● You can use **standard form** to estimate calculations involving very large or very small numbers. For example,

$$5130 \times 0.000\,178 \approx (5 \times 10^3) \times (2 \times 10^{-4})$$
$$= 5 \times 2 \times 10^3 \times 10^{-4}$$
$$= 10 \times 10^{-1} = 1$$

Example

Estimate the value of $\dfrac{4217 \times 0.0625}{23\,563}$.

Writing the calculation in standard form.

$$(4.217 \times 10^3) \times (6.25 \times 10^{-2}) \div (2.3563 \times 10^4)$$
$$\approx (4 \times 10^3) \times (6 \times 10^{-2}) \div (2 \times 10^4)$$
$$= 12 \times 10^{-3}$$
$$= 1.2 \times 10^{-2}$$
$$= 0.012$$

For standard form, the multiplier must be between 1 and 10.

1 Estimate answers to these calculations.

   a $4.88 + 3.07$ **b** $216 + 339$ **c** $0.0049 + 0.003\ 02$

   d $43.89 - 28.83$ **e** 3.77×0.85 **f** $44.66 \div 0.89$

2 Estimate these square roots mentally, to 1 decimal place.

   a $\sqrt{2}$ **b** $\sqrt{8}$ **c** $\sqrt{10}$ **d** $\sqrt{15}$ **e** $\sqrt{20}$

   f $\sqrt{26}$ **g** $\sqrt{32}$ **h** $\sqrt{45}$ **i** $\sqrt{70}$ **j** $\sqrt{85}$

   Use a calculator to check your estimates.

3 Explain why approximating the numbers in these calculations to 1 significant figure would *not* be an appropriate method for estimating the results of the calculations.

   a $\dfrac{5.39 + 4.72}{0.53 - 0.46}$ **b** $(2.45 - 0.96)^8$ **c** $(1.52 - 1.49)^2$

4 Use approximations to estimate the value of each of these calculations. You should show all your working.

   a $\dfrac{317 \times 4.22}{0.197}$ **b** $\dfrac{4.37 \times 689}{0.793}$ **c** $\dfrac{4.75 \times 122}{522 \times 0.38}$

   d $4.8^3 - 8.5^2$ **e** $\dfrac{9.32 - 3.85}{0.043 - 0.021}$ **f** $7.73 \times \left(\dfrac{0.17 \times 234}{53.8 - 24.9} \right)$

5 Find approximate values for these. Show your working.

   a $\dfrac{48.75 \times 4.97}{10.13^2}$ **b** $\sqrt{\dfrac{305.3^2}{913}}$ **c** $\dfrac{\sqrt{9.67 \times 8.83}}{0.087}$

   d $\dfrac{6.8^2 + 11.8^2}{\sqrt{47.8 \times 52.1}}$ **e** $\dfrac{(23.4 - 18.2)^2}{3.2 + 1.8}$ **f** $\sqrt{\dfrac{2.85 + 5.91}{0.17^2}}$

6 Use standard form approximations to find an estimate for each of these calculations. Show your working.

   a $4800 \div 465$ **b** $7326 \div 0.069$ **c** $\dfrac{83\ 550 \times 0.039}{4378}$

   d $\dfrac{653 \times 0.415}{0.07 \times 0.38}$ **e** $\dfrac{735 + 863}{0.06 \times 0.85}$ **f** $\dfrac{3400 \times 475}{(28.5 + 36.9)^2}$

Exact calculations

This spread will show you how to:

● Use π and surds in exact calculations

Keywords
Irrational
π (pi)
Recurring
Surds
Terminating

● A fraction will have a terminating decimal equivalent if the only prime factors of the denominator are 2 or 5. $\frac{3}{20} = \frac{3}{(2 \times 2 \times 5)}$

● All other fractions give **recurring** decimals. $\frac{1}{18} = \frac{1}{(2 \times 3 \times 3)} = 0.05555... = 0.0\dot{5}$

When you calculate with fractions, you should work with the numbers in fraction form as far as possible.

Example

Calculate each of these.

a $\frac{1}{2} + \frac{2}{3}$ **b** $\frac{3}{4} \times \frac{2}{9}$ **c** $\frac{4}{5} \div \frac{3}{10}$

a $\frac{1}{2} + \frac{2}{3} = \frac{3}{6} + \frac{4}{6} = \frac{7}{6} = 1\frac{1}{6}$ **b** $\frac{{}^1\cancel{3}}{{}_2\cancel{4}} \times \frac{\cancel{2}^1}{\cancel{9}_3} = \frac{1}{2} \times \frac{1}{3} = \frac{1}{6}$

c $\frac{4}{5} \div \frac{3}{10} = \frac{4}{\cancel{5}_1} \times \frac{\cancel{10}^2}{3} = \frac{4}{1} \times \frac{2}{3} = \frac{8}{3} = 2\frac{2}{3}$

● Some numbers cannot be written as fractions. These **irrational** numbers have no repeating decimal patterns.

● Irrational numbers such as $\sqrt{2}$ are called **surds**.

$\sqrt{2} = 1.414213562...$

You can work with combinations of rational numbers and surds.

$$\frac{1 + \sqrt{5}}{2} \quad \text{or} \quad \frac{1}{2\pi}\sqrt{\frac{7}{9.8}}$$

Again, work with the numbers in surd form as far as possible.

Example

Calculate each of these.

a $\frac{7 + \sqrt{20}}{4} - \frac{3 + \sqrt{5}}{5}$ **b** $(1 + \sqrt{5}) \div (2 + \sqrt{3})$

Rationalise the denominator.

a $\frac{7 + \sqrt{20}}{4} - \frac{3 + \sqrt{5}}{5} = \frac{35 + 5\sqrt{20} - 12 - 4\sqrt{5}}{20}$

                    $= \frac{35 + 5\sqrt{20} - 12 - 4\sqrt{5}}{20}$

                    $= \frac{23 + 10\sqrt{5} - 4\sqrt{5}}{20}$

                    $= \frac{23 + 6\sqrt{5}}{20}$

b $(1 + \sqrt{5}) \div (2 + \sqrt{3})$

       $= \frac{1 + \sqrt{5}}{2 + \sqrt{3}}$

       $= \frac{1 + \sqrt{5}}{2 + \sqrt{3}} \times \frac{2 - \sqrt{3}}{2 - \sqrt{3}}$

       $= 2 - \sqrt{3} + 2\sqrt{5} - \sqrt{15}$

1 Find the exact solution.

a $\frac{1}{3} + \frac{1}{5}$ **b** $\frac{2}{5} + \frac{3}{7}$ **c** $\frac{3}{8} - \frac{2}{7}$ **d** $\frac{8}{15} + \frac{4}{9}$

e $6\frac{1}{2} - 1\frac{5}{8}$ **f** $\frac{2}{5} + \frac{1}{3} + \frac{1}{4}$ **g** $\frac{3}{8} + \frac{1}{2} - \frac{2}{5}$ **h** $7\frac{2}{9} + 2\frac{1}{4}$

2 Find the exact solution.

a $\frac{2}{3} \times \frac{3}{4}$ **b** $\frac{5}{9} \div \frac{1}{3}$ **c** $2\frac{1}{2} \times \frac{5}{8}$ **d** $\frac{8}{9} \div 1\frac{2}{3}$

e $3\frac{1}{2} \div 2\frac{1}{4}$ **f** $5\frac{1}{5} \times 2\frac{3}{4}$ **g** $8\frac{2}{5} \div 3\frac{1}{7}$ **h** $3\frac{1}{2} \times 7\frac{5}{9}$

3 For each answer from questions **1** and **2**, either give an exact decimal equivalent of the answer, or explain why it is not possible to do so.

4 Find the exact solution.

a $\frac{2}{5} \times (\frac{3}{4} + \frac{2}{3})$ **b** $(\frac{2}{3} + \frac{4}{5}) \div (\frac{2}{5} + \frac{3}{7})$ **c** $\frac{5}{6} \div \frac{3^2 + 4^2}{10}$

d $(\frac{3}{7})^2 \times (\frac{4}{5} - \frac{1}{7})$ **e** $(\frac{1}{2} + \frac{5}{9})^2 + (\frac{2}{3})^3$ **f** $(\frac{5}{6} + \frac{1}{2}) - (\frac{4}{7} \times 3)$

5 Simplify these expressions.

a $\sqrt{20} + 3\sqrt{5}$ **b** $(5 + 3\sqrt{3}) - (2 + 4\sqrt{3})$ **c** $\frac{6 + \sqrt{27}}{3} - \frac{2 + \sqrt{3}}{4}$

d $\frac{20 + 3\sqrt{7}}{3} - (5 + \sqrt{28})$ **e** $\frac{6 + \sqrt{8}}{2} + \frac{3 + \sqrt{2}}{9}$ **f** $\frac{8 + \sqrt{45}}{3} - \frac{4 + \sqrt{20}}{9}$

6 Evaluate, giving your answers in surd form.

a $\sqrt{5}(1 + \sqrt{5})$ **b** $\sqrt{3}(5 - \sqrt{3})$ **c** $(2 + \sqrt{3})(3 + \sqrt{3})$

d $(3 - \sqrt{2})^2$ **e** $(5 - \sqrt{7})(5 + \sqrt{7})$ **f** $(4 + 3\sqrt{5})(6 - \sqrt{5})$

7 Write each expression without irrational numbers in the denominator.

a $\frac{1}{\sqrt{2}}$ **b** $\frac{1 + \sqrt{3}}{\sqrt{3}}$ **c** $\frac{5 - \sqrt{20}}{\sqrt{5}}$

d $\frac{3}{1 + \sqrt{7}}$ **e** $\frac{8}{5 - 2\sqrt{3}}$ **f** $\frac{3 - \sqrt{3}}{4 + \sqrt{5}}$

8 Simplify these expressions, writing your solutions without irrational numbers in the denominators.

a $(1 + \sqrt{5}) \div (1 + \sqrt{3})$ **b** $(2 + \sqrt{7}) \div (2 + \sqrt{3})$

c $(5 - \sqrt{11}) \div (2 + \sqrt{11})$ **d** $(1 + 2\sqrt{3}) \div (2 - \sqrt{3})$

e $(5 + 2\sqrt{3}) \div (3 - 2\sqrt{3})$ **f** $(7 - 2\sqrt{5}) \div (8 - 2\sqrt{5})$

Limits of accuracy

This spread will show you how to:

- Recognise limitations on the accuracy of data and measurements and calculate their upper and lower bounds
- Give answers to an appropriate degree of accuracy

Keywords

Implied accuracy
Lower bound
Upper bound

In calculations, give your answers to an appropriate degree of accuracy.

A measurement of 2.3 cm has an implied accuracy of 1 decimal place.

A measurement given as 3.50 m has an implied accuracy of 2 decimal places.

If all measurements are to 1 dp, give your answers to 1 or 2 dp.

$2.25 \leqslant x < 2.35$

2.2 2.25 2.3 2.35 2.4

All values in the range
$2.25 \leqslant x < 2.35$
round to 2.3 to
1 dp.
The upper bound is
2.35 cm and the
lower bound is
2.25 cm.

Example

A pile of 16 sheets of card is 7.3 mm thick.
Calculate the thickness of one sheet.

7.3 mm ÷ 16 = 0.456 25 mm = 0.46 mm (2 sf)

Use upper and lower bounds to calculate precise limits of accuracy.

- The **upper bound** is the smallest value that is greater than or equal to any possible value of the measurement.
- The **lower bound** is the greatest value that is smaller than or equal to any possible value of the measurement.

0.456 25 mm
implies that you
could measure the
thickness of a
piece of card to the
nearest
0.000 01 mm –
unlikely!

For continuous
data, the upper
bound is *not* a
possible value of
the data.

Example

In a science experiment, a trolley travelled 8.4 m in 3.6 seconds. Calculate the upper and lower bound of the average speed of the trolley in metres per second (ms^{-1}), writing down all the digits on your calculator display.

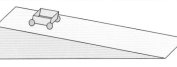

Speed = distance ÷ time
Speed$_{Upper}$ = Distance$_{Upper}$ ÷ Time$_{Lower}$ = 8.45 ÷ 3.55 = 2.380 281 7 m s^{-1}
Speed$_{Lower}$ = Distance$_{Lower}$ ÷ Time$_{Upper}$ = 8.35 ÷ 3.65 = 2.287 671 2 m s^{-1}

Distance is in the
range
$8.35 \leqslant d < 8.45$ m.
Time is in the
range
$3.55 \leqslant t < 3.65$
seconds.

Example

A measurement is given as 3.8 cm, correct to the nearest 0.1 cm.
Pritesh writes:

Lower bound = 3.75 cm
Upper bound is a bit less than 3.85 cm,
say 3.849 99...

Is he correct?

The lower bound is correct; 3.75 is the largest value that is less than or equal to any possible value of the measurement.
The upper bound is incorrect. It should be 3.85.

1 The length of a rod is 0.45 m, correct to the nearest centimetre.
   Write the upper and lower bounds for the length of the rod.

2 Use the implied accuracy of these measurements to write down the
   upper and lower bound for each one.
   a 6.4 mm **b** 4.72 m **c** 18 s **d** 0.388 kg **e** 6.5 volts

3 A car travels a distance of 32 m (to the nearest metre) in a time of
   1.6 seconds (to the nearest 0.1 s). Find the upper and lower bounds
   of the average speed of the car.

4 A rectangle has a length of 4.3 m and a width of 3.1 m, both
   measured to the nearest 0.1 m.
   Find
   a the lower bound of the perimeter of the rectangle
   b the upper bound of the area of the rectangle.

5 The maximum load a van can carry is 450 kg.
   The van is used to carry boxes that weigh 30 kg to the nearest 1 kg.
   Find the maximum number of boxes that the van can safely carry.
   Show your working.

6 A crane has a maximum working load of 670 kg to 2 sf.
   It is used to lift crates that weigh 85 kg, to the nearest 5 kg.
   What is the greatest number of crates that the crane can safely lift at
   one time?

7 A lift can carry a maximum of five people, and the total load must
   not exceed 440 kg. Five members of a judo team enter the lift. Each
   person weighs 87 kg to the nearest kilogram. Is it possible that the
   total weight of the group exceeds the maximum load of the lift?
   Show your working.

8 This trapezium has area 450 cm^2 to 2 sf. It has parallel
   sides 18 cm and 22 cm, each to the nearest centimetre.

   Calculate the lower bound of the height, h, of the
   trapezium.

9 The length of Eva's stride is 86 cm, to the nearest centimetre.
   a Write the upper and lower bounds of Eva's stride.
   b The length of a path is 28 m, to the nearest metre.
   Starting at the beginning of the path, Eva takes 32 strides in a
   straight line along the path. Explain, showing all your working,
   why Eva may not reach the end of the path.

This spread will show you how to:

- Use written calculation methods

- You can make an **estimate** before starting a written calculation. You can use it to check your answer or to adjust place value.

- In a multi-stage calculation, you need to remember the **order of operations**, **BIDMAS**.

BIDMAS
Brackets
Indices (or powers)
Division or **M**ultiplication
Addition or **S**ubtraction

Example

Evaluate these without a calculator.

a $\dfrac{2.3 \times (4^2 + 5^2)}{7}$ **b** $-2.1 + (-9 \times 2.4)$ **c** $\dfrac{2}{7} + \dfrac{3}{5}$ **d** 23% of 760

a
23 × 41 = 943
943 ÷ 7 = 134.7 to 4 sf
÷ 10 to adjust place value: answer = 13.5 to 3 sf

Estimate
$4^2 + 5^2 = 16 + 25 = 41$
$41 \div 7 \approx 6$, and $6 \times 2.3 \approx 14$.

b
9 × 24 = 216 ⇒ −9 × 2.4 = −21.6
−2.1 − 21.6 = −23.7

Estimate
$-9 \times 2.4 \approx -10 \times 2 = -20$,
and $-2.1 - 20 \approx -22$

c
$\dfrac{2}{7} + \dfrac{3}{5} = \dfrac{10}{35} + \dfrac{21}{35} = \dfrac{31}{35}$

Estimate
$\dfrac{2}{7} \approx \dfrac{1}{3}$ and $\dfrac{3}{5} \approx \dfrac{2}{3}$, so the answer should be about 1.

d
23% of 760 ⇒ $\dfrac{23}{100} \times 760$
23 × 76 = 1748
adjust place value: answer = 174.8

Estimate
23% of 760 $\approx \dfrac{1}{4}$ of 760 =
$\dfrac{1}{2}$ of 380 = 190

Use your estimate to adjust place value.

Example

Evaluate these without a calculator, giving your answers in standard form to 2 significant figures.

a $(2.1 \times 10^4) \times (7.3 \times 10^{-6})$ **b** $(1.512 \times 10^{-2}) \div (9 \times 10^3)$

a 21 × 73 = 1533
adjust place value: 1.5×10^{-1}
b 1512 ÷ 9 = 168
adjust place value: answer = 1.68×10^{-6}

Estimate
$(2 \times 10^4) \times (7 \times 10^{-6}) = 14 \times 10^{-2}$
$= 1.4 \times 10^{-1}$

Estimate
$9 \times 10^3 \approx 10^4$
$\Rightarrow (1.512 \times 10^{-2}) \div (9 \times 10^3)$
$\approx (1.5 \times 10^{-2}) \div 10^4$
$= 1.5 \times 10^{-6}$

1 Use a written method to calculate these.

   a 836×46 b 774×38 c 397×171 d 617×259

2 Use a written method to calculate these.

   a $2496 \div 16$ b $4071 \div 23$ c $7942 \div 38$ d $21\,373 \div 67$

3 Evaluate these without using a calculator.

   a $\dfrac{-7 + (-18 \times +45)}{+3 \times -3}$ b $-6.7 + \dfrac{4.3 \times -2.1}{-2 \times +5}$

   c $\dfrac{8.51 - 30.77}{1.8 + 0.3}$ d $\dfrac{25.19}{11} - 3.6^2$

4 Evaluate these without a calculator, giving your answers to 3 significant figures.

   a $\dfrac{4.7 \times 3.8}{7}$ b $5.8^2 - \dfrac{3.2 \times 1.8}{9}$ c $\dfrac{5.2^2}{8} - \dfrac{2.1^2}{5}$ d $\dfrac{4.8 \times 6.3}{2.3 + 5.1}$

5 Use a written method to evaluate these. Show your working.

   a 52% of 416   b 38% of 98   c 63% of 881

   d 119% of 77   e 28.5% of 515   f 106.5% of 24.5

   g 3.8% of 46.2   h 0.35% of 17   i 0.07% of 309.5

6 Use a written method for these calculations. Show your working.

   a $\frac{2}{7} + \frac{1}{3}$ b $\frac{3}{4} - \frac{2}{5}$ c $\frac{3}{8} \times \frac{4}{9}$ d $\frac{7}{8} \div \frac{3}{4}$

   e $3\frac{2}{3} - 1\frac{5}{8}$ f $15\frac{3}{4} + 7\frac{5}{9}$ g $4\frac{1}{2} \times 9\frac{3}{4}$ h $3\frac{2}{5} \div 2\frac{1}{10}$

7 Evaluate these without using a calculator. Give your answers in standard form.

   a $(2.3 \times 10^3) + (4.2 \times 10^3)$ b $(2.7 \times 10^5) - (6.3 \times 10^4)$

   c $(8.75 \times 10^{-8}) + (5.66 \times 10^{-8})$ d $(6.31 \times 10^7) + (6.09 \times 10^8)$

   e $(1.82 \times 10^2) - (7.3 \times 10^1)$ f $(3.19 \times 10^{-6}) - (8.4 \times 10^{-7})$

8 Evaluate these without using a calculator. Give your answers in standard form, correct to 3 significant figures.

   a $(4.5 \times 10^5) \times (3.6 \times 10^2)$ b $(8.9 \times 10^{-4}) \times (7.7 \times 10^8)$

   c $(5.85 \times 10^4) \div (5 \times 10^{-6})$ d $(2.2 \times 10^7)^2$

   e $(3.77 \times 10^{-3}) \times (6.08 \times 10^{11})$ f $(4.82 \times 10^8) \div (2 \times 10^{-4})$

9 Use a calculator to check your answers to questions **1** to **8**.

Efficient use of a calculator

This spread will show you how to:

Keywords
Function
Mode
Standard form

- Use calculators effectively and efficiently for complex calculations
- Use calculators to calculate in standard form

You need a calculator to find the results of complex calculations.

In such calculations you have to follow the order of operations. Make sure you know how to enter complex calculations in your calculator.

- You need to know how to use the **function** keys on your calculator.
- You need to know how to enter and interpret numbers in **standard form**.

Not all calculators are the same!
For example, to calculate $\sqrt{74}$:

- in some you press
  $\sqrt{}$ then [7] [4]
- in some you press
  [7] [4] then $\sqrt{}$.

Example

Use your calculator to work out $\sqrt{\dfrac{42\,389}{31.6^2}}$.

Give your answer to 3 significant figures.

$$\boxed{6.515376289}$$

$$\sqrt{\frac{42389}{31.6^2}} = 6.52 \ (3 \ \text{sf})$$

Example

Use your calculator to work out $(6.43 \times 10^6) \div (4.21 \times 10^{-2})$.

$$\boxed{1.527315914 \ ^{08}}$$

$$(6.43 \times 10^6) \div (4.21 \times 10^{-2}) = 1.53 \times 10^8 \ (\text{to 3 sf})$$

Example

Calculate $\sqrt{7.9^2 - 8.8 \cos 35°}$.
Rashima calculated the answer 8.39.
Is she correct?

No. The correct answer is 7.43 to 3 sf.

Rashima had her calculator in the wrong **mode** for the angle 35°. When calculating with angles, make sure your calculator is in degree mode.

1   Use a calculator to find the value of these in standard form.

   a $(6.4 \times 10^{-4}) + (7.1 \times 10^{-3})$ **b** $(9.9 \times 10^{5}) - (2.7 \times 10^{4})$

   c $(4.8 \times 10^{-6}) + (3.9 \times 10^{-5})$ **d** $(3.3 \times 10^{2}) - (7.5 \times 10^{1})$

   e $(9.8 \times 10^{5}) - (6.4 \times 10^{5})$ **f** $(3.5 \times 10^{-2}) + (9.7 \times 10^{-3})$

2   Use a calculator to find the value of these in standard form.

   a $(5.3 \times 10^{-4}) \times (4.1 \times 10^{-7})$ **b** $(5.4 \times 10^{-5}) \div (3.1 \times 10^{2})$

   c $(8.9 \times 10^{-6}) \div (6.5 \times 10^{-4})$ **d** $(4.7 \times 10^{-2}) \times (9.2 \times 10^{-8})$

   e $(3.8 \times 10^{3}) \times (1.7 \times 10^{5})$ **f** $(2.4 \times 10^{-2}) \div (3.8 \times 10^{-6})$

3   Use your calculator to evaluate these.

   a $6.34 + \sqrt{\dfrac{8.79}{0.35}}$ (Give your answer correct to 2 significant figures.)

   b $11\frac{3}{4} - 2\frac{4}{7}$ (Give your answer as a mixed number.)

   c $(4.78 \times 10^{-4}) \times (6.1 \times 10^{6})$ (Give your answer in standard form.)

4   Evaluate these.

   a $3 \times 4.7^{3} + 2 \times 4.7 - 3$ **b** $\left(1 + \dfrac{4.5}{0.65}\right)^{3}$ **c** $\left(\dfrac{1.54 - 0.79}{0.03}\right)^{3}$

   d $105\left(1 + \dfrac{5.6}{100}\right)^{8}$ **e** $4.9^{2} \times \left(\dfrac{3.2 - 0.75}{2.8 - 0.75}\right)^{2}$ **f** $\dfrac{3.2^{2}}{1.7} \times \dfrac{2.9}{1.6^{2}}$

5   Evaluate these, giving your results to 3 significant figures.

   a $\sqrt{31.9^{2} - 8.77^{2}}$ **b** $6.75 + \sqrt{\dfrac{3.92}{4.15}}$ **c** $\sqrt{\dfrac{5.68^{4}}{4.75^{2} - 2.59^{2}}}$

   d $\sqrt[3]{3.87 \times 4.36^{2}}$ **e** $\dfrac{1}{2\pi}\sqrt{\dfrac{1.55}{9.81}}$ **f** $\left(\dfrac{1 + \sqrt{5}}{2}\right)^{-\frac{3}{2}}$

6   Use the fraction facility on your calculator to evaluate these.

   a $\frac{2}{3} + \frac{4}{5}$ **b** $\frac{3}{4} + \frac{5}{6}$ **c** $\frac{7}{9} - \frac{2}{3}$ **d** $\frac{3}{8} \times \frac{2}{9}$

   e $\frac{4}{5} \div \frac{2}{3}$ **f** $2\frac{1}{2} \times \frac{3}{4}$ **g** $4\frac{3}{4} + 3\frac{1}{3}$ **h** $4\frac{1}{2} \div 1\frac{2}{3}$

   i $1\frac{5}{8} - \frac{11}{16}$ **j** $5\frac{1}{4} \times 2\frac{3}{5}$

7   Evaluate these expressions, giving your answers to 3 significant figures.

   a $\dfrac{2.5 \sin 60°}{4.6^{2} - 3.8^{2}}$ **b** $\sqrt{19.7^{2} + 14.5^{2} - 2 \times 19.7 \times 14.5 \times \cos 63°}$

   c $\dfrac{\left(\frac{4.7}{9.8}\right)^{2} \times \tan 84°}{\sqrt{4.7^{2} + 5.2^{2}}}$ **d** $\dfrac{\sqrt{3.65 \times 1.93^{2} - 8.75 \cos 45°}}{8.75^{2} + 3.65^{2}}$

Key objectives

- Estimate answers to problems involving decimals, round to a given number of significant figures
- Recognise limitations on the accuracy of data and measurements
- Select and justify appropriate degrees of accuracy for answers to problems
- Use calculators effectively and efficiently, knowing how to enter complex calculations
- Use standard index form display and know how to enter numbers in standard index form

1 Use your calculator to find the value of x where
$$x = \frac{2.67 \times 5.41^3 - \sin 45°}{\sqrt{(5.91 + 2.83)}}$$
giving your answer to 4 decimal places. (3)

2 Elliot did an experiment to find the value of g m/s², the acceleration due to gravity. He measured the time, T seconds, that a block took to slide L m down a smooth slope of angle $x°$. He then used the formula

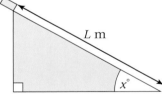

$$g = \frac{2L}{T^2 \sin x°}$$

to calculate an estimate for g.
$T = 1.3$ correct to 1 decimal place.
$L = 4.50$ correct to 2 decimal places.
$x = 30$ correct to the nearest integer.
a Calculate the lower bound and the upper bound for the value of g. Give your answers correct to 3 decimal places. (4)
b Use your answers to part a to write down the value of g to a suitable degree of accuracy.
Explain your reasoning. (1)

(Edexcel Ltd., 2003)

This unit will show you how to

- Solve quadratic equations by factorisation and using the quadratic formula
- Set up simple equations
- Solve simultaneous equations in two unknowns by eliminating a variable
- Use substitution to solve simultaneous equations where one equation is linear and one is quadratic
- Use Pythagoras's theorem to find the equation of a circle centred at the origin
- Use simultaneous equations to find where a line intersects a circle centred at the origin
- Solve simple linear inequalities in one variable, representing the solution set on a number line
- Solve simple quadratic inequalities

Before you start ...

You should be able to answer these questions.

Review

1 Solve these using factorisation or the quadratic formula.

Unit A3

a $x^2 - 7x + 12 = 0$ **b** $y^2 - 8y = 0$
c $x^2 + 3x - 2 = 0$ **d** $2x^2 - 7x + 3 = 0$
e $y^2 = 11y + 24$ **f** $3x^2 - 2x - 1 = 0$

2 Use inspection to find

Unit A5

a two numbers that differ by three and have a product of 40
b two numbers that have sum 9 and difference 5
c two values which add to 7 and whose squares add to 25.

3 Given that $2x + 3y = 9$, find each value.

Unit N4

a $4x + 6y$ **b** $x + 1.5y$
c $20x + 30y$ **d** $14x + 21y$

4 On a graph with both axes from −10 to 10, shade these regions.

Unit A4

a $x > 7$ **b** $y \leqslant -2$ **c** $y < 2x + 1$

This spread will show you how to:

- Solve quadratic equations by factorisation and using the quadratic formula
- Set up simple equations

- A **quadratic** equation has a squared term as its highest power.
- To solve a quadratic equation, you:

Always try to factorise first.

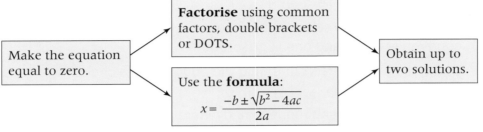

Make the equation equal to zero. → **Factorise** using common factors, double brackets or DOTS. → Obtain up to two solutions.

Use the **formula**: $x = \dfrac{-b \pm \sqrt{b^2 - 4ac}}{2a}$ → Obtain up to two solutions.

To solve a problem, you may be able to set up an equation and solve it.

For quadratics you may need to reject one solution, depending on the context.

For example, if the solutions are $x = -3$ or $x = 5$, and x is a length, then reject $x = -3$.

Example

Two numbers have a product of 105 and a difference of 8.
If the larger number is x:
a show that $x^2 - 8x - 105 = 0$
b solve this equation to find the two numbers.

a The two numbers are x and $x - 8$.
So $x(x - 8) = 105$
$x^2 - 8x - 105 = 0$ (as required)
b $x^2 - 8x - 105 = 0$
$(x + 7)(x - 15) = 0$
Either $x + 7 = 0$ or $x - 15 = 0$
So $x = -7$ and $x - 8 = -15$ or $x = 15$ and $x - 8 = 7$.
The two numbers are -7 and -15 or 7 and 15.

Two answers for x lead to two answers for $x - 8$.

Exam Question

If the area of the trapezium is 400 cm², show that $x^2 + 20x = 400$ and find the value of x correct to 3 dp.

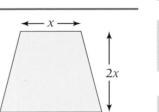

Area of trapezium $= \dfrac{(a + b)}{2} h$

$A = \frac{1}{2}(x + 20) \times 2x = x(x + 20)$
$400 = x(x + 20)$
$x^2 + 20x = 400$ (as required)
$x^2 + 20x - 400 = 0$

$x = \dfrac{-20 \pm \sqrt{20^2 - 4 \times 1 \times -400}}{2}$

$= 12.360\,67\ldots$ or $-32.360\,67\ldots$

The equation does not factorise, so use $x = \dfrac{-b \pm \sqrt{b^2 - 4ac}}{2a}$

Since x is a length, it must be positive, so $x = 12.361$ to 3 dp.

(Edexcel Ltd., 2005)

1 Solve these quadratic equations by factorisation.

a $x^2 - 9x + 18 = 0$ **b** $x^2 - 100 = 0$ **c** $7x = x^2$

2 Solve these quadratic equations using the formula, giving your answers to 3 significant figures.

a $x^2 - 8x - 2 = 0$ **b** $6x^2 - 5x - 5 = 0$ **c** $7x^2 = 10 - 3x$

3 In each question

   i write a quadratic equation to represent the information given

   ii solve the equation to find the necessary information.

   Give answers to 2 decimal places where appropriate.

   a Two numbers which differ by 4 have a product of 117.
   Find the numbers.

   b The length of a rectangle exceeds its width by 4 cm.
   The area of the rectangle is 357 cm².
   Find the dimensions of the rectangle.

   c The square of three less than a number is ten.
   What is the number?

   d Three times the reciprocal of a number is one less than ten times
   the number. What is the number?

   e The diagonal of a rectangle is 17 mm. The length is 7 mm more
   than the width. Find the dimensions of the rectangle.

   f If this square and rectangle have equal areas, find the side length
   of the square.

$3x - 2$

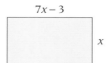
$7x - 3$
x

4 a The height of a closed cylinder is 5 cm and its surface area
   is 100 cm².

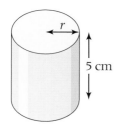
r
5 cm

   Given that the radius is r, show that
   $$\pi r^2 + 5\pi r - 50 = 0$$

   b Hence, find the diameter of the base of the cylinder.

Solving simultaneous linear equations

This spread will show you how to:

- Solve simultaneous equations in two unknowns by eliminating a variable

Keywords

Eliminate
Simultaneous
Variable

- **Simultaneous** equations are true at the same time.
  They share a solution.
  For example $x + y = 10$ x and y add to make 10 and their difference is 4.
   $x - y = 4$ The solution must be $x = 7$ and $y = 3$

- You can solve simultaneous equations by **eliminating** one of the two variables.
  - Multiply one or both equations to get equal numbers of one variable in both equations.
  - Add or subtract the equations to eliminate the variable.

$3x + 2y = 16$ $x - 6y = 2$	$3x + 2y = 16$ ① $3x - 18y = 6$ ②	$20y = 10$ $y = \frac{1}{2}$	$3x + 1 = 16$ $x = 5$

Obtain equal numbers of x's by multiplying the second equation by 3

Subtract ① − ② to eliminate x

Substitute for y in ① to find x

Example

Solve the simultaneous equations $2x - 4y = 8$
 $3x + 3y = -15$

$2x - 4y = 8$ ①   Multiply by 3: $6x - 12y = 24$ ③
$3x + 3y = -15$ ②   Multiply by 4: $12x + 12y = -60$ ④
Add ③ + ④: $18x = -36$
$x = -2$
Substituting in ②: $-6 + 3y = -15$ so $3y = -9$
$y = -3$

Check the solution by substituting in one of the original equations.

Example

The perimeter of this isosceles triangle is 50 cm. Find the length of its base.

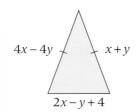

$4x - 4y$ $\qquad$ $x + y$

$2x - y + 4$

Isosceles: $4x - 4y = x + y \rightarrow 3x - 5y = 0$ ①
Perimeter: $7x - 4y + 4 = 50 \rightarrow 7x - 4y = 46$ ②
Multiply ① by 7 and ② by 3:
$21x - 35y = 0$ ③
$21x - 12y = 138$ ④
Subtract ③ from ④:
$23y = 138$
$y = 6$

Substituting in ①: $3x - 30 = 0$
$3x = 30$
$x = 10$
Hence, the base is $(2 \times 10) - 6 + 4 = 18$ cm.

There are two unknowns, so you need to set up two equations to find them.

Equal terms have the **S**ame **S**igns so **S**ubtract.

Finish by answering the question.

1 Solve these simultaneous equations.

 a $2x + y = 18$
  $x - 2y = -1$

 b $5x + 2y = -30$
  $3x + 4y = -32$

 c $14(c + d) = 14$
  $5d - 3c = -11$

 d $5x = 7 + 6y$
  $8y = x + 2$

 e $24a + 12b + 7 = 0$
  $6a + 12b - 5 = 0$

 f $4p + 1\frac{1}{2}q = 5\frac{1}{2}$
  $6p - 2q = 21$

2 For each question, set up a pair of simultaneous equations and solve them to find the required information.

 a Two numbers have a sum of 23 and a difference of 5. What numbers are they?

 b Two numbers have a difference of 6. Twice the larger plus the smaller number also equals 6. What numbers are they?

3 Use simultaneous equations to find the value of each symbol in the puzzle.

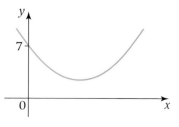

4 A straight-line graph passes through the points (3, 7) and (21, −9). Apply $y = mx + c$ and use simultaneous equations to find the equation of the line.

5 Tickets for a theatre production cost £3.50 per child and £5.25 per adult. 94 tickets were sold for a total of £365.75. How many children attended the production?

6 The equation of a parabola is $y = ax^2 + bx + c$.
A parabola crosses the y-axis at (0, 7) and passes through (2, 5) and (5, 42).

Find the values of a, b and c.

This spread will show you how to:

- Use substitution to solve simultaneous equations where one equation is linear and one is quadratic

Keywords

Linear
Quadratic
Simultaneous

- You can use substitution to solve **simultaneous** equations where one is **linear** and one **quadratic**.

- You rearrange the linear equation, if necessary, to make one unknown the subject. Then substitute this expression into the quadratic equation and solve.

A linear equation contains no square or higher terms. A quadratic equation contains a square term, but no higher powers.

Example

Solve the simultaneous equations $x + y = 7$
$$x^2 + y = 13$$

$x + y = 7$ ①
$x^2 + y = 13$ ②
Rearranging ①: $y = 7 - x$
Substitute in ②: $x^2 + (7 - x) = 13$
$$x^2 - x - 6 = 0$$
$$(x + 2)(x - 3) = 0$$
Either $x = -2$ and $y = 7 - (-2) = 9$
or $x = 3$ and $y = 7 - 3 = 4$

Equation ① is linear. Rearrange it to make y the subject. Equation ② is quadratic.

Example

The line $x + 3y = 5$ crosses the parabola $y = x^2 - 25$ at two points. Find the coordinates of these points.

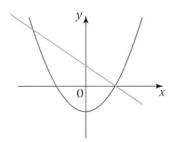

$y = x^2 - 25$ ①

$x + 3y = 5$ ②

Rearranging ②: $y = \frac{5 - x}{3}$

Substitute in ①: $\frac{5 - x}{3} = x^2 - 25$
$$5 - x = 3x^2 - 75$$
$$3x^2 + x - 80 = 0$$
$$(3x + 16)(x - 5) = 0$$
Either $x = -\frac{16}{3}$ or $x = 5$

When $x = -\frac{16}{3}$, $y = \frac{31}{9}$ and when $x = 5$, $y = 0$.

The line crosses the parabola at $\left(-\frac{16}{3}, \frac{31}{9}\right)$ and $(5, 0)$.

Check that you have paired the x- and y-values correctly.

1 Solve these pairs of simultaneous equations.

   a $x^2 + y = 55$
      $y = 6$

   b $x + y^2 = 32$
      $x = 7$

   c $x^2 - 3y = 73$
      $y = 9$

2 Solve these simultaneously equations.

   a $y = x^2 - 2x$
      $y = x + 4$

   b $y = x^2 - 1$
      $y = 2x - 2$

   c $x = 2y^2$
      $x = 9y - 4$

3 Solve these simultaneous equations.

   a $y = x^2 - 3x + 7$
      $y - 5x + 8 = 0$

   b $p^2 + 3pq = 10$
      $p = 2q$

> The equation of a parabola is $y = ax^2 + bx + c$.

4 The graph shows a parabola and a line.

   a The parabola crosses the x-axis at $(2, 0)$ and $(-3, 0)$. Find its equation.

   b The line intersects the y-axis at $(0, 10)$. Find its equation.

   c Use the graph to find the coordinates of the points where the graphs intersect.
   Check this solutions using simultaneous equations.

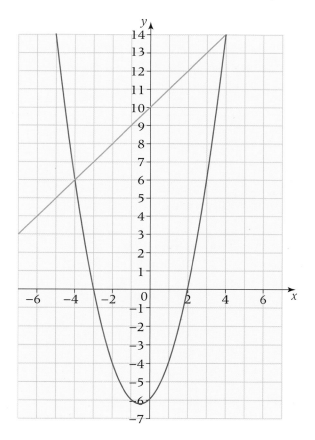

5 Solve simultaneously $xy^5 = -96$
                        $2xy^3 = -48$

6 Solve simultaneously $2^{p+q} = 32$

                           $\dfrac{3^q}{3^{2p}} = 6561$

> Write 32 as a power of 2 and 6561 as a power of 3

This spread will show you how to:

- Use Pythagoras's theorem to find the equation of a circle centred at the origin
- Use simultaneous equations to find where a line intersects a circle centred at the origin

Keywords

Circle
Origin
Radius

- You can use Pythagoras's theorem to find the equation of a **circle** of **radius** 4 and centre the **origin**.
  Choose a point (x, y) on the circumference of the circle.
  Draw in a right-angled triangle.

$$x^2 + y^2 = 4^2$$
$$x^2 + y^2 = 16$$

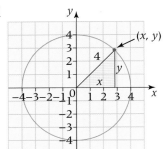

- You can generalise to give the equation of any circle with radius r and centre (0, 0): $x^2 + y^2 = r^2$

Example

What is the equation of the circle, centre (0, 0) with radius $\frac{3}{4}$?

$$x^2 + y^2 = \left(\tfrac{3}{4}\right)^2$$
$$x^2 + y^2 = \tfrac{9}{16}$$

- You can use simultaneous equations to find where a line intersects a circle with centre at the origin.

Example

Find the points of intersection of the line $y = x + 1$ and the circle $x^2 + y^2 = 25$.

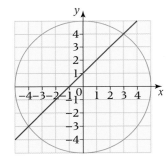

Substituting $y = x + 1$ into $x^2 + y^2 = 25$ gives:
$$x^2 + (x + 1)^2 = 25$$
$$2x^2 + 2x + 1 = 25$$
$$2x^2 + 2x - 24 = 0$$
$$x^2 + x - 12 = 0$$
$(x + 4)(x - 3) = 0$ so either $x = -4$ or $x = 3$
When $x = -4$, $y = (-4) + 1 = -3$
When $x = 3$, $y = 3 + 1 = 4$
Hence, the circle and line intersect at $(-4, -3)$ and $(3, 4)$.

Simplify by dividing each term by 2.
Solve the quadratic by factorising.

1 Write the equations of these circles.

   a centre origin, radius 6   **b** centre origin, radius $\frac{1}{2}$

   c centre origin, radius 0.4   **d** centre origin, radius $\sqrt{5}$

2 Find the equations of these circles.

   a 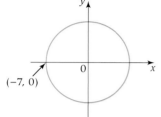   **b**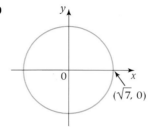

3 The diagram shows the circle, $x^2 + y^2 = 169$ and the line $y = 2x + 2$.
Find the coordinates of the two points of intersection of the line and the circle.

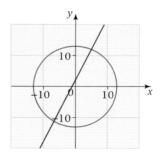

4 Where do the circles $x^2 + y^2 = 25$ and $x^2 + y^2 = 49$ intersect?

Imagine/draw a diagram.

5 **a** How many times do these lines intersect with the circle $x^2 + y^2 = 25$?

     i $y = -5$ **ii** $y = 3$ **iii** $4y + 3x = 25$

   b Hence, what word could you use to describe the line in **a** part **i**?

Imagine/draw a diagram if necessary.

6 Find, using algebraic methods, the points of intersection of these lines and circles.

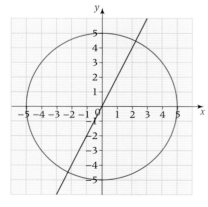 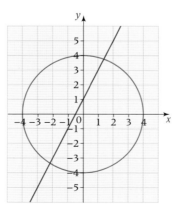

7 **a** Use Pythagoras's theorem to find the equation of a circle, centre (3, 5) and radius 6.

   b By imagining a diagram, find where the circles $x^2 + y^2 = 36$ and $(x - 12)^2 + y^2 = 36$ intersect.

This spread will show you how to:

● Solve simple linear inequalities in one variable, and represent the solution set on a number line

Keywords
Inequality
Integer

● An **inequality** is a mathematical statement including one of these symbols:

$<$	$>$	$\leqslant$	$\geqslant$
less than	more than	less than or equal to	more than or equal to

For example $3 < 5$.

● You can solve an inequality by rearranging and using inverse operations, in a similar way to solving an equation.

$$3x + 2 > 5x - 1$$
$$2 > 2x - 1$$
$$3 > 2x$$
$$1.5 > x$$

Compare with…

$$3x + 2 = 5x - 1$$
$$2 = 2x - 1$$
$$3 = 2x$$
$$1.5 = x$$

● If you multiply or divide an inequality by a positive number, the inequality remains true.

$4 < 6$ and $8 < 12$ and $2 < 3$

● If you multiply or divide an inequality by a negative number you need to reverse the inequality sign to keep it true.

$4 < 6$ but $-2 > -3$
$5 > 2$ but $-15 < -6$

● The solution to an inequality can be a range of values, which you can show on a number line:

$x \geqslant 1$

$x < -2$

Use an 'empty' circle for $<$ and $>$.
Use a 'filled' circle for $\leqslant$ and $\geqslant$.

Example

a Find the range of values of x that satisfy both $3x \geqslant 2x - 1$ and $10 - 3x > 6$.
Represent the solution set on a number line.
b List the **integer** values of x that satisfy both inequalities.

Integers: positive and negative whole numbers and zero.

a $3x \geqslant 2(x - 1)$
$3x \geqslant 2x - 2$
$x \geqslant -2$

$10 - 3x > 6$
$10 > 6 + 3x$
$4 > 3x$
$1\frac{1}{3} > x$

So $-2 \leqslant x < 1\frac{1}{3}$.

Combine the two inequalities. $1\frac{1}{3} > x$ is the same as $x < 1\frac{1}{3}$.

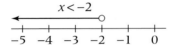

$1\frac{1}{3}$

b The integer values of x that satisfy both inequalities are -2, -1, 0 and 1.

1 **a** What is the smallest prime number p such that $p > 30$?

b What is the smallest value of x such that $2^x \geqslant 1024$?

Use trial and error.

2 True or false: if x is a number, then $x^2 > x$?
If false, write the range of values of x for which it is false, as an inequality.

3 Solve these inequalities, representing each solution on a number line.

a $3x - 5 > 18$

b $\dfrac{p}{4} + 6 \leqslant -2$

c $6x + 3 \leqslant 2x - 8$

d $-3y > 12$

e $\dfrac{q}{-5} \geqslant -2$

f $4z - 3 \leqslant 3(z - 2)$

g $3(y - 2) < 8(y + 6)$

4 **a** The area of this rectangle exceeds its perimeter.

Write an inequality and solve it to find the range of values of x.

b Given that x is an integer, find the smallest possible value that x can take.

8

$x - 5$

5 Find the range of values of x that satisfy *both* inequalities.

a $3x + 6 < 18$ and $-2x < 2$

b $10 > 5 - x$ and $3(x - 9) < 27$

6 Explain why it is not possible to find a value of y such that $3y \leqslant 18$ and $2y + 3 > 15$.

7 Solve, by treating the two inequalities separately.

a $y \leqslant 3y + 2 \leqslant 8 + 2y$

b $z - 8 < 2(z - 3) < z$

c $4p + 1 < 7p < 5(p + 2)$

In part **a**, solve $y \leqslant 3y + 2$ and $3y + 2 \leqslant 8 + 2y$.

8 **a** Explain why the solution to $x^2 \leqslant 25$ is not simply $x \leqslant 5$.

b Hence, solve:

i $3x^2 + 1 > 49$

ii $10 - p^2 < 6$

9 By drawing a graph or otherwise, and using a calculator, find the range of values between $0°$ and $360°$ such that

a $\sin x > 0.5$

b $-1 \leqslant \tan x \leqslant 1$

c $2 \cos x > \sqrt{3}$

This spread will show you how to:

● Solve quadratic inequalities

Keywords
Quadratic
  inequality
Critical values

To solve $x^2 < a^2$ you need to take into account the fact that $(-a)^2 = (a)^2 = a^2$.

Look at the following number line

$$-a < x < a$$

Any number between $-a$ and a satisfies $x^2 < a^2$. The set of values that satisfies $x^2 < a^2$ is thus $-a < x < a$. The boundary values $-a$ and a are called the **critical values**.

To solve $x^2 \geq a^2$ you need to concentrate on the critical values.

The solution is: $x \geq a$ and $x \leq -a$

$$x \leq -a \qquad\qquad x \geq a$$

Example

Solve $\dfrac{(x-3)^2}{4} \leq 9$

$(x-3)^2 \leq 36$	multiply both sides by 4 to remove the fraction
$-6 \leq x - 3 \leq 6$	the solution of $x^2 \leq a^2$ is $-a \leq x \leq a$
	in this case $x \to x - 3$
$-3 \leq x \leq 9$	add 3, to isolate x

Example

Solve $x^2 + 2x - 6 > 3x$

$x^2 - x - 6 > 0$	rearrange so that the right hand side is 0
$(x-3)(x+2) > 0$	factorise or use the quadratic formula to find the critical values

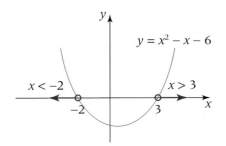

draw a sketch of $y = x^2 - x - 6$

identify the values of x for which the curve is above the x-axis

the solution is: $x > 3$ and $x < -2$

1 Find the set of values of x for which

a $x^2 < 169$ **b** $x^2 \geqslant 169$

c $x^2 < 5$ **d** $x^2 - 5 \leqslant 4$

2 Find the set of values of x for which

a $(x - 3)^2 < 169$ **b** $(x - 3)^2 \geqslant 169$

c $(x - 3)^2 < 5$ **d** $(x - 3)^2 - 5 \leqslant 4$

3 Find the set of values of x for which

a $\dfrac{(x - 5)^2}{5} \leqslant 5$ **b** $\dfrac{(2x - 7)^2}{4} \geqslant 9$

4 Find the set of values of x for which

a $(x - 3)(x + 2) \leqslant 0$ **b** $(2x + 5)(x + 8) > 0$

5 Find the set of values of x for which

a $x^2 + 4x - 5 < 0$ **b** $6x^2 + 7x - 3 \geqslant 0$

c $2x^2 + 9x < 5$ **d** $x(2x - 4) < 12 + x$

6 List the set of integers n for which

a $n^2 + 4n - 5 < 0$

b $\dfrac{(n - 2)^2}{3} \leqslant 3$

c $(2n + 1)(n - 3) < 4(3 + 2n)$

7 Find the set of values of x for which both $x^2 + 4x - 5 < 0$ and $2x > x$ are true.

Exam review

Key objectives

- Solve quadratic equations by factorisation, completing the square and using the quadratic formula
- Solve exactly, by elimination of an unknown, two simultaneous equations in two unknowns, one of which is linear in each unknown, and the other is linear in one unknown and quadratic in the other, or where the second is of the form $x^2 + y^2 = r^2$
- Solve linear inequalities in one variable, and represent the solution set on a number line
- Solve simple quadratic inequalities

1 Find the range of integers that satisfy both the following inequalities.

Represent the solution set on a number line.

$$x - 1 < 2(x + 3) \qquad x + 1 \leqslant \tfrac{3}{2}$$

(3)

2 Bill said that the line $y = 6$ cuts the curve $x^2 + y^2 = 25$ at two points.

a By eliminating y show that Bill is incorrect. (2)

b By eliminating y, find the solutions to the simultaneous equations

$$x^2 + y^2 = 25$$

$$y = 2x - 2$$

(6)

(Edexcel Ltd., 2004)

This unit will show you how to

● Solve problems involving bearings

● Use a ruler and compasses to draw standard constructions

● Understand, recall and use Pythagoras's theorem

● Understand, recall and use trigonometrical relationships in right-angled triangles and triangles that are not right-angled

● Use coordinates to identify a point on a 3-D grid

● Use Pythagoras's theorem and the rules of trigonometry to solve 2-D and 3-D problems

Before you start ...

You should be able to answer these questions.

Review

1 Use Pythagoras's theorem to find the missing side in these triangles:

Unit S4

a

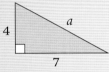

b

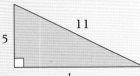

2 Solve these equations.

Unit A2

a $\frac{x}{4} = \frac{5}{7}$ **b** $\frac{2}{11} = \frac{x}{9}$ **c** $\frac{5}{4} = \frac{7}{x}$

3 Use the order of operations and calculate:

Unit N5

a $4^2 + 3 - 2 \times 7$ **b** $5 - 7^2 + 2 \times 3$

c $\dfrac{8^2 + 3^2 - 5^2}{2 \times 4}$

4 Copy this cuboid and mark **planes** of symmetry.

CD – S5

Bearings and scale drawings

This spread will show you how to:

- Use geometry to solve problems involving bearings
- Use and interpret maps and scale drawings

Keywords
Angle
Bearing

- A bearing is an **angle** measured in a clockwise direction from north.

To find the bearing of A from B

- Imagine you are standing at B, facing north.
- Turn clockwise until you face A.

The angle you have turned through is the bearing of A from B.

You always write bearings with 3 digits, for example 070°, 190°, 230°

The bearing of A from B is 256°.

Example

The bearing of G from B is 028°. Find the bearing of B from G.

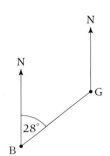

Bearing of B from G is the angle at G measured clockwise from north to B.

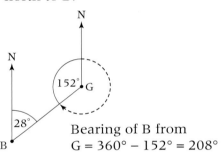

Bearing of B from G = 360° − 152° = 208°

Interior angles are supplementary.

Example

A church, C, is 10 km due west of a school, S.
Joe is 6 km from the school on a bearing of 320°.
He wants to walk directly to the church.

Draw a diagram to show the positions of Joe, the church and the school, and use it to find the bearing Joe should take.
Use a scale of 1 cm to 2 km.

Label point S.
Draw C 5 cm west of S.
Draw the north line at S.
Measure and draw the 320° bearing from S and, 3 cm from S, mark a point, J, to show Joe's position.

Draw the line JC.
Draw the north line at J.
Measure the clockwise angle between the north line and JC.
The bearing Joe needs is 233°.

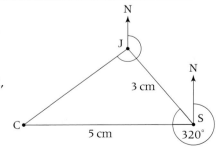

Scale 1 cm to 2 km, so represent 10 km by a line 5 cm long.

Represent 6 km by a line 3 cm long.

Draw and measure lengths and angles carefully or your answer will be inaccurate.

1 These diagrams are drawn accurately.
Measure the bearing of T from S in each.

a N

T

S

b T N

S

2 P and Q are points 2 cm apart. Draw diagrams to show the position of points P and Q where the bearing of Q from P is

a 070°   b 155°   c 340°   d 260°

3 These diagrams have *not* been drawn accurately.
Find the bearing of X from Y in each case.

a N

136°

X

Y

b N

95°

X

Y

c N

X

248°

Y

4 Find these bearings.

a The bearing of A from B is 104°. Work out the bearing of B from A.

b The bearing of E from F is 083°. Work out the bearing of F from E.

c The bearing of J from K is 297°. Work out the bearing of K from J.

> Draw a sketch to help you.

5 A youth club (Y) is 4 km due east of a school (S).
Hazel leaves school and walks 5 km on a bearing of 042° to her house (H).

a Make a scale drawing to show the position of Y, S and H.
Use a scale of 1 cm to 1 km.

b Hazel walks directly from her house to the youth club.
What bearing does she take?

6 A lighthouse, L, is 6 km on a bearing of 160° from a point H at the harbour.
A boat, B, is 3 km from L on a bearing of 125°.

a i Make a scale drawing to show the positions of L, H and B.

ii What bearing should B travel to go directly to H?

The boat moves 4 km due west.

b i Mark on your drawing the new position of B.

ii What bearing should B now travel to go directly to H?

Constructing triangles

This spread will show you how to:

● Use a ruler and compasses to draw standard constructions

Keywords
Arc
Compasses
Radius
Triangle

You can **construct** a unique triangle when you know

| Two sides and the angle between them (SAS) | or | Two angles and a side (ASA) | or | Right angle, the hypotenuse and a side (RHS) | or | Three sides (SSS) |

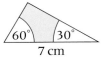
5 cm 30° 6 cm

60° 30° 7 cm

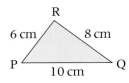
5 cm 3 cm

R 6 cm 8 cm P 10 cm Q

You will need a ruler and a protractor for SAS, ASA and RHS triangles.

You will need a ruler and compasses for SSS triangles.

Example

a Construct this equilateral triangle ABC with side length 4 cm.
b Construct another equilateral triangle with base AB and side length 4 cm.
c What special quadrilateral have you drawn?

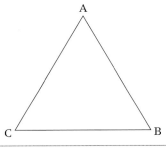

---

a

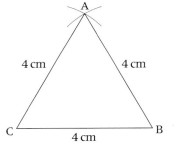

Draw BC 4 cm long.

Draw arcs of **radius** 4 cm from B and C to intersect at A.

Join AC and BC.

The construction **arcs** show your method, so do not erase them.

You can use this method to construct an angle of 60° (angles in an equilateral triangle = 60°).

b

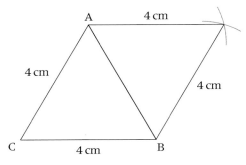

Draw arcs of radius 4 cm from A and B to intersect at D.
Join AD and DB.

Four equal sides, two pairs of parallel sides, opposite angles equal.

c A rhombus

1 Use a straight edge and compasses or a protractor to construct these triangles.

It helps to draw a rough sketch first.

  a Sides 8 cm, 4 cm, 7 cm (SSS)   **b** 3 cm, 30°, 4 cm (SAS)

  c Sides 10 cm, 7.5 cm, 6 cm (SSS) **d** 8 cm, 2 cm, 90° (RHS)

  e Sides 6 cm, 9 cm, 5 cm (SSS)   **f** 45°, 4 cm, 45° (ASA)

2 **a** Explain why you cannot construct a triangle with sides 9 cm, 4 cm, 3 cm.

Try to construct the triangles to see what happens.

  b Explain what happens when you try and construct a triangle with sides 9 cm, 4 cm, 5 cm.

3 Without drawing the construction, write whether these sets of 3 sides will make a triangle.

  a Sides 5 cm, 5 cm, 9 cm   **b** Sides 2 cm, 2 cm, 2 cm

  c Sides 29 cm, 26 cm, 4 cm   **d** Sides 22 cm, 12 cm, 10 cm

  e Sides 20 cm, 7 cm, 9 cm   **f** Sides 14 cm, 8 cm, 6 cm

  g Sides 15 cm, 60 mm, 100 mm **h** Sides 120 mm, 8 cm, 9 cm

4 Construct isosceles triangles with sides

  a 7 cm, 7 cm, 5 cm   **b** 5 cm, 5 cm, 7 cm

5 **a** Construct an equilateral triangle ABC with sides 5 cm.

  b Construct a second equilateral triangle with base AB and sides 5 cm to get a rhombus.

  c Follow the steps in **a** and **b** to construct a rhombus with sides 3.5 cm.

6 **a** Construct a triangle with sides 5 cm, 12 cm, 13 cm.

  b What type of triangle is this?

7 **a** Construct a triangle ABC with sides AB = 3 cm, BC = 4 cm, CA = 5 cm.

  b Construct triangle ADC with side CA from the triangle in part **a**, and side AD = 4 cm and side DC = 3 cm.

  c What special type of quadrilateral is this?

8 **a** Construct a triangle ABC such that BC = 12 cm, AC = 7 cm and ∠B = 30°.

This triangle is called SSA because you have two sides and the non-included angle.

  b Now try to draw a second, **different** triangle with the same measurements. (Move the position of A.)

  c Are SSA triangles unique?

Constructing bisectors

This spread will show you how to:

- Use a ruler and compasses to draw standard constructions

Keywords
Bisect
Equidistant
Perpendicular

- **Bisect** means cut into two equal parts.

You can use a straight edge and compasses to construct an angle bisector.

Use the same compass radius throughout the construction. Start at the red dots.

- To bisect angle ABC

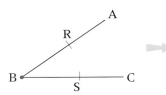 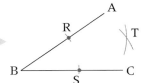 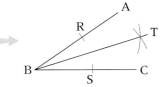

BRTS is a rhombus.

- All points on the angle bisector are equidistant from the arms of the angle.

Equidistant means equal distance from.

- The **perpendicular** bisector of a line bisects the line at right angles.

- To construct the perpendicular bisector of line AB

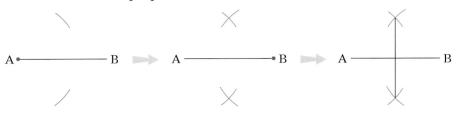

Use the same compass radius throughout the construction.

Start at the red dots.

- All points on the perpendicular bisector of AB are equidistant from A and B.

Example

Construct an angle of 45°.

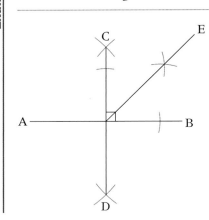

Draw a line AB
Construct the perpendicular bisector CD.
Construct the angle bisector of ∠BCD.
∠BCE = 45°

$45°$ is $\frac{1}{2}$ of $90°$.
Construct a perpendicular bisector to AB (90°) and then bisect the angle.

1 Trace these angles.
Construct the angle bisector of each angle, using ruler and compasses.

a

b

c

You will need to extend the lines.

2 a Construct an equilateral triangle with sides 5 cm.

b Construct the angle bisector of each angle of the triangle.

c What do you notice about the three angle bisectors?

See the example in S5.2 for help with constructing an equilateral triangle.

3 Follow these steps to construct an angle of 30°.

a Construct an equilateral triangle with sides 4 cm.

b Bisect one of the base angles.

Angles in an equilateral triangle = 60°.

4 Draw lines AB for these lengths and construct their perpendicular bisectors.

a 6 cm   **b** 9 cm   **c** 5.6 cm   **d** 10 cm   **e** 11.2 cm

Check by measuring that each bisector intersects the line AB at its midpoint.

5 a Construct an equilateral triangle with sides 5 cm.

b Construct the perpendicular bisectors of each side of the triangle.

c Compare your diagram with the one for question **2**. Write down what you notice.

6 a Construct two triangles with sides 8 cm, 5 cm, 7 cm. Label them triangle A and triangle L.

b On triangle A construct the angle bisector of each internal angle of the triangle.

c On triangle L construct the perpendicular bisectors of each side of the triangle.

d Compare and comment on your answers to **b** and **c**.

7 a Draw a line AB, 8 cm long, and construct its perpendicular bisector.

b Construct an angle of 45° where the perpendicular bisector intersects AB.

c What other angles have you created in this construction?

Further constructions

This spread will show you how to:

- Use a ruler and compasses to draw standard constructions

You can construct a **perpendicular** from a point to a line or from a point on a line.

- To construct a perpendicular from a point X to a line YZ

Start at the red dots.

Keep the same compass radius throughout the construction.

- To construct a perpendicular from a point E on a line DF

Start at the red dots.

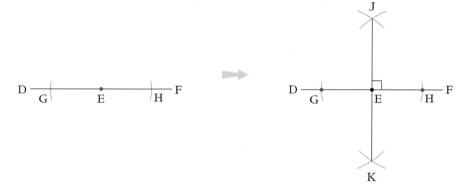

Change your radius for the second part of the construction to a larger one.

- The shortest distance from a point to a line is the perpendicular distance.

Example

Construct a right-angled triangle with sides 3 cm, 4 cm and 5 cm.

Draw a line longer than 4 cm.
Mark the points A and B, 4 cm apart.

Construct a line perpendicular to A.
Mark point C at 3 cm above A on this line.

Draw the third side of the triangle.

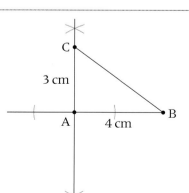

Start by drawing one of the shorter sides.

In a right-angled triangle, the hypotenuse is the longest side.
So the other two sides (3 cm and 4 cm here) meet at right angles.

1 Trace these lines and the points marked X.
For each, use ruler and compasses to construct
a perpendicular from the point X to the line.
Check your constructions using a protractor.

Leave in the
construction lines
and arcs.

a

b X•

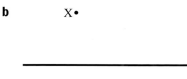

c

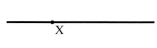 • X

d X•

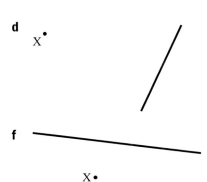

e • X

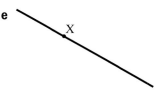

f

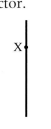

X•

2 Trace these lines.
For each, use ruler and compasses to construct the perpendicular
from the point X on the line. Show all construction lines.
Check your constructions using a protractor.

a

X

b

X•

c

X

d

X

e

X

f

X

261

This spread will show you how to:

- Find loci by reasoning and using diagrams

Keywords
Bisector
Equidistant
Locus
Perpendicular

Loci is the plural of locus.

A **locus** is the path traced out by a moving point.

- The locus of a point which is a constant distance from another point is a circle.

- The locus of a point at a constant distance from a fixed line is a parallel line.

- The locus of a point that is **equidistant** from two other fixed points is the **perpendicular bisector** of the line joining the fixed points.

- The locus of a point equidistant from two intersecting lines is the angle bisector of the lines.

Example

P and Q are two points 2.5 cm apart.

P   Q

Shade in the region that satisfies all these conditions:

- Right of the perpendicular to the line at point P.
- Closer to P than to Q.
- More than 1 cm from P.

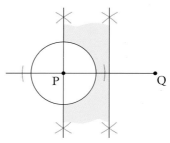

Construct the perpendicular to the line at point P.

Construct the perpendicular bisector of PQ. Points to the left are nearer to P than Q.

Draw a circle radius 1 cm, centre P. Points outside are more than 1 cm from P.

Example

ABCD is the plan of a garden.

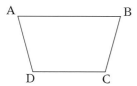

A tree is to be planted in the garden so that it is

- nearer to BC than to BA
- nearer to AD than AB.

Shade the region where the tree may be planted.

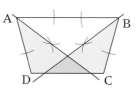

Construct the angle bisector of ∠ABC. Shade the region between this line and BC.

Construct the angle bisector of ∠BAD. Shade the region between this line and AD.

The tree can be planted in the region where the shadings overlap.

1 a Draw points A and B, 6 cm apart.

  b Shade in the region that satisfies both these conditions.

    i Closer to A than to B

    ii Less than 4 cm from B

2 a Draw points J and K, 5 cm apart.

  b Shade the region that satisfies both these conditions.

    i More than 4 cm from J

    ii More than 3 cm from K

3 a Trace the points X, Y and Z.   X•

  b Shade the region that satisfies all three of these conditions.

    i Closer to X than to Y

    ii Closer to the line XZ than to the line XY

    iii More than 1 cm from X   Z•   Y•

4 a Draw a rectangle PQRS where PQ = 5 cm and QR = 3 cm.

  b Shade the region of the rectangle that is within 4 cm of P and within 2.5 cm of R.

5 a Construct a right-angled triangle ABC where angle ABC = 90°, AB = 6 cm, BC = 4.5 cm.

  b Shade the region that satisfies all three of these conditions.

    i Closer to A than to B

    ii Less than 4 cm from A

    iii Less than 4 cm from C

6 The diagram shows the rectangular garden of a house.

There are two trees, *T*, in the garden.

A radio mast is to be placed in the garden.

It must be more than 5 m from the rear of the house.

It must be more than 3 m from a tree.

Using a scale of 1 cm : 2 m, draw a scale diagram and shade the possible site for the radio mast.

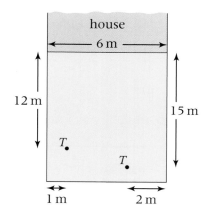

More Pythagoras's theorem and trigonometry

This spread will show you how to:

- Understand, recall and use Pythagoras' theorem in 2-D problems
- Understand, recall and use trigonometrical relationships in right-angled triangles

Keywords
Cosine
Pythagoras's
  theorem
Sine
Tangent

You can use **Pythagoras's theorem** to find a side in a right-angled triangle if you know the other two sides.

You can use Pythagoras's theorem and trigonometry to solve problems involving right-angled triangles.

- Pythagoras's theorem for right-angled triangles:

  In a right-angled triangle, the square on the hypotenuse equals the sum of the squares on the other two sides.

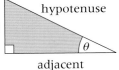

$$c^2 = a^2 + b^2$$
c is the hypotenuse

The hypotenuse is always opposite the right-angle.

You can use trigonometry to find an unknown side if you know one side and an angle (in addition to the right angle).

- $\sin \theta = \dfrac{\text{opposite side}}{\text{hypotenuse}}$ $\quad \cos \theta = \dfrac{\text{adjacent side}}{\text{hypotenuse}}$ $\quad \tan \theta = \dfrac{\text{opposite side}}{\text{adjacent side}}$

Example

Find the missing sides and angles in these triangles.

a

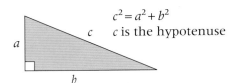

3.2 cm

43°

x

b

12.5 cm

9.8 cm

y

Give your answer to the same degree of accuracy as the measurements given in the question.

a $\tan 43° = \dfrac{3.2}{x}$

$x = \dfrac{3.2}{\tan 43°}$

$x = 3.4$ cm (to 1 dp)

b $9.8^2 + y^2 = 12.5^2$

$y = \sqrt{12.5^2 - 9.8^2}$

$y = 7.8$ cm

You can find an unknown angle if you know two sides.

You use the inverse trig functions $\sin^{-1}$, $\cos^{-1}$ and $\tan^{-1}$.

Example

Find the missing angle.

Give your answer to the nearest degree.

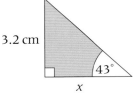

8 cm

4.5 cm

z

$\cos z = \dfrac{4.5}{8}$

$z = \cos^{-1}\left(\dfrac{4.5}{8}\right)$

$z = 56°$ to nearest degree

Make sure your calculator is in degree mode.

1 Find the missing sides in these triangles.

a

7.3 cm
a
38°

b
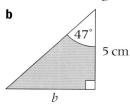
47°
5 cm
b

c
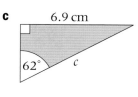
6.9 cm
62°
c

d
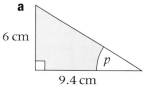
12 cm
d
32°

e
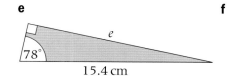
e
78°
15.4 cm

f
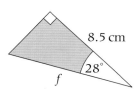
8.5 cm
28°
f

2 Work out the missing angles.

a
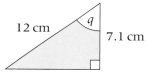
6 cm
p
9.4 cm

b
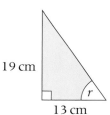
12 cm
q
7.1 cm

c
19 cm
r
13 cm

d
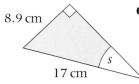
8.9 cm
s
17 cm

e
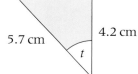
5.7 cm
4.2 cm
t

3 ABC and ACD are right-angled triangles.
AB = 12 cm
BC = 3 cm
CD = 4 cm

a Show that AD = 13 cm.

b Work out AD if the lengths are
changed so that AB = 16 cm,
BC = 5 cm and CD = 7 cm.

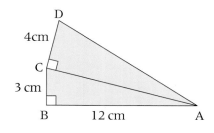
D
4 cm
C
3 cm
B   12 cm   A

Find AC first.

4 Work out PQ.

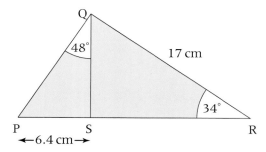
Q
48°
17 cm
34°
P   S   R
←6.4 cm→

265

The sine rule

This spread will show you how to:

● Understand, recall and use trigonometrical relationships in triangles that are not right-angled

You can split any triangle into two right-angled triangles.

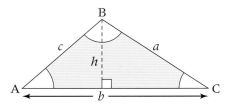

The angles are A,B and C.
Label the side opposite angle A as a,
the side opposite angle B as b, etc.

In the left-hand triangle:

$\sin A = \frac{h}{c}$

so $\quad h = c \times \sin A$

In the right-hand triangle:

$\sin C = \frac{h}{a}$

so $\quad h = a \times \sin C$

$c \times \sin A = a \times \sin C$

so $\frac{\sin A}{a} = \frac{\sin C}{c}$ or $\frac{a}{\sin A} = \frac{c}{\sin C}$

You can extend the rule to include the third side:

● This is the **sine rule**: $\frac{a}{\sin A} = \frac{b}{\sin B} = \frac{c}{\sin C}$

You use the sine rule when a problem involves two sides and an angle or two angles and a side.

The sine rule works in triangles whether they are right-angled or not.

a Find the lengths PQ and PR.

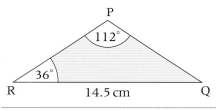

b Find the angles X and Y.

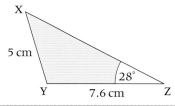

a $\frac{PQ}{\sin 36°} = \frac{14.5}{\sin 112°}$

$PQ = \sin 36° \times \frac{14.5}{\sin 112°}$

$PQ = 9.2$ cm

$\frac{PR}{\sin 32°} = \frac{14.5}{\sin 112°}$

$PR = \sin 32° \times \frac{14.5}{\sin 112°}$

$PR = 8.3$ cm

b $\frac{\sin X}{7.6} = \frac{\sin 28°}{5}$

$\sin X = 7.6 \times \frac{\sin 28°}{5}$

$X = \sin^{-1}\left(7.6 \times \frac{\sin 28°}{5}\right)$

$X = 46°$

$Y = 180° - (28° + 46°)$
$\quad = 106°$

To find a side, use
$\frac{a}{\sin A} = \frac{c}{\sin C}$

To find an angle, use
$\frac{\sin A}{a} = \frac{\sin C}{c}$

1 Find the sides marked x.

Use $\frac{a}{\sin A} = \frac{b}{\sin B} = \frac{c}{\sin C}$

a
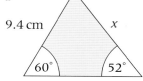
8 cm · x · 49° · 38°

b
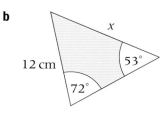
x · 12 cm · 53° · 72°

c
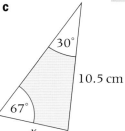
30° · 10.5 cm · 67° · x

d
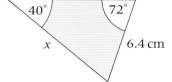
9.4 cm · x · 60° · 52°

e
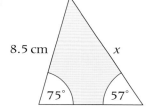
40° · 72° · x · 6.4 cm

f
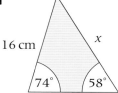
16 cm · x · 74° · 58°

g
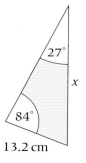
27° · x · 84° · 13.2 cm

h
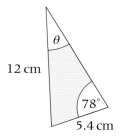
8.5 cm · x · 75° · 57°

2 Find the angles marked θ.

Use $\frac{\sin A}{a} = \frac{\sin B}{b} = \frac{\sin C}{c}$

a
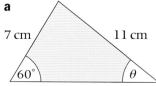
7 cm · 11 cm · 60° · θ

b
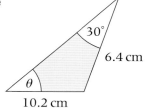
θ · 12 cm · 78° · 5.4 cm

c

13 cm · 4.2 cm · θ · 44°

d
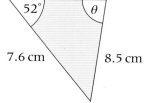
52° · θ · 7.6 cm · 8.5 cm

e
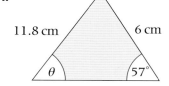
30° · 6.4 cm · θ · 10.2 cm

f
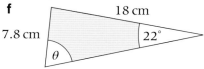
18 cm · 7.8 cm · 22° · θ

g
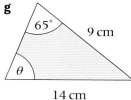
65° · 9 cm · θ · 14 cm

h
11.8 cm · 6 cm · θ · 57°

267

The cosine rule

This spread will show you how to:

- Understand, recall and use trigonometrical relationships in triangles that are not right-angled

Keywords
Cosine rule
Pythagoras's theorem

You need the **cosine rule** if a problem involves all three sides and one angle of a triangle.

Split side b as shown into lengths x and $b - x$.

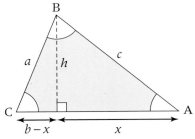

Use **Pythagoras's theorem**

In the left-hand triangle

$$h^2 = a^2 - (b - x)^2$$

In the right-hand triangle

$$h^2 = c^2 - x^2$$

Eliminate h.

$$a^2 - (b - x)^2 = c^2 - x^2$$
$$a^2 - (b^2 - 2bx + x^2) = c^2 - x^2$$
$$a^2 = b^2 + c^2 - 2bx \qquad ①$$

In the right-hand triangle

$$\cos A = \frac{x}{c}$$

so $\qquad x = c \cos A$

Substitute this expression for x in equation ①

$$a^2 = b^2 + c^2 - 2bc \cos A$$

- This is the cosine rule:
  $$a^2 = b^2 + c^2 - 2bc \cos A$$

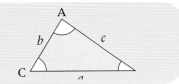

You can rearrange the rule to

$$\cos A = \frac{b^2 + c^2 - a^2}{2bc}$$

The cosine of an obtuse angle is negative. You will solve problems involving the cosine of the negative angles in S8.1.

a Work out the length EF.

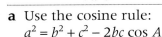

b Work out the angle P.

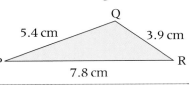

a Use the cosine rule:
$$a^2 = b^2 + c^2 - 2bc \cos A$$
$$EF^2 = 9.5^2 + 4.2^2 - 2 \times 9.5 \times 4.2 \times \cos 37°$$
$$= 90.25 + 17.64 - 63.73...$$
$$EF = 6.6 \text{ cm}$$

b Use the rearranged cosine rule:
$$\cos A = \frac{b^2 + c^2 - a^2}{2bc}$$
$$\cos P = \frac{7.8^2 + 5.4^2 - 3.9^2}{2 \times 7.8 \times 5.4}$$
$$P = \cos^{-1}\left(\frac{7.8^2 + 5.4^2 - 3.9^2}{2 \times 7.8 \times 5.4}\right)$$
$$= 27.4° \text{ (nearest degree)}$$

1 Find the sides marked x.

Use $a^2 = b^2 + c^2 - 2bc \cos A$.

a

6 cm
x
50°
8 cm

b

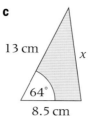

9.4 cm
80°
12 cm
x

c

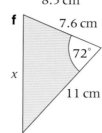

13 cm
x
64°
8.5 cm

d

7 cm
x
38°
10.5 cm

e

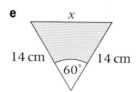

x
14 cm
14 cm
60°

f

7.6 cm
72°
x
11 cm

g

9 cm
105°
6 cm
x

h

13.2 cm
116°
8.8 cm
x

2 Find the angles marked θ.

Use $\cos A = \dfrac{b^2 + c^2 - a^2}{2bc}$.

a

9 cm
6 cm
θ
11 cm

b

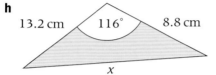

7 cm
θ
5.5 cm
4 cm

c

9.2 cm
θ
8.5 cm
12 cm

d

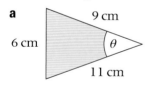

12 cm
θ
8 cm
10 cm

e

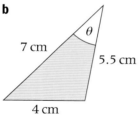

8.4 cm
3 cm
θ
7 cm

f

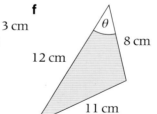

θ
8 cm
12 cm
11 cm

g

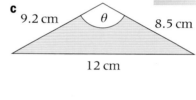

8.2 cm
6.6 cm
θ
10 cm

h

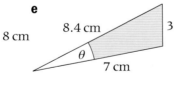

12.7 cm
6.3 cm
9 cm
θ

3 Use the sine rule to find angles B and C.
Then use the cosine rule to find the side x.

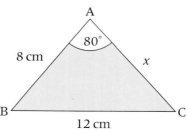

A
80°
8 cm
x
B
12 cm
C

Solving problems using the sine and cosine rules

This spread will show you how to:

● Understand, recall and use trigonometrical relationships in triangles that are not right-angled

Keywords

Bearing
Cosine rule
Sine rule

You use the **sine rule** and the **cosine rule** to solve problems in triangles that are not right-angled.

You may need to use a combination of rules to solve a problem.

● To use the sine rule you need
  – an angle
  – the side opposite to it
  – one other angle or side.
● To use the cosine rule you need
  – all three sides
  – or two sides and the angle between them.

Always start by sketching a diagram.

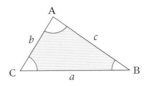

$$\frac{a}{\sin A} = \frac{b}{\sin B} = \frac{c}{\sin C}$$

$$a^2 = b^2 + c^2 - 2bc \cos A$$

Example

Peter walks 5 km, from S, on a **bearing** of 063°.
At C he changes direction and walks a further 3.2 km on a bearing of 138°, to F.
Find the distance, SF, from where he began.

Make a sketch.

Shaded angle = 180° − 63° = 117°   (angles in parallel lines)
∠SCF = 360° − 117° − 138° = 105° (angles at a point)

Use the cosine rule.

$SF^2 = 5^2 + 3.2^2 - 2 \times 5 \times 3.2 \cos 105°$
  $SF = 6.6$ km

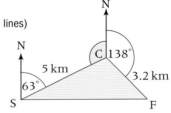

You know two sides and the included angle.

Example

The diagram shows a bicycle frame.
PQ is parallel to SR.

Work out the length QR.

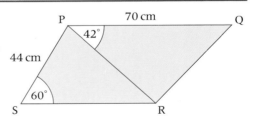

PQ is parallel to RS so ∠SRP = ∠QPR = 42°.
Use the sine rule to find the length PR.

$\frac{PR}{\sin 60°} = \frac{44}{\sin 42°}$ $PR = \frac{44 \times \sin 60°}{\sin 42°}$ $PR = 56.947...$

Use the cosine rule to find QR.

$QR^2 = 56.947...^2 + 70^2 - 2 \times 56.947... \times 70 \cos 42°$
  $QR = 47$ cm

In triangle PRS you know an angle, the side opposite it, one other angle and one other side.

In triangle PQR you know two sides and the included angle.

1 Debbie runs 8 km on a bearing of 310°.
She stops, changes direction and continues running for 10 km on
a bearing of 055°.
Find the distance and bearing on which Debbie should run to return
to her starting point.

2 Clare cycles 12 km on a bearing of 050°.
She stops, changes direction and continues cycling 10 km on a
bearing of 120°.
Find the distance and bearing on which Clare should cycle to return
to her starting point.

3 AB is parallel to DC.
Work out the length BC.

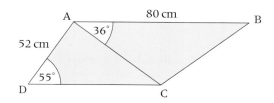

4 JK is parallel to ML.
Work out the length JM.

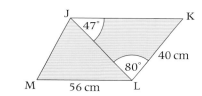

5 PQ is parallel to SR.
Work out the length PS.

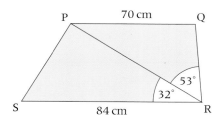

6 a Find x.

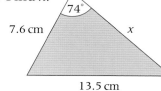

b Find y.

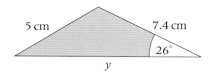

7 Two sides of a triangle are 15.4 cm and 12 cm.
The angle between them is 72°.
Work out the perimeter of the triangle.

8 Adjacent sides of a parallelogram are 6 cm and 8.3 cm.
The shorter diagonal is 7 cm.
Work out the length of the other diagonal.

9 Adjacent sides of a parallelogram are 8 cm and 11 cm.
The longer diagonal is 15.2 cm.
Work out the length of the other diagonal.

Pythagoras's theorem and trigonometry in 3-D

This spread will show you how to:

- Use coordinates to identify a point on a 3-D grid
- Use Pythagoras's theorem and the rules of trigonometry to solve 2-D and 3-D problems

In 3-D, a point has x-, y- and z-coordinates.

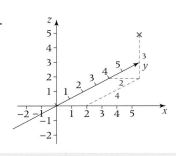

A is the point $(2, 4, 3)$

- You can use **Pythagoras's theorem** to find the distance d between two points in 3-D.

$$d^2 = x^2 + y^2 + z^2$$

Example

Find the distance from A $(3, 4, 1)$ to B $(1, 3, 5)$.

$$AB^2 = (3 - 1)^2 + (4 - 3)^2 + (1 - 5)^2$$
$$= 2^2 + 1^2 + (-4)^2 = 21$$
$$AB = \sqrt{21} = 4.6$$

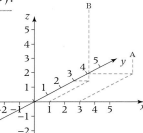

Example

Work out the angle θ between the diagonal and the base in this cuboid.

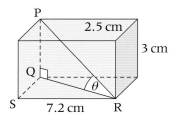

SRQ and PQR are right-angled triangles:

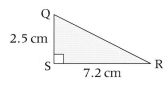

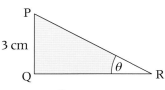

Use Pythagoras in SRQ to find QR.

Use trigonometry to find θ in triangle PQR.

$$QR^2 = 7.2^2 + 2.5^2$$
$$QR = \sqrt{58.09}$$
$$= 7.62 \text{ cm}$$

$$\tan \theta = \frac{3}{7.62}$$
$$\theta = 21.49 \approx 21°$$

1 Find the distances between these points.

   a (2, 4, 5) and (3, 7, 10)   **b** (1, 5, 3) and (6, 2, 6)

   c (−1, 9, 2) and (−3, 0, 5)   **d** (4, −2, 3) and (2, −4, 7)

2 Work out the length of the diagonal *d* in these cuboids.

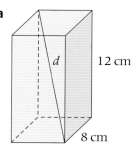 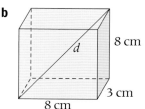 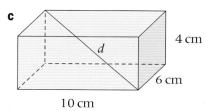

a   12 cm   8 cm   5 cm

b   8 cm   3 cm   8 cm

c   4 cm   6 cm   10 cm

3 For each cuboid in question **2**, work out the angle between the base and the diagonal *d*.

4 **a** Find the angle between the base and the diagonal *d* in this cube.

   b Repeat part **a** for a cube with the side length 10 cm.

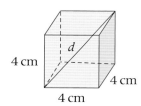

4 cm   4 cm   4 cm

5 A vertical pole is placed at the corner of a horizontal, rectangular garden.

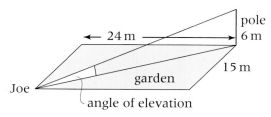

pole 6 m   24 m   15 m   garden   Joe   angle of elevation

The garden is 15 m by 24 m.
The pole is 6 m high.
Joe is at a corner of the garden diagonally opposite the pole.
Work out the angle of elevation from Joe to the top of the pole.

6 The diagram shows a wedge.
The rectangular base ABCD is
perpendicular to the side CDEF.
AB = 8 cm, BC = 6 cm and CF = 3 cm.
Work out

   a AF   **b** angle FAC.

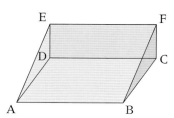

E   F   D   C   A   B

Key objectives

- Solve problems involving bearings, constructions and loci
- Understand, recall and use Pythagoras's theorem in 2-D and 3-D problems
- Understand, recall and use trigonometrical relationships in right-angled triangles, and use these to solve problems, including those involving bearings, then use these relationships in 3-D contexts, including finding the angles between a line and a plane
- Locate points in 3-D with given coordinates

1 Work out the distances between these points.

    a (2, 1, 0) and (7, 3, 2)   (2)

    b (−1, 0, 4) and (−3, 5, 2)   (2)

2 The diagram represents a cuboid ABCDEFGH.

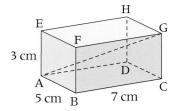

Diagram **NOT** accurately drawn

AB = 5 cm

BC = 7 cm

AE = 3 cm

    a Calculate the length of AG.
       Give your answer correct to 3 significant figures.   (2)

    b Calculate the size of the angle between AG and the face ABCD.
       Give your answer correct to 1 decimal place.   (2)

(Edexcel Ltd., 2004)

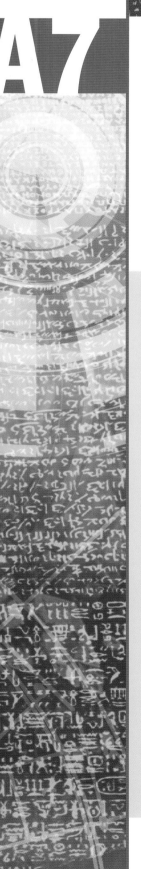

This unit will show you how to

- Solve quadratic equations by factorisation and using the quadratic formula
- Recognise the shape of a graph of a quadratic function
- Generate points and plot graphs of general quadratic functions
- Select and use appropriate and efficient techniques and strategies to solve problems

Before you start ...

You should be able to answer these questions.

Review

1 a Solve these using the quadratic formula.
   i $x^2 - 14x + 33 = 0$ **ii** $x^2 + 5x - 2 = 0$
   b Using your working from **a**, explain why one equation results in integer solutions and the other in decimal solutions.
   c Could either equation in **a** have been solved by factorisation instead?

Unit A3, A6

2 Explain why, by using the quadratic equation formula, the equation $x^2 - 4x + 10 = 0$ has no solutions.

Unit A3, A6

3 a Given that $y = x^2 + 5x + 6$, find the value of y when x is zero and the values of x when y is zero.
   b Repeat part **a** for $y = 2x^2 - 7x + 6$.

Unit A3, A6

4 A rectangle has length $3x + 2$, width $x - 1$ and area 40 units^2.
By forming an equation and solving it, find the value of x.
Hence find the dimensions of the rectangle.

Unit A3, A6

More quadratic equations

This spread will show you how to:

● Solve quadratic equations by factorisation and using the quadratic formula

Keywords

Discriminant
Quadratic

● You can solve some **quadratic** equations by making them equal to zero, then either factorising or using the quadratic equation formula.

● Quadratic equations can have one, two or no solutions.

Quadratic	Does it factorise?	Solution	Comment
$x^2 + 8x + 12 = 0$	Yes: $(x+6)(x+2)$	$x = -2$ or -6	● Factorises ● Two solutions ● Integer solutions
$x^2 + 8x + 11 = 0$	No	$x = \dfrac{-8 \pm \sqrt{8^2 - 4 \times 1 \times 11}}{2 \times 1}$ $x = \dfrac{-8 \pm \sqrt{20}}{2}$ $x = -6.236\ldots$ or $-1.7639\ldots$	● Does not factorise ● Two solutions ● Non-integer solutions
$x^2 + 8x + 16 = 0$	Yes: $(x+4)(x+4)$ or $(x+4)^2$	$x = -4$ twice	● Factorises ● One repeated solution ● Integer solution
$x^2 + 8x + 20 = 0$	No	$x = \dfrac{-8 \pm \sqrt{8^2 - 4 \times 1 \times 20}}{2 \times 1}$ $x = \dfrac{-8 \pm \sqrt{-16}}{2}$	● Does not factorise ● No solutions You cannot find the square root of -16

● The value in the square root of the formula $b^2 - 4ac$ is called the **discriminant**. It tells you how to solve the equation and how many solutions there will be.

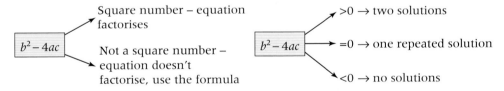

Square number – equation factorises

Not a square number – equation doesn't factorise, use the formula

$b^2 - 4ac$ ──→ $>0 \to$ two solutions

$b^2 - 4ac$ ──→ $=0 \to$ one repeated solution

──→ $<0 \to$ no solutions

Example

By calculating the discriminant, decide how many solutions $2x^2 - 3x - 7 = 0$ has and whether to solve the equation by factorising or by formula.

$a = 2$, $b = -3$ and $c = -7$ so $b^2 - 4ac = (-3)^2 - 4 \times 2 \times (-7) = 65$
Discriminant > 0, so there are two solutions. Discriminant is not a square, so the equation will not factorise – use the formula.

$x = \dfrac{+3 \pm \sqrt{65}}{4}$

$x = 2.77$ or -1.27 (both to 3 sf.)

1 Solve these quadratic equations by factorisation.

a $x^2 + 7x + 10 = 0$ **b** $x^2 + 4x - 12 = 0$ **c** $x^2 - 49 = 0$

d $x^2 - 8x = 0$ **e** $(x + 2)^2 = 16$ **f** $3y^2 - 7y + 2 = 0$

2 Solve these quadratic equations using the formula.

a $x^2 + 8x + 6 = 0$ **b** $7x^2 + 6x + 1 = 0$ **c** $x^2 - 2x - 1 = 0$

d $10y^2 - 2y - 3 = 0$ **e** $4x^2 + 3x = 2$ **f** $(2x - 3)^2 = 2x$

3 For each equation, use the discriminant test to decide

i if it factorises

ii how many solutions it has.

a $x^2 + 2x - 15 = 0$ **b** $x^2 - 6x + 9 = 0$ **c** $2x^2 - 3x - 4 = 0$

d $3y^2 + 11x + 6 = 0$ **e** $10x^2 - x - 3 = 0$ **f** $6x^2 - 11x + 3 = 0$

4 Solve these quadratics by the most efficient method.

a $y(y + 3) = 88$ **b** $(2x)^2 = 25$

c $(7 + 2x)^2 + 4x^2 = 37$ **d** $(x + 2)^2 + 45 = 2x + 1$

e $(y + 1)^2 = 2y(y - 2) + 10$ **f** $\dfrac{15}{p} = p + 22$

g $\dfrac{3}{x} + \dfrac{3}{x + 1} = 7$

> If you are not sure a quadratic will factorise, use the discriminant test.

5 True or false? $x^2 - 5x + 10 = 0$ can be solved by factorisation, producing two solutions. Explain your answer.

6 The perimeter of a rectangle is 46 cm and its diagonal is 17 cm.

17 cm

> You should not need to use the formula.

Set up a quadratic equation and solve it to find the dimensions of the rectangle.

7 You can solve the equation $x^2 - 5x = 0$ by factorisation, $x(x - 5) = 0$, and then letting either $x = 0$ or $x - 5 = 0$. Extend this method to find three solutions for each of these equations.

a $x^3 - 5x^2 + 6x = 0$ **b** $2x^3 + 5x^2 - x = 0$ **c** $x^3 = 6x - x^2$

Sketching quadratic graphs

This spread will show you how to:

● Recognise the shape of a graph of a quadratic function

● Generate points and plot graphs of general quadratic functions

Keywords

Function
Intercept
Maximum
Minimum
Parabola

● You can write a quadratic expression as a **function**, for example $f(x) = x^2 + 2$.

● The graph of a quadratic function is always a **parabola**.

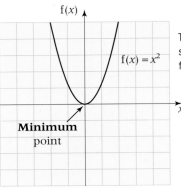

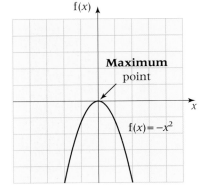

The graphs are symmetrical.
$f(a) = a^2 = f(-a)$

● To sketch a quadratic graph, you need to know
  – where it intersects the axes
  – the coordinates of its turning point (minimum or maximum point).

You can work these out from the quadratic function.
For example, for $f(x) = x^2 - 6x + 8$ (or $y = x^2 - 6x + 8$)

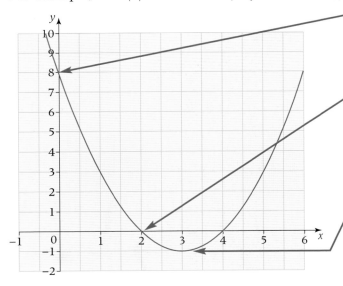

y-axis intercept
On the y-axis the x-coordinate is always zero.
Substitute $x = 0$ into the function:
$f(0) = 0^2 - 6 \times 0 + 8 = 8$
y-intercept is $(0, 8)$.

x-axis intercept(s)
On the x-axis the y-coordinate is always zero.
Substitute $y = 0$ (or $f(x) = 0$) and solve the quadratic:
$0 = x^2 - 6x + 8$
$0 = (x - 2)(x - 4)$
Either $x - 2 = 0$ so $x = 2$ or $x - 4 = 0$ so $x = 4$
Hence, the intercepts are $(2, 0)$ and $(4, 0)$.

Turning point
Find the minimum value by symmetry.
The minimum value must be halfway
between $x = 2$ and $x = 4$, so $x = 3$.
The minimum value is $(3, -1)$

Example

Which quadratic function crosses the x-axis at $(-3, 0)$ and $(5, 0)$?

$x = -3$ and $x = 5$ are the solutions to $0 = (x + 3)(x - 5)$. So the function is
$f(x) = (x + 3)(x - 5) = x^2 - 2x - 15$.

1 Find the y-intercept of each of these quadratic functions.

a $y = x^2 + 8x + 12$ **b** $f(x) = x^2 - 8x + 15$ **c** $y = x^2 + 6x + 5$

d $f(x) = x^2 + 5x + 6$ **e** $y = x^2 + 10x + 25$

2 Find the coordinates where each of the quadratic graphs in question **1** intercept the x-axis.

3 Find the coordinates of the minimum point of each of the quadratic graphs in question **1**.

4 Explain why $y = x^2 + 8x + 20$ does not intercept the x-axis.

5 Sketch a graph of the quadratic function $y = x^2 + 7x + 12$, indicating the coordinates of any points of intersection with the axes and of the minimum point.

6 Match these sketch graphs with the equations given. Sketch the remaining function, labelling its points of intersection with the axes.

a 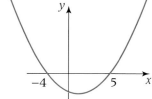 **b** 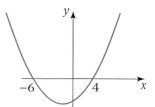 **c**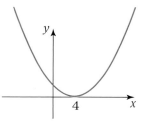

i $y = (x-4)^2$ **ii** $y = x^2 - x - 20$ **iii** $y = (x-6)(x+4)$ **iv** $y = x^2 + 2x - 24$

7 State the equation of these quadratic functions, using the information given in their sketch graphs.

a 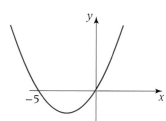 **b** 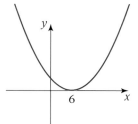 **c**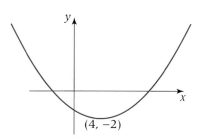

8 Give the equation of a quadratic function that

a intersects the y-axis at $(0, 5)$ but that does not intersect the x-axis

b intersects the x-axis just once, at $(-12, 0)$

c has a maximum value of $(4, 9)$.

9 a Explain why $f(x) = (x-2)(x+3)(x+4)$ is not a parabola. What shape is it?

b Sketch the graph of $f(x) = (x-2)(x+3)(x+4)$.

> Where does $f(x)$ intersect with the x-axis?

Solving problems involving quadratics

This spread will show you how to:

- Solve quadratic equations by factorisation or using the quadratic formula.
- Select and use appropriate and efficient techniques and strategies to solve problems.

Keywords
Quadratic
Solve

- When a problem involves a **quadratic** equation:

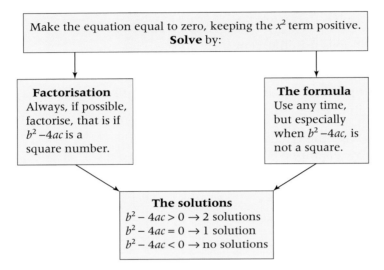

Make the equation equal to zero, keeping the x^2 term positive.
Solve by:

Factorisation
Always, if possible, factorise, that is if $b^2 - 4ac$ is a square number.

The formula
Use any time, but especially when $b^2 - 4ac$, is not a square.

The solutions
$b^2 - 4ac > 0 \rightarrow$ 2 solutions
$b^2 - 4ac = 0 \rightarrow$ 1 solution
$b^2 - 4ac < 0 \rightarrow$ no solutions

Spot a quadratic by its x^2 term.

Examiner's tip
You could also use a graph to find the number of solutions to a quadratic equation.

Example

Find the radius of this circle.

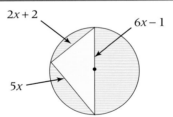
2x + 2
6x − 1
5x

Using Pythagoras's theorem:
$$a^2 + b^2 = c^2$$
$$(5x)^2 + (2x + 2)^2 = (6x - 1)^2$$
$$25x^2 + (2x + 2)(2x + 2) = (6x - 1)(6x - 1)$$
$$25x^2 + 4x^2 + 4x + 4x + 4 = 36x^2 - 6x - 6x + 1$$
$$29x^2 + 8x + 4 = 36x^2 - 12x + 1$$
$$0 = 7x^2 - 20x - 3$$
$$0 = 7x^2 - 21x + x - 3$$
$$0 = 7x(x - 3) + 1(x - 3)$$
$$0 = (7x + 1)(x - 3)$$
Either $\qquad 7x + 1 = 0 \quad or \quad x - 3 = 0$
$$x = -\tfrac{1}{7} \qquad\qquad x = 3$$

In this context, x cannot be negative, hence, $x = 3$.
Diameter $= 6 \times 3 - 1 = 17$, so radius $= 8.5$.

Angle in a semicircle is 90°.

The diameter is the hypotenuse of the right-angled triangle.

Keep the x^2 term positive.

$b^2 - 4ac = (-20)^2 - 4 \times 7 \times (-3) = 484$
= a square number, so it does factorise.

1 Solve these, by writing them as quadratic equations.

a $2x^2 = 5x - 1$ **b** $(y - 5)^2 = 20$ **c** $p(p + 10) + 21 = 0$

d $10x + 7 = \dfrac{3}{x}$ **e** $\dfrac{x^2 + 3}{4} + \dfrac{2x - 1}{5} = 1$

2 In parts **a–d**, form an equation and solve it to answer the problem.

a The diagonal of this rectangle is 13 cm. What is the perimeter of the rectangle?

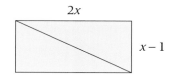
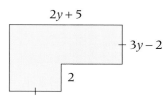

b The area of this hexagon is 25 m². Find the perimeter of the hexagon.

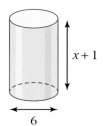

c A circle with radius 5 and centre (0, 0) intersects the line $y = 2(x - 1)$ in two places. Where do they intersect?

d The surface area of the sphere is equal to the curved surface area of the cylinder. Which has the largest volume?

3 **a** Sketch the graphs $y = 3 - 2x$ and $y = x^2 - 4x + 3$ on the same axes.

b Hence write down one point of intersection of these graphs.

c Use algebra to find their second point of intersection.

4 For what value of x is the *total* surface area of the cylinder equal to three times the area of the *curved* surface of the cone?

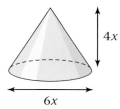
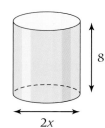

A7 Exam review

Key objectives

- Solve quadratic equations by factorisation or using the quadratic formula
- Generate points and plot graphs of simple quadratic functions, then more general quadratic functions

1 a Copy and complete the table of values for $y = x^2 - x - 2$. (2)

x	-2	-1	0	1	2	3	4
y	4			-2	0		10

b Copy the grid and draw the graph of $y = x^2 - x - 2$. (2)

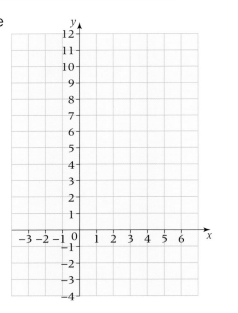

2 For all values of x and m,
$x^2 - 2mx = (x - m)^2 - k$.

a Express k in terms of m. (2)

The expression $x^2 - 2mx$ has a minimum value as x varies.

b i Find the minimum value of $x^2 - 2mx$.
Give your answer in terms of m.

ii State the value of x for which this minimum value occurs.
Give your answer in terms of m. (3)

(Edexcel Ltd., 2003)

D4

This unit will show you how to

- Represent grouped data on a histogram
- Understand frequency density
- Calculate frequencies from a histogram
- Use the information given in tables and histograms to deal with problems such as missing data
- Use histograms to compare two or more data sets, considering the modal class, range and skewness

Before you start ...

You should be able to answer these questions.

1 Work out each of these.

    a $(20 \times 0.2) + (10 \times 1.4) + (5 \times 0.8)$

    b $(30 \times 0.6) + (20 \times 1.2) + (5 \times 0.6)$

2 Work out each of these.

    a $13 \div (50 - 40)$

    b $12 \div (10 - 5)$

    c $9 \div (40 - 15)$

    d $9 \div (0.5 - 0.2)$

Review

Unit N2

Unit N2

This spread will show you how to:

- Represent grouped data on a histogram
- Understand frequency density

Keywords
Class width
Frequency
Frequency
    density
Histogram

You can represent grouped continuous data in a **histogram**.

- In a histogram, the area of each bar represents the **frequency**.

The data can be in equal or unequal sized class intervals.

The vertical axis represents **frequency density**.

- Frequency density = $\dfrac{\text{frequency}}{\text{class width}}$

Area = frequency =
class width × bar height

So bar height = $\dfrac{\text{frequency}}{\text{class width}}$

Example

Ursula collected data on the time taken to complete a simple jigsaw.
Draw a histogram to represent these data.

Time, t seconds	$40 \leqslant t < 60$	$60 \leqslant t < 70$	$70 \leqslant t < 80$	$80 \leqslant t < 90$	$90 \leqslant t < 120$
Frequency	6	6	10	7	6

Time, t seconds	$40 \leqslant t < 60$	$60 \leqslant t < 70$	$70 \leqslant t < 80$	$80 \leqslant t < 90$	$90 \leqslant t < 120$
Class width	20	10	10	10	30
Frequency	6	6	10	7	6
Frequency density	0.3	0.6	1	0.7	0.2

Add rows to the
table to calculate
class width and
frequency density.

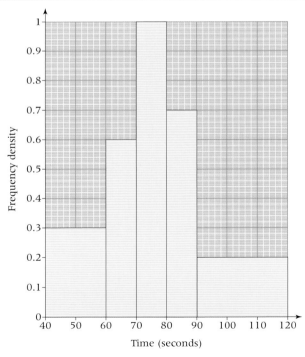

For each set of data

a Copy and complete the table to calculate frequency density.

b Draw a histogram to represent the data.

1 Reaction times of a sample of students.

Time, t seconds	$1 \leqslant t < 3$	$3 \leqslant t < 4$	$4 \leqslant t < 5$	$5 \leqslant t < 6$	$6 \leqslant t < 9$
Class width					
Frequency	12	17	19	11	18
Frequency density					

2 Amounts spent by the first 100 customers in a shop one Saturday.

Amount spent, £a	$0 \leqslant a < 5$	$5 \leqslant a < 10$	$10 \leqslant a < 20$	$20 \leqslant a < 40$	$40 \leqslant a < 60$	$60 \leqslant a < 100$
Class width						
Frequency	6	10	23	29	24	8
Frequency density						

3 Distance travelled to work by 100 office workers.

Distance, d miles	$0 \leqslant d < 2$	$2 \leqslant d < 5$	$5 \leqslant d < 10$	$10 \leqslant d < 20$	$20 \leqslant d < 30$
Class width					
Frequency	8	15	27	44	6
Frequency density					

4 Times of goals scored in Premiership football matches one Saturday.

Time, t minutes	$0 \leqslant t < 10$	$10 \leqslant t < 40$	$40 \leqslant t < 45$	$45 \leqslant t < 55$	$55 \leqslant t < 85$	$85 \leqslant t < 90$
Class width						
Frequency	12	48	18	11	30	22
Frequency density						

5 Distances swum by children in a sponsored swim.

Distance, d km	$0.1 \leqslant d < 0.2$	$0.2 \leqslant d < 0.5$	$0.5 \leqslant d < 1$	$1 \leqslant d < 2$	$2 \leqslant d < 5$
Class width					
Frequency	3	12	22	25	18
Frequency density					

6 Times dog owners spend on daily walks.

Time, t minutes	$10 \leqslant t < 20$	$20 \leqslant t < 40$	$40 \leqslant t < 60$	$60 \leqslant t < 90$	$90 \leqslant t < 120$
Class width					
Frequency	8	16	28	39	9
Frequency density					

Interpreting histograms

This spread will show you how to:

- Understand frequency density
- Calculate frequencies from a histogram

Keywords

Class width
Frequency
Frequency
density
Histogram

You can calculate frequencies from a **histogram**.

For each bar, the area represents the **frequency**.

- Frequency = frequency density × class width

Frequency density = $\dfrac{\text{frequency}}{\text{class width}}$

Example

The histogram shows the times a sample of students spent on the internet one evening.

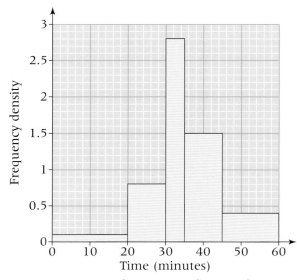

a How many students spent longer than 50 minutes on the internet?
b Complete the frequency table for these data.

Time, t minutes	$0 \leqslant t < 20$	$20 \leqslant t < 30$	$30 \leqslant t < 35$	$35 \leqslant t < 45$	$45 \leqslant t < 60$
Frequency					

c How many students were included in the sample?

a The area in the histogram that represents >50 minutes is only part of the last bar.

Area $50 \leqslant t < 60$ = height × width of 50 – 60 class interval
$= 0.4 \times 10 = 4$

Four students spent longer than 50 minutes on the internet.

b

Time, t minutes	$0 \leqslant t < 20$	$20 \leqslant t < 30$	$30 \leqslant t < 35$	$35 \leqslant t < 45$	$45 \leqslant t < 60$
Frequency	$0.1 \times 20 =$ 2	$0.8 \times 10 =$ 8	$2.8 \times 5 =$ 14	$1.5 \times 10 =$ 15	$0.4 \times 15 =$ 6

The area of each bar gives the frequency.

c $2 + 8 + 14 + 15 + 6 = 45$
45 students were included in the sample.

1 The histogram shows the times a sample of students spent watching TV one evening.

   a How many students spent longer than $2\frac{1}{2}$ hours watching TV?

   b Copy and complete the frequency table for these data.

   c How many students were in the sample?

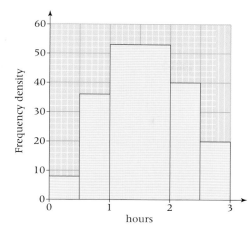

Time, t hours	$0 \leqslant t < 0.5$	$0.5 \leqslant t < 1$	$1 \leqslant t < 2$	$2 \leqslant t < 2.5$	$2.5 \leqslant t < 3$
Frequency					

2 The histogram shows the distances a sample of teachers travel to work each day.

   a How many teachers travel between 10 and 30 kilometres?

   b Copy and complete the frequency table for these data.

   c How many teachers were in the sample?

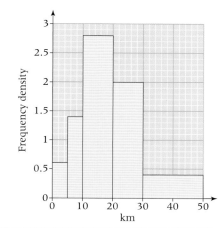

Distance, d km	$0 \leqslant d < 5$	$5 \leqslant d < 10$	$10 \leqslant d < 20$	$20 \leqslant d < 30$	$30 \leqslant d < 50$
Frequency					

3 The histograms show the heights of some boys aged 11 and 16. For each histogram, draw a frequency table and calculate the number of boys in each sample   **a** aged 11   **b** aged 16.

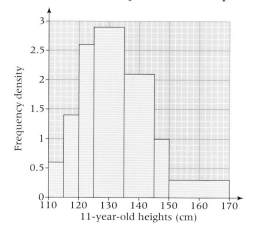

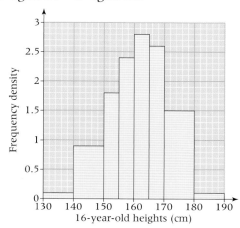

More histograms

This spread will show you how to:

● Use the information given in tables and histograms to deal with problems such as missing data

Keywords
Class width
Frequency
Frequency
   density
Histogram

You can use information from tables and **histograms** to fill in gaps in data.

Example

The incomplete table and histogram give some information about the lengths of phone calls Wendy made at work one day.

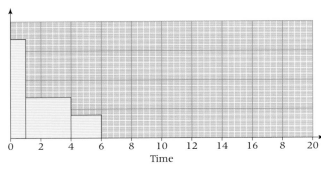

Time t, minutes	Frequency
$0 \leqslant t < 1$	17
$1 \leqslant t < 4$	
$4 \leqslant t < 6$	
$6 \leqslant t < 10$	12
$10 \leqslant t < 20$	10

a Use the information in the histogram to complete the table.
b Complete the histogram.

a The class $0 \leqslant t < 1$ has **frequency density** $= \dfrac{\text{frequency}}{\text{class width}} = \dfrac{17}{1} = 17$

Use the information in the table to work out the scale on the vertical axis.

On the histogram, $0 \leqslant t < 1$ bar has height 3.4 cm, so vertical scale = 17 ÷ 3.4 = 5 per cm

From the histogram the class, $1 \leqslant t < h$ has **frequency** = frequency density × class width = 7 × 3 = 21 calls.
The class $4 \leqslant t < 6$ has frequency = 4 × 2 = 8 calls.
The completed table is:

Time t, min	Frequency	Class width	Frequency density
$0 \leqslant t < 1$	17	1	17
$1 \leqslant t < 4$	21	3	7
$4 \leqslant t < 6$	8	2	4
$6 \leqslant t < 10$	12	4	3
$10 \leqslant t < 20$	10	10	1

Calculate the frequency densities:

$6 \leqslant t < 10$:
fd = 12/4 = 3
$10 \leqslant t < 20$:
fd = 10/10 = 1

b

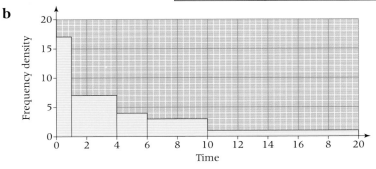

Remember to scale and label the vertical axis.

1 The incomplete table and histogram give some information about the weights, in grams, of a sample of apples.

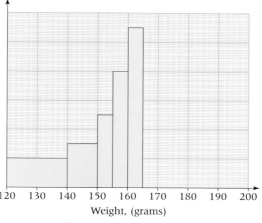

Weight, g grams	Frequency
$120 \leqslant g < 140$	8
$140 \leqslant g < 150$	6
$150 \leqslant g < 155$	
$155 \leqslant g < 160$	
$160 \leqslant g < 165$	
$165 \leqslant g < 175$	16
$175 \leqslant g < 185$	12
$185 \leqslant g < 200$	6

a Use the information in the histogram to work out the missing frequencies in the table.

b Copy and complete the histogram.

2 The incomplete table and histogram give some information about distances travelled by sales representatives on one day.

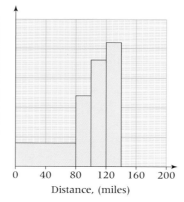

Miles travelled, m	Frequency
$0 \leqslant m < 80$	32
$80 \leqslant m < 100$	24
$100 \leqslant m < 120$	
$120 \leqslant m < 140$	
$140 \leqslant m < 160$	44
$160 \leqslant m < 200$	28

a Copy and complete the table.

b Copy and complete the histogram.

3 Copy the tables and work out the height, in terms of f, for the second bar in each of these histograms.

a

Time, t minutes	Frequency	Bar height	Frequency density
$0 \leqslant t < 10$	8	3.2 cm	$8 \div 10 = 0.8$
$10 \leqslant t < 40$	f		

b

Time, t minutes	Frequency	Bar height	Frequency density
$0 \leqslant t < 25$	10	2cm	
$25 \leqslant t < 30$	f		

Using histograms to compare data sets

This spread will show you how to:

● Use histograms to compare two or more data sets, considering the modal class, range and skewness

Keywords
Frequency density
Histogram
Modal class
Skewness

You can use **histograms** to compare data sets.

● The highest bar on a histogram represents the **modal class**.

● You can estimate the range.

The shape of a histogram shows whether the data is skewed.

This is the class with the highest **frequency density**. It **may not** be the class with the highest frequency.

The range is an estimate, as you do not have the actual data values.

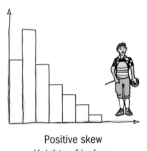

Positive skew
Heights of jockeys

No skew – symmetrical
Heights of random sample of men

Negative skew
Heights of basketball players

● To compare histograms for two or more data sets, consider the modal class, range and **skewness**.

Example

The histograms show the times taken by a sample of boys and a sample of girls to complete the same puzzle.
Compare the times taken by the boys and the girls.

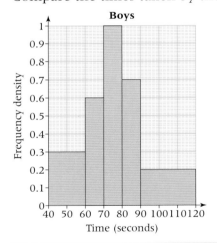

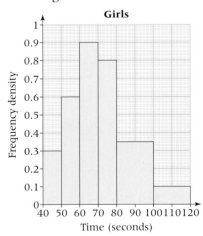

The range of times is the same for both boys and girls $120 - 40 = 80$

The modal class for boys (70–80 seconds) is a slower time than the modal class for girls (60–70 seconds), so the boys were generally slower.

The girls' times are more positively skewed than the boys' times indicating that girls' times were shorter, that is, the girls were quicker.

1 Compare the heights of two samples of boys of different ages summarised in these histograms.

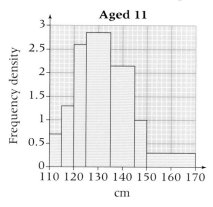

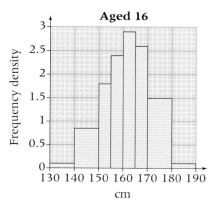

2 Compare the weights of samples of apples and pears summarised in these histograms.

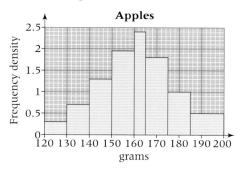

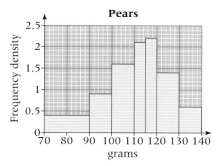

3 Compare the reaction times of girls and boys summarised in these histograms.

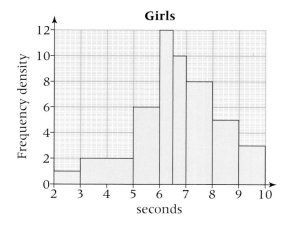

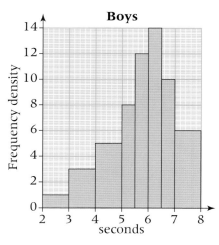

Exam review

Key objectives

- Draw and produce cumulative frequency tables and diagrams, and histograms for grouped continuous data
- Understand frequency density
- Compare distributions and make inferences, using shapes of distributions and measures of average and spread, including median and quartiles

1 The tables show the race times for a sample of boys and girls.

Boys:

Time, t seconds	$11 < t \leqslant 12$	$12 < t \leqslant 13$	$13 < t \leqslant 14$	$14 < t \leqslant 15$	$15 < t \leqslant 16$
Frequency	1	3	6	4	2
Frequency density					

Girls:

Time, t seconds	$11 < t \leqslant 12$	$12 < t \leqslant 13$	$13 < t \leqslant 14$	$14 < t \leqslant 15$	$15 < t \leqslant 16$
Frequency	0	2	4	6	4
Frequency density					

   a Draw two histograms to summarise the data. (4)

   b Use your histograms from part **a** to compare the two data sets. (3)

2 The incomplete table and histogram give some information about the ages of the people who live in a village.

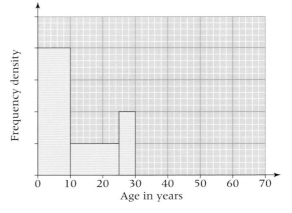

Age, x years	Frequency
$0 < x \leqslant 10$	160
$10 < x \leqslant 25$	
$25 < x \leqslant 30$	
$30 < x \leqslant 40$	100
$40 < x \leqslant 70$	120

   a Use the information in the histogram to copy and complete the frequency table. (2)

   b Copy and complete the histogram. (2)

(Edexcel Ltd., 2003)

This unit will show you how to

- Identify dimensions by looking at units of measurement
- Convert between length measures, area measures and volume measures
- Understand the difference between formulae for perimeter, area and volume by considering dimensions
- Calculate the area of any triangle using $\frac{1}{2}ab \sin C$
- Use the sine and cosine rules to solve 2-D problems
- Solve problems involving more complex shapes, including segments of circles and frustums of cones
- Use the formulae for surface area and volume of a cone
- Solve 3-D problems involving surface areas and volumes of cones and spheres
- Improve the accuracy of solutions to multi-step problems

Before you start ...

You should be able to answer these questions.

1 Solve these equations.

  a $\frac{x+2}{x} = \frac{5}{4}$ b $\frac{x+3}{x} = \frac{7}{5}$

  c $\frac{x}{x+5} = \frac{2}{7}$ d $\frac{x}{x+1} = \frac{4}{5}$

2 For the triangles drawn,

  i   Use trigonometry to find the perpendicular height.

  ii   Use the height to find the area of the triangles.

a   b

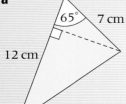

Review

Unit A2

Unit S5

Measures and dimensions

This spread will show you how to:

● Identify dimensions by looking at units of measurement

● Convert between length measures, area measures and volume measures

● Understand the difference between formulae for perimeter, area and volume by considering dimensions

You can identify **dimensions** by looking at units of measurement.

● **Perimeter** is the distance drawn around a shape.
It is a length.
Perimeter is measured in mm, cm and m.

● **Area** is the space covered by a two-dimensional shape.
It is the product of length × length.
Area is measured in square units: mm^2, cm^2 and m^2.

$1 m^2 = 100\ cm × 100\ cm$

● **Volume** is the space inside a three-dimensional object.
It is the product of length × length × length.
Volume is measured in cubic units: mm^3, cm^3 and m^3.

$1 m^3 = 100\ cm × 100\ cm × 100\ cm$

Example

Change **a** 5 400 000 cm to m **b** 5 400 000 cm^2 to m^2
c 5 400 000 cm^3 to m^3.

a 5 400 000 ÷ 100 = 54 000 m **b** 5 400 000 ÷ (100 × 100) = 540 m^2
c 5 400 000 ÷ (100 × 100 × 100) = 5.4 m^3

Larger unit means smaller number
→ divide.

Example

Change **a** 0.026 m to mm **b** 0.041 m^2 to cm^2 **c** 8 cm^3 to mm^3.

a 0.026 × 1000 = 26 mm **b** 0.041 × (100 × 100) = 410 cm^2
c 8 × (10 × 10 × 10) = 8000 mm^3

Smaller unit means larger number
→ multiply.

Example

Change 0.042 m^3 to **a** cm^3 **b** litres.

a 0.042 × (100 × 100 × 100) = 42 000 cm^3
b 42 000 ÷ 1000 = 4.2 litres

1 litre = 1000 cm^3

You can identify dimensions in expressions.

Constants, such as π and numbers, have no dimensions.

Example

u, v and w represent lengths.
Decide whether each expression could represent a length, area, volume or none of these.

$3u + 2v + w$	uvw	uv^2	$uv - v$	$uv + vw$
length + length + length	length × length × length	length × length^2 = length^3	length × length − length	length × length + length × length
= length	= volume	= volume	none of these	= area

1 Change each amount to the unit given.

a 320 000 cm² to m² **b** 0.004 m to mm

c 0.02 m³ to cm³ **d** 1900 cm³ to litres

e 5100 m to km **f** 580 mm to m

g 6 300 000 mm to km **h** 24 cm³ to mm³

i 630 000 000 mm³ to cm³ **j** 900 000 000 cm² to m²

k 0.0007 km² to m² **l** 10 cm² to mm²

2 m, n and l are all lengths.
Explain why the expression

a $m + n + l$ represents a length

b mnl can represent a volume.

3 Jade said the volume of this shape is
$\frac{5}{8}\pi r l^3$

Explain why the expression cannot be correct.

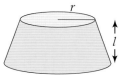

4 In these expressions, the letters a, b, d, h and r all represent lengths.
For each expression write down if it represents a length, area,
volume or none of these.

a abh **b** $\frac{1}{2}dh$ **c** $\frac{1}{3}\pi r^2 - \frac{1}{2}ab$ **d** $ab + bh$

e $2\pi r$ **f** $h - r$ **g** $a^2 - b^2$ **h** $\frac{1}{3}d + br$

i $hr^2 + ab$ **j** $a^2h + b^3$ **k** $a^2b + h$ **l** $\frac{1}{2}h + rd$

5 The diagram shows a pool that is
rectangular with semicircular ends.
Write an expression for

a the perimeter **b** the area.

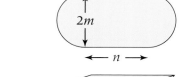

6 The diagram shows a prism with a
cross-section that is a parallelogram.
Write an expression for

a the surface area **b** the volume.

7 An unusually shaped window has area given by the expression

$4ab + b^2$ cm² where a and b are lengths in cm.

The glass fitted in the window has thickness 6 mm.
Write an expression for the volume of the glass.

8 The cross-sectional area of a prism is given by the expression

wl where w and l are lengths in cm.

The height of the prism is h where h is in mm.
Write an expression for the volume of the prism, stating the
correct unit.

The area of a triangle

This spread will show you how to:

- Calculate the area of a triangle using $\frac{1}{2}ab \sin C$

You can use the **sine** ratio to find a formula for the area of any triangle.

Label the angles A, B and C.
Label the sides opposite these angles a, b and c.
Divide the triangle into two right-angled triangles.
h is perpendicular to side b.

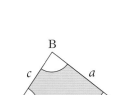

Area of triangle ABC $= \frac{1}{2} \times b \times h$

In the right-hand triangle, $\sin C = \frac{h}{a}$ so $h = a \sin C$.

Area of triangle ABC $= \frac{1}{2} \times b \times a \sin C$

- **The area of a triangle $= \frac{1}{2}ab \sin C$**

You can use this formula when you know two sides
and the angle between them.

The formula can
also be written as:

Area $= \frac{1}{2}bc \sin A$

Area $= \frac{1}{2}ac \sin B$

Example

Work out the areas of these triangles.

a
3.6 cm 108° 4.9 cm

b
5.2 cm 42° 10.7 cm

a Area $= \frac{1}{2} \times 3.6 \times 4.9 \times \sin 108°$
   $= 8.4 \text{ cm}^2$

b Area $= \frac{1}{2} \times 5.2 \times 10.7 \times \sin 42°$
   $= 18.6 \text{ cm}^2$

You can split other shapes into triangles to work out their areas.

Example

Work out the area of the parallelogram.

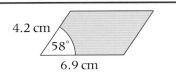
4.2 cm 58° 6.9 cm

Draw a diagonal to divide the parallelogram
into two congruent triangles.

4.2 cm 58° 6.9 cm

Area of one triangle:
   $\frac{1}{2} \times 6.9 \times 4.2 \times \sin 58 = 12.288$
Area of parallelogram:
   $2 \times 12.288 \ldots = 24.6 \text{ cm}^2$

Congruent
triangles are
identical.

Don't round until
the end.

1 Work out the area of each triangle.
Give your answers correct to 3 significant figures.

a 7 cm 65° 8 cm

b 9 cm 42° 12 cm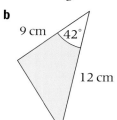

c 8.5 cm 72° 6.4 cm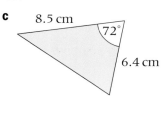

d 4.7 cm 36° 9 cm

e 58° 10.2 cm 9.4 cm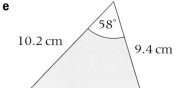

f 5.6 cm 118° 8.8 cm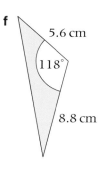

g 9.5 cm 125° 6.8 cm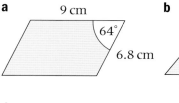

h 7.5 cm 98° 10 cm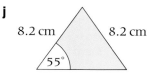

i 6 cm 80° 6 cm

j 8.2 cm 8.2 cm 55°

2 Work out the area of each parallelogram or rhombus.
Give your answers correct to 3 significant figures.

a 9 cm 64° 6.8 cm

b 52° 8.2 cm 5.4 cm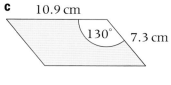

c 10.9 cm 130° 7.3 cm

d 16 cm 142° 11.6 cm

e 8 cm 50°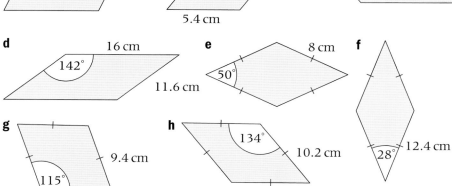

f 12.4 cm 28°

g 9.4 cm 115°

h 134° 10.2 cm

Solving problems using trigonometry

This spread will show you how to:

- Use the sine and cosine rules to solve 2-D problems
- Solve problems involving more complex shapes, including segments of circles
- Improve the accuracy of solutions to multi-step problems

Keywords
Area of triangle
Cosine rule
Segment
Sine rule

In multi-stage problems you often need to calculate extra information.

Your answers will be more accurate if you do not round numbers at intermediate steps.

For any triangle

- **Cosine rule:** $a^2 = b^2 + c^2 - 2bc \cos A$
- **Sine rule:** $\frac{a}{\sin A} = \frac{b}{\sin B} = \frac{c}{\sin C}$
- **Area of triangle** $= \frac{1}{2}ab \sin C$

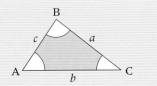

For sine rule see page 244.
For cosine rule see page 246.

Example

Work out the area of the shaded **segment**.

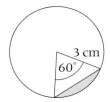

Area of sector
$$\frac{60}{360}\pi \times 3^2 = 4.712 \ldots$$
Area of triangle
$$\frac{1}{2} \times 3 \times 3 \sin 60 = 3.897 \ldots$$
Area of segment
$$4.712 \ldots - 3.897 \ldots$$
$$= 0.815 \text{ cm}^2$$

360° at centre of circle, so area of sector $= \frac{60}{360}$ area of circle.

Area of segment = area of sector − area of triangle

Example

The diagram shows a four-sided field. Find the area of the field.

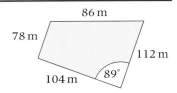

Divide the shape into two triangles by drawing in the diagonal DB.

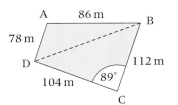

$DB^2 = 112^2 + 104^2 - 2 \times 112 \times 104 \times \cos 89°$
$DB^2 = 22953.42\ldots$
$\cos DAB = \frac{78^2 + 86^2 - 22953.42 \ldots}{2 \times 78 \times 86}$
$\angle DAB = \cos^{-1}(-0.7061292 \ldots)$
$\quad = 134.920 \ldots°$

Area of triangle ADB $= \frac{1}{2} \times 78 \times 86 \times \sin 137.405° \ldots$
$\quad = 2374.9106 \ldots \text{ m}^2$

Area of triangle BCD $= \frac{1}{2} \times 104 \times 112 \times \sin 89°$
$\quad = 5823.112 \ldots \text{ m}^2$

Area of field $= 2374.9106 \ldots + 5823.112$
$\quad = 8198 \text{ m}^2$

Use the cosine rule in triangle DCB to find DB.

Use the cosine rule in triangle ADB to work out angle DAB.

Use $\frac{1}{2}ab \sin C$.

1 Work out the area of the shaded segment of each circle.

a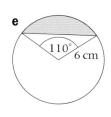

b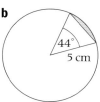

c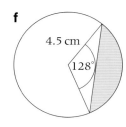

d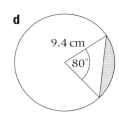

e

f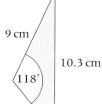

g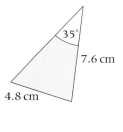

h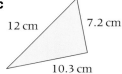

2 **a** Use the sine rule to find angle Q.

   b What is angle R?

   c Work out the area of triangle PQR.

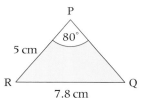

3 Work out the area of each of these triangles.

a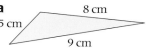

b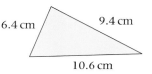

c

4 **a** Use the cosine rule to find angle B.

   b Work out the area of triangle ABC.

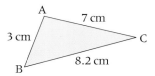

5 Work out the area of these triangles.

a   8 cm   **b**   **c**

6 Work out the area of quadrilateral PQRS.

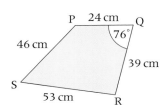

This spread will show you how to:

● Use the formulae for surface area and volume of a cone
● Solve problems involving more complex shapes, including frustums of cones

Keywords
Cone
Frustum

When you cut the top off a **cone** with a cut parallel to the base, the part left is called the **frustum**.

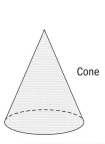

Cone

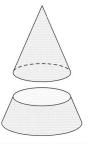

Frustum

The cone removed is similar to the whole cone.

● Volume of a frustum =
  volume of the whole cone − volume of the smaller cone

● Volume of a cone = $\frac{1}{3}\pi r^2 h$
● Curved surface area of a cone = $\pi r l$

Example

Find **a** the volume
       b the surface area of this frustum.

8 cm
5 cm
12 cm

Use similar triangles to find the height of the whole cone.

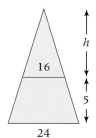

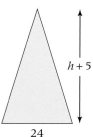

$\frac{h}{16} = \frac{h+5}{24}$

$24h = 16(h+5)$

$8h = 80$

$h = 10$

The triangles are similar, so the ratio of the sides is constant.

You can leave π in your working to avoid rounding. It is more accurate.

a Volume of whole cone: $\frac{1}{3}\pi \times 12^2 \times 15 = 720\pi$

Volume of small cone: $\frac{1}{3}\pi \times 8^2 \times 10 = 213\frac{1}{3}\pi$

Volume of frustum: $720\pi - 213\frac{1}{3}\pi = 1592$ cm^3

b Curved surface area of whole cone

$\pi \times 12 \times \sqrt{369} = 12\pi\sqrt{369}$

Curved surface area of small cone

$\pi \times 8 \times \sqrt{164} = 8\pi\sqrt{164}$

Surface area of frustum

= section of curved surface + area top circle + area bottom circle

= $(12\pi\sqrt{369} - 8\pi\sqrt{164}) + \pi \times 8^2 + \pi \times 12^2 = 1056$ cm^2

Use Pythagoras to find the sloping length l for the whole cone:
$l^2 = 12^2 + (h+5)^2$
$l^2 = 12^2 + 15^2$
$l = \sqrt{369}$
For the small cone:
$l^2 = 8^2 + h^2$
$l^2 = 8^2 + 10^2$
$l = \sqrt{164}$

1 Work out

i the volume

ii the surface area of each of these frustums.

a
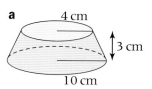
4 cm
3 cm
10 cm

b
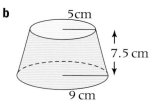
5 cm
7.5 cm
9 cm

c
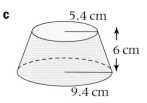
5.4 cm
6 cm
9.4 cm

d
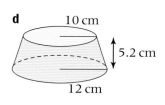
10 cm
5.2 cm
12 cm

e
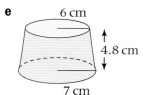
6 cm
4.8 cm
7 cm

f
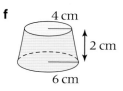
4 cm
2 cm
6 cm

g
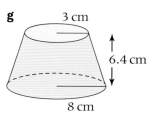
3 cm
6.4 cm
8 cm

h
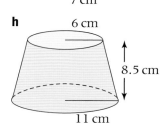
6 cm
8.5 cm
11 cm

i
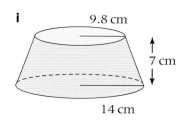
9.8 cm
7 cm
14 cm

j
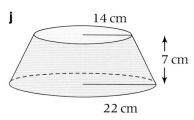
14 cm
7 cm
22 cm

k
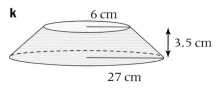
6 cm
3.5 cm
27 cm

l
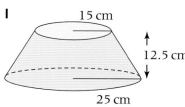
15 cm
12.5 cm
25 cm

2 The diagram shows a frustum.
The base radius, $2r$ cm, is twice the radius
of the top of the frustum, r cm.
The height of the frustum is h cm.
Write down an expression for

a the volume

b the surface area of this frustum.

r

h

$2r$

Solving 3-D problems

This spread will show you how to:

- Use the formulae for surface area and volume of a cone
- Solve 3-D problems involving surface areas and volumes of cones and spheres

You can make answers to multi-stage problems exact by

- leaving **constants** such as π in your answer
- giving your answer as a fraction
- giving your answer in **surd** form.

Example

This is a **net** for a cone.
Show that the volume of the cone is $36\pi\sqrt{5}$.

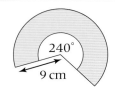
240°
9 cm

The made-up cone will look like this:

Volume of a cone = $\frac{1}{3}\pi r^2 h$
The arc length of the **sector** is the circumference of the base of the cone.

Circumference of base of cone = $2\pi r$
So $\frac{240}{360} \times 2\pi \times 9 = 2\pi r$
$12\pi = 2\pi r$
$r = 6$

$h^2 = 9^2 - 6^2$
$h = \sqrt{45} = 3\sqrt{5}$

Volume of cone: $\frac{1}{3}\pi r^2 h = \frac{1}{3}\pi \times 6^2 \times 3\sqrt{5} = 36\pi\sqrt{5}$

h
9 cm
r

You need to find:
- r, the radius of the base of the cone
- h, the height of the cone.

Arc length = $\frac{\theta}{360} \times 2\pi r$
(see S1)

Use Pythagoras to find h.

- Volume of a sphere = $\frac{4}{3}\pi r^3$
- Surface area of a sphere = $4\pi r^2$

Example

A sphere of radius $2r$ has the same volume as a cylinder with base radius r and height 8 cm.
Work out the surface area of the sphere. Give your answer in terms of π.

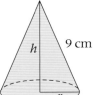
2r

r
8 cm

Volume of sphere = $\frac{4}{3}\pi(2r)^3 = \frac{32}{3}\pi r^3$
So $\frac{32}{3}\pi r^3 = 8\pi r^2$
$r = \frac{3}{4}$

Volume of cylinder = $8\pi r^2$

Surface area of sphere = $4\pi r^2 = 4\pi(\frac{3}{4})^2 = \frac{9\pi}{4}$ cm^2

1 This is a net for a cone.

Show that the volume of the cone is $\frac{200}{3}\pi\sqrt{11}$.

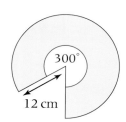

300°

12 cm

2 This is a net for a cone.

Show that the volume of the cone is $\frac{250}{3}\pi\sqrt{2}$.

120°

15 cm

3 The surface area of a sphere of radius r is the same as the curved surface area of a cylinder with base radius 3 cm and height 2 cm.

Show that the volume of the sphere is $4\pi\sqrt{3}$.

4 The radius of a sphere is 5 cm.
The radius of the base of a cone is 10 cm.
The volume of the sphere is twice the volume of the cone.

5 cm

10 cm

Work out the curved surface area of the cone, give your answer as a multiple of π.

5 The radius of a sphere is three times the radius of the base of a cone.
The base radius of the cone is 2 cm.
The volume of the sphere is nine times the volume of the cone.

Work out the curved surface area of the cone.

6 The curved surface area of a cylinder is equal to the curved surface area of a cone.
The base radius of the cylinder is twice the base radius of the cone.

Show that the height of the cylinder is a quarter of the slant height of the cone.

7 A solid metal sphere with radius 3 cm is melted down and re-formed as a cylinder.
The base radius of the cylinder is 3 cm.

Work out the height of the cylinder.

S6

Exam review

Key objectives

- Convert measurements from one unit to another
- Calculate the area of a triangle using $\frac{1}{2}ab \sin C$
- Solve problems involving more complex shapes and solids, including segments of circles and frustums of cones
- Understand how errors are compounded in certain calculations

1 In these expressions x, y and z represent lengths.

   i xyz **ii** $x + y$

   iii $2y^2 - xz$ **iv** $x + yz$

Which of these expressions represent

   a length   **b** area

   c volume   **d** none of these?   (4)

2 In triangle ABC,

   AC = 8 cm

   CB = 15 cm

   Angle ACB = 70°

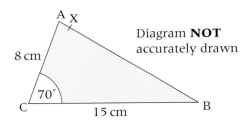

Diagram **NOT** accurately drawn

   a Calculate the area of triangle ABC.
      Give your answer correct to 3 significant figures.   (2)

X is the point on AB such that angle $CXB = 90°$.

   b Calculate the length of CX.
      Give your answer correct to 3 significant figures.   (4)

(Edexcel Ltd., 2003)

A8

This unit will show you how to

- Recognise the shapes of graphs of quadratic, cubic and exponential functions and functions involving reciprocals
- Draw graphs of quadratic, cubic and exponential functions and functions involving reciprocals by plotting points
- Find the intersection points of graphs of a linear and quadratic function
- Solve simultaneous equations graphically by drawing a graph to represent each equation
- Solve equations graphically by splitting them into two equations
- Recognise the equation of a circle
- Solve two simultaneous equations representing a line and a circle graphically

Before you start ...

You should be able to answer these questions.

Review

1 Choose which word best describes each line.

Unit A4

horizontal   vertical   diagonal

a $y = 9$ **b** $y = 6x - 5$ **c** $x = -2$

d $y + 9 = 0$ **e** $x + 2y = 11$

2 Evaluate each expression for the given value of x.

Unit A5

a $3x^2 - 2x$ when $x = 4$

b $x^2 - x$ when $x = -2$

c $x^2(2x + 3)$ when $x = -4$

3 a Write the radius of each circle, centred at the origin.

Unit A6

a $x^2 + y^2 = 25$ **b** $x^2 + y^2 = 121$

c $x^2 + y^2 = 400$.

b Give the equation of each circle.

i   A circle, centre the origin, radius 2

ii A circle, centre the origin, radius $\frac{3}{8}$

Graphs of quadratic and cubic functions

This spread will show you how to:

- Recognise the graphs of quadratic and cubic functions
- Draw graphs of quadratic and cubic functions by plotting points

Keywords

Cubic
Function
Parabola
Quadratic
S-shaped

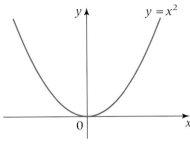

- The graph of a **quadratic function** is a U-shaped curve called a parabola.
  $$f(x) = x^2 \quad \text{or} \quad y = x^2$$

- The graph of a **cubic** function is an **'S'-shaped** curve.
  $$f(x) = x^3 \quad \text{or} \quad y = x^3$$

f(2) is positive, since $2^3 = +8$

f(−2) is negative, since $-2^3 = -8$

- You draw graphs of quadratic and cubic functions by plotting points.

Example

Plot the graph of $f(x) = x^2 - 3x + 5$ for $-3 \leqslant x \leqslant 3$.

From your graph, find the coordinates of the minimum point and confirm this using an algebraic technique.

$f(x) = x^2 - 3x + 5$ is the same as $y = x^2 - 3x + 5$.

x	−3	−2	−1	0	1	2	3
x^2	9	4	1	0	1	4	9
$-3x$	9	6	3	0	−3	−6	−9
$+5$	5	5	5	5	5	5	5
$f(x)$	23	15	9	5	3	3	5

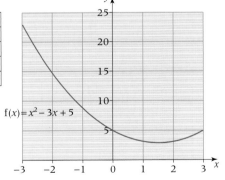

Draw up a table of values. Keep signs with the terms. Add down a column for each value of $f(x)$.

Plot the points (−3, 23), (−2, 15), (−1, 9), …, (3, 5), joining them with a smooth curve. Make sure the bottom of the curve is 'rounded'.

From the graph, the minimum point is at approximately $(1\frac{1}{2}, 2\frac{3}{4})$.

Complete the square on $x^2 - 3x + 5$.

$$x^2 - 3x + 5 = (x - \tfrac{3}{2})^2 + 5 - \tfrac{9}{4}$$
$$= (x - 1\tfrac{1}{2})^2 + 2\tfrac{3}{4} \qquad \text{So the minimum point is } (1\tfrac{1}{2}, 2\tfrac{3}{4}).$$

For the minimum value of $f(x)$, $x - 1\frac{1}{2} = 0$.

1 Match the graphs with the equations.

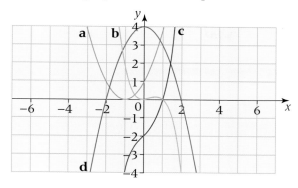

i $y = x^2 + 2x + 1$

ii $f(x) = x^3 + x - 2$

iii $y = 4 - x^2$

iv $g(x) = x^2 - x^3$

2 **a** Copy and complete this table of values for the graph of
$y = 2x^2 + 3x - 6$.

x	−4	−3	−2	−1	0	1	2	3	4
$2x^2$		18							
$3x$		−9							
−6	−6	−6	−6	−6	−6	−6	−6	−6	−6
y		3							

  b Draw suitable axes and plot the graph of $y = 2x^2 + 3x - 6$, joining
the points with a smooth curve.

  c From your graph, write down the coordinates of the minimum
point of $y = 2x^2 + 3x - 6$.

3 Draw graphs of these functions for the range of x-values given.

  a $y = x^2 + 3x$, for $-3 \leqslant x \leqslant 3$ **b** $y = x^2 + x - 2$, for $-3 \leqslant x \leqslant 3$

  c $y = 2x^2 - 3x$, for $-2 \leqslant x \leqslant 5$ **d** $f(x) = 3 - x^2$, for $-3 \leqslant x \leqslant 3$

  e $y = x^3 + x - 4$, for $-2 \leqslant x \leqslant 3$ **f** $f(x) = x^3 - x^2 + 3x$, for $-3 \leqslant x \leqslant 3$

4 **a** Plot the graph $f(x) = x^3 - 2x^2 + x + 4$ for $-3 \leqslant x \leqslant 3$.

  b Use your graph to find

    i the value of x when $y = -20$ **ii** the value of y when $x = 1.7$.

5 A farmer has 100 metres of fencing with which
to make a chicken pen. He wants to build the
pen against the side of his barn, as shown.
The farmer wants to enclose the maximum
possible area in the pen.

BARN

CHICKEN PEN

  a Let the width of the pen be w. Write a formula in terms of w for

    i the length   **ii** the area of the pen.

  b Plot a graph of area against width and use it to find the dimensions
the farmer should use.

This spread will show you how to:

- Recognise the shapes of graphs of quadratic and cubic functions
- Draw graphs of quadratic and cubic functions by plotting points

- Graphs of **functions** involving **reciprocals** have a characteristic shape, called a **hyperbola**.

For example, $f(x) = \frac{1}{x}$

As x gets larger, y gets smaller and vice versa.

If x is 1000, y is $\frac{1}{1000}$.
If x is $\frac{1}{1\,000\,000}$, y is 1 000 000.

x cannot take the value zero, since you cannot evaluate $\frac{1}{0}$. The same is true for y. So the x- and y-axes are **asymptotes** – the graphs will never touch them.

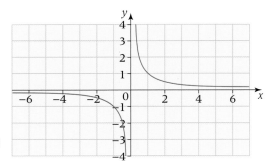

- **Exponential** functions include a term with a variable **index**, for example $f(x) = 2^x$.
  Graphs of exponential functions have a characteristic shape.

As x gets larger, y gets larger and vice versa.

If x is 10, y is 2^{10} or 1024.
If x is -4, y is 2^{-4} or $\frac{1}{16}$.

Since $2^x (= y)$ can never be zero for any value of x, the x-axis is an asymptote, the curve gets near but never actually touches it.

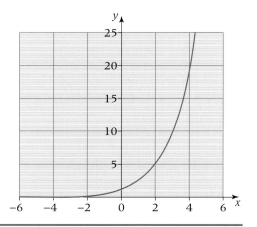

Example

A piece of paper is 1 unit thick. It is folded in half, then in half again, then again repeatedly.

If x is the number of folds and y is the thickness of the paper, form an equation connecting x and y and draw the graph of this equation.

x	0	1	2	3	4	5
y	1	2	4	8	16	32

The thickness doubles each time. After 2 folds, it is 2^2 thick, after 5 folds it is 2^5. Hence the equation is $y = 2^x$.

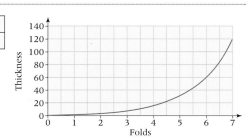

Write the information in a table and spot the pattern.

1 Match each graph with its equation.

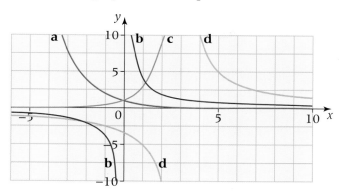

i $\quad y = \frac{4}{x}$

ii $\quad y = 3^x$

iii $\quad y = \frac{10}{x-3}$

iv $\quad y = 2^{-x}$

2 a Copy and complete this table of values for the function $f(x) = \frac{12}{x}$

x	−6	−5	−4	−3	−2	−1	0	1	2	3	4	5	6
f(x)		−2.4					Asymptote				3		

b Draw suitable axes and plot the graph of $f(x) = \frac{12}{x}$, joining your points with a smooth curve.

c What shape is your graph and where are its asymptotes?

d Use your graph to estimate the value of $f(2.5)$.

3 a Without using a calculator, copy and complete this table of values for the function $g(x) = 2^{x+1}$

x	−4	−3	−2	−1	0	1	2	3	4
g(x)		$\frac{1}{4}$					8		

b Draw suitable axes and plot the graph of $g(x) = 2^{x+1}$, joining your points with a smooth curve.

c Approximate the value of x for which $g(x) = 25$.

4 Copy and complete the table of values below and use it to plot the graph of $y = \frac{20}{x} + x - 5$.

x	−4	−3	−2	−1	0	1	2	3	4	5
$\frac{20}{x}$										
−5										
y										

5 Plot these functions for the range of x-values given.

a $y = \frac{12}{x-2}$ for $-2 \leqslant x \leqslant 6$

b $f(x) = 4^{x-2}$ for $-2 \leqslant x \leqslant 6$

c $y = \frac{x}{x+4}$ for $-4 \leqslant x \leqslant 4$

d $f(x) = \frac{6}{x} + x - 2$ for $-3 \leqslant x \leqslant 3$

e $y = 3^{-x} - 1$ for $-4 \leqslant x \leqslant 4$

Solving equations using graphs

This spread will show you how to:

Keywords
Intersection
Simultaneous

- Find the intersection points of graphs of a linear and a quadratic function
- Solve simultaneous equations graphically by drawing a graph to represent each equation

- You can solve **simultaneous** equations graphically by drawing a graph to represent each one. A solution to the simultaneous equations is at a point of **intersection** of the graphs.

For example, for the equations $3x - y = 2$ and $2x + y = 8$, the lines intersect at $(2, 4)$ so the solution is $x = 2$ and $y = 4$.

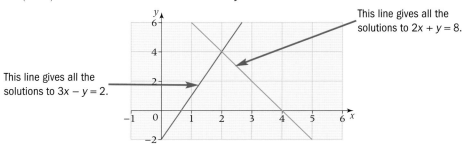

This line gives all the solutions to $2x + y = 8$.

This line gives all the solutions to $3x - y = 2$.

- You can solve some complex equations graphically by splitting them into two equations and treating these as simultaneous equations.

For example, $x^2 - 5x + 6 = 3$ splits into the two equations $y = x^2 - 5x + 6$ and $y = 3$.

Example

Solve $x^2 - 5x + 6 = 3$ graphically.

Plot the graphs of $y = x^2 - 5x + 6$ and $y = 3$ on the same axes.

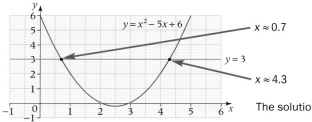

$x \approx 0.7$

$x \approx 4.3$

The solutions are $x \approx 0.7$ and $x \approx 4.3$.

Draw up a table of values to plot $y = x^2 - 5x + 6$.

To solve the equation you need to find the values of x. Read these off the graph.

Example

Solve the equation $5x - x^2 = 2x - 1$ graphically.

Plot the graphs of $y = 5x - x^2$ and $y = 2x - 1$.

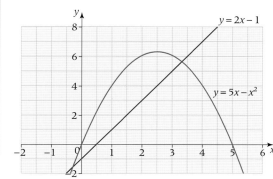

The solutions are $x \approx -0.3$ and $x \approx 3.3$.

The answers are approximate, since you read them off the graph, rather than finding them by a direct algebraic approach.

1 Solve these simultaneous equations graphically.

   a $y = 2x + 1$
   $x + y = 10$

   b $y = 3x - 2$
   $x + y = 2$

   c $2x + y = 5$
   $x - y = 4$

   Confirm your solution to each pair algebraically.

2 **a** Use these graphs to find the approximate solutions of
   these equations.

   i $x^2 - x - 2 = 2$ **ii** $x^2 - x - 2 = -1$ **iii** $x^2 - x - 2 = x + 1$

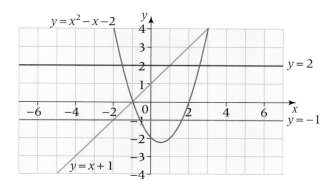

   b Where would you find the solutions to $x^2 - x - 2 = 0$?
   Approximately what are these? Confirm your answer algebraically.

   c Which graph would you need to add to the diagram to
   solve $2 - x = x^2 - x - 2$?

3 Draw graphs to find the approximate solutions of each of these
   equations. Draw all the graphs on one pair of axes for the
   range $-3 \leqslant x \leqslant 4$.

   a $x^2 - 2x - 2 = 0$ **b** $x^2 - 2x - 2 = 2$ **c** $x^2 - 2x - 2 = x + 1$

   d $x^2 - 2x - 2 = 3 - x$ **e** $x^2 - 2x - 2 = \frac{1}{2}x + 1$ **f** $y - x = 1$ and $x + y = 3$

4 Draw appropriate graphs to solve each of these equations.
   For each part of the question draw a new pair of axes for the
   range $-3 \leqslant x \leqslant 4$.

   a $\frac{12}{x} = 2.5$ **b** $2^x = 5$ **c** $3^x = 3x - 2$ **d** $\frac{1}{x - 3} = 5$

5 Using *sketch* graphs only, decide how many solutions each of these
   equations will have.

   a $2^x = x - 1$ **b** $2^{-x} = -3$ **c** $\frac{2}{x} = 4$ **d** $\frac{1}{x} = 2^x$

This spread will show you how to:

- Find the intersection points of graphs of a linear and quadratic function
- Solve complex equations graphically by splitting them into two equations

Keyword
Function
Intersection
Transform

You can solve $x^2 - 5x + 7 = 2$ by drawing the graphs of the **functions** $y = x^2 - 5x + 7$ and $y = 2$ and finding the x-value(s) at their point(s) of **intersection**.

You can use the same graphs to solve other equations, such as $x^2 - 8x + 1 = 0$.

To **transform** $x^2 - 8x + 1$ to $x^2 - 5x + 7$, you need to add $3x + 6$.

Adding $3x + 6$ to both sides of $x^2 - 8x + 1 = 0$ gives $x^2 - 5x + 7 = 3x + 6$.

Plot the graph of $y = 3x + 6$ and see where the two graphs intersect.

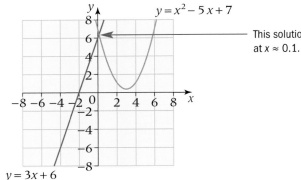

This solution is at $x \approx 0.1$.

Example

Given the graph of $y = 3x^2 + 2x - 4$, what graph do you need to draw to solve

a $3x^2 + 2x - 4 = 3x - 1$ **b** $3x^2 + 6x = 0$?

By drawing the graphs, solve the equations.

a Draw the graph of $y = 3x - 1$.
b To transform $3x^2 + 6x$ to $3x^2 + 2x - 4$ you subtract $4x$ and subtract 4.
  Doing the same to both sides: ① $3x^2 + 6x - 4x - 4 = 0 - 4x - 4$
                                    ② $3x^2 + 2x - 4 = -4x - 4$

Draw the graph of $y = -4x - 4$.

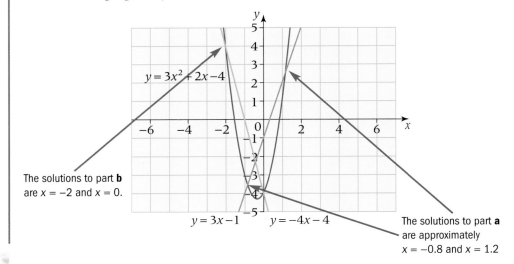

The solutions to part **b** are $x = -2$ and $x = 0$.

The solutions to part **a** are approximately $x = -0.8$ and $x = 1.2$.

1 Use the graphs to find approximate solutions of the these equations.

a $x^2 + 2x - 3 = 0$ **b** $x^2 + 2x - 3 = x + 1$

c $x^2 + 2x - 5 = 0$ **d** $x^2 + x = 0$

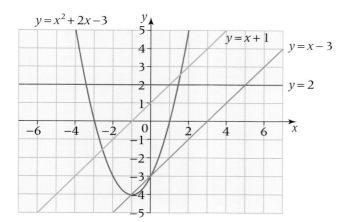

2 Which graph, if any, would you need to add to the grid in question **1** in order to solve each of these equations?

a $x^2 + 2x = 3$ **b** $x^2 + x = 5$ **c** $2x - 3 = 0$

3 If you have the graph of $y = x^2 + 4x - 2$, which one line would you need to draw in order to solve each of these equations?

> You do not need to draw them.

a $x^2 + 4x - 2 = 3$ **b** $x^2 + 4x - 2 = 0$ **c** $x^2 + 4x - 2 = 2x + 1$

d $x^2 + 4x = 6$ **e** $x^2 + 5x = x + 4$ **f** $x^2 + 2x - 3 = 6x$

4 If you have the graph of $y = x^3 + 2x^2 + 5x - 1$, which graph would you need to draw in order to solve each of these equations?

a $x^3 + 2x^2 + 5x - 1 = 2x - 1$ **b** $x^3 + 2x^2 + 2 = 0$

c $x^3 + 2x^2 + 3x = 4$ **d** $x^3 + 2x^2 + 5x - 1 = 0$

e $x^3 + x^2 + 5x = 0$ **f** $x(x + 5) = 0$

5 **a** Draw the graph $f(x) = 2^x$ for $-4 \leqslant x \leqslant 4$.

     b Draw suitable graphs to solve

       i $2^x = 3$ **ii** $2^x + 4x = 2$ **iii** $2^x - x^2 = 0$

6 **a** Draw the graph $f(x) = \frac{24}{x} + 1$ for $-4 \leqslant x \leqslant 4$.

     b Draw suitable graphs to find the approximate solutions of

       i $\frac{24}{x} = 3$ **ii** $\frac{24}{x} = x^2$

     c What graph could you draw in order to find the approximate value of $\sqrt{24}$?

This spread will show you how to:

- Recognise the equation of a circle
- Solve two simultaneous equations representing a line and a circle graphically

Keywords
Intersect
Origin
Radius
Simultaneous

- A circle with centre (0, 0) and **radius** r has equation $x^2 + y^2 = r^2$
- You can solve two **simultaneous** equations representing a line and a circle graphically.

Example

Solve $x^2 + y^2 = 16$ and $y = 2x - 1$ graphically.

$x^2 + y^2 = 16$ is a circle, centre the **origin**, radius 4.
$y = 2x - 1$ is a straight line with gradient 2 and y-intercept $(0, -1)$.

Compare with $x^2 + y^2 = r^2$.

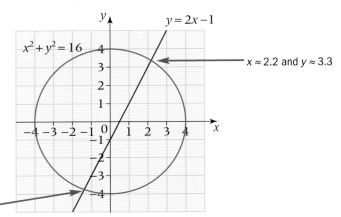

$x \approx 2.2$ and $y \approx 3.3$

$x \approx -1.2$ and $y \approx -3.8$

Read off the approximate x- and y-values where the graphs **intersect**. Give them in pairs as they include x and y.

Example

Solve $x^2 + y^2 = 25$ and $y = x^2 - 2$ **a** graphically   **b** algebraically.

a The solutions are $x \approx 2.5$, $y \approx 4.3$ and $x \approx -2.5$, $y \approx 4.3$.

b $y = x^2 - 2$, so $x^2 = y + 2$

Substitute $x^2 = y + 2$ into the equation of the circle

$y + 2 + y^2 = 25 \implies y^2 + y - 23 = 0$

$y = \dfrac{-1 \pm \sqrt{1^2 + 4 \times 1 \times 23}}{2 \times 1} = \dfrac{-1 \pm \sqrt{93}}{2}$

$y = 4.321\,825\,381\ldots$ or $-5.321\,825\,381\ldots$

From the diagram, $y = -5.321\,825\,381$ is impossible in this case.
Hence, $y = 4.3$ (to 1 dp).

Since $x^2 = y + 2$, then $x^2 = 6.3218\ldots$,
so $x = \pm 2.5$ (to 1 dp).

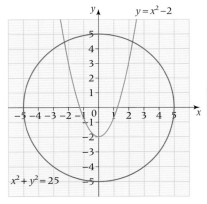

Solve the quadratic using the formula.

1 The diagram shows the circle $x^2 + y^2 = 9$ and the line $y = x + 1$.
Use the diagram to find the approximate solution of the
simultaneous equations

$x^2 + y^2 = 9$
$y = x + 1$

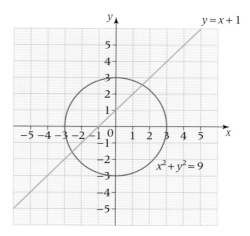

2 Determine the number of solutions
to the simultaneous equations by
imagining an extra graph on the
diagram that currently shows
$x^2 + y^2 = 16$.

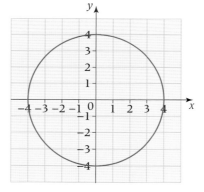

a $x^2 + y^2 = 16$ **b** $x^2 + y^2 = 16$
    $y = 2$ $y = x$

c $x^2 + y^2 = 16$ **d** $x^2 + y^2 = 16$
    $x = 4$ $y = x^2$

e $x^2 + y^2 = 16$ **f** $x^2 + y^2 = 16$
    $y = 5$ $y = x^3$

g $x^2 + y^2 = 16$ **h** $x^2 + y^2 = 16$
    $x + y = 4$ $y = 4 - x^2$

3 By drawing a suitable graph in each case, find approximate solutions
to these simultaneous equations.

   a $x^2 + y^2 = 25$ **b** $x^2 + y^2 = 4$ **c** $x^2 + y^2 = 36$ **d** $x^2 + y^2 = 1$
        $y = 3$ $y = x$ $y = 3x - 1$ $y = x^2$

4 For each pair of simultaneous equations, find their solutions
graphically and then confirm them algebraically.

   a $x^2 + y^2 = 16$ **b** $x^2 + y^2 = 25$ **c** $x^2 + y^2 = 9$ **d** $x^2 + y^2 = 49$
        $y = 2x$ $y = x$ $y = 3x - 1$ $y = x^2$

5 Is it possible to have three intersections between a circle and another
function? Give an example to support your findings.

Key objectives

- Generate points and plot graphs of general quadratic functions, simple cubic functions, the reciprocal function $y = \frac{1}{x}$ with $x \neq 0$, the exponential function $y = k^x$ for integer values of x and simple positive values of k and the circular functions

- Find the intersection points of the graphs of a linear and quadratic function, knowing that these are the approximate solutions of the corresponding simultaneous equations representing the linear and quadratic functions

- Find graphically the intersection points of a given straight line with this circle and know that this corresponds to solving the two simultaneous equations representing the line and the circle

1 Match the equations to the shape of graph they represent. (4)

   a $y = x^3 - x + 2$ **b** $4 = 9x^2 + 9y^2$ **c** $xy = 2$ **d** $x^2 = 2 + y$

   i parabola   **ii** cubic   **iii** hyperbola   **iv** circle

2 a Find the equation of the straight line which passes through the point (0, 3) and is perpendicular to the straight line with equation $y = 2x$. (2)

The graphs of $y = 2x^2$ and $y = mx - 2$ intersect at the points A and B. The point B has coordinates (2, 8).

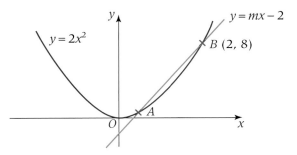

b Find the coordinates of the point A. (2)

(Edexcel Ltd., 2003)

This unit will show you how to

- Understand direct and inverse proportion
- Set up and use equations to solve problems involving direct and inverse proportion
- Relate algebraic solutions to graphical representations of the equations of the original problem
- Recognise the characteristic shape of quadratic, cubic, exponential and reciprocal functions
- Discuss and interpret graphs modelling real situations

Before you start ...

You should be able to answer these questions.

Review

1 a If 7 litres of petrol cost £13, how much will 9 litres cost?

Unit N4

b It takes 9 hours to paint 4 rooms. How long would it take to paint 7 rooms?

2 a If it takes 3 men 4 days to build a new home, how long would it take 2 men?

Unit N4

b It takes 3 minutes for 8 women to eat a large bag of chocolate. How long would it take 5 women to eat the same amount of chocolate, assuming they eat at the same rate?

3 a Given that $y = 4x$, find
 i y when $x = 2.5$ **ii** x when $y = 22$.

Unit A5

b Given that $y = 3x^2$, find
 i y when $x = 4$ **ii** x when $y = 75$.

c Given that $y = \frac{12}{x}$, find
 i y when $x = 8$ **ii** x when $y = 10$.

4 Sketch a graph to show each situation.

Unit A8

a Your height as you age from a baby to a twenty year old.
 (*x*-axis time, *y*-axis height.)

b The temperature of a cup of tea as it is left to stand for half an hour.
 (*x*-axis time, *y*-axis temperature.)

Distance–time graphs

This spread will show you how to:

● Draw and interpret distance–time graphs

● Understand and use compound measures, including speed

Keywords

Distance
Gradient
Speed

● You can represent a journey on a distance–time graph.

● Time is always plotted on the horizontal axis.

● Distance is plotted on the vertical axis.

This graph shows Dan's journey on his bike.

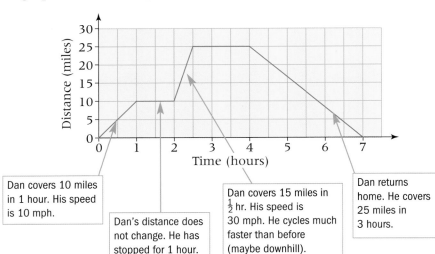

Dan covers 10 miles in 1 hour. His speed is 10 mph.

Dan's distance does not change. He has stopped for 1 hour.

Dan covers 15 miles in $\frac{1}{2}$ hr. His speed is 30 mph. He cycles much faster than before (maybe downhill).

Dan returns home. He covers 25 miles in 3 hours.

Example

Janine leaves home at 1 p.m. and cycles to her friend's house, 30 km away, at a speed of 20 km/h. She stays for 2 hours, then cycles home, arriving at 6 o'clock.

Draw a distance-time graph to represent the journey and determine her speed on the way home and her average speed for the entire journey.

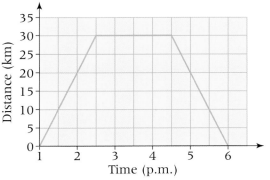

Janine returning home, is shown by the graph going down, back to the x-axis

On the journey home, Janine covers the 30 km in $1\frac{1}{2}$ hours. This means that she has covered 10 km in each half hour and, hence, 20 km in each hour. Her speed is 20 km/h.

Since she covers 60 km in 5 hours, her average speed for the whole journey, including the stop, is 12 km/h.

average speed = $\dfrac{\text{total distance}}{\text{total time}}$

1 The distance–time graph shows the journey of a car between Birmingham and Stoke-on-Trent.

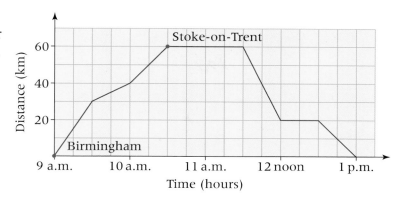

  a How far is it from Birmingham to Stoke-on-Trent?

  b For how long did the car stop?

  c What was the speed of the car for the first part of the journey?

  d Between which two times was the car travelling fastest?

  e What was the average speed of the car for the whole journey?

2 Three students have drawn distance–time graphs. Two have made mistakes. Which two students have made a mistake and what mistake is it?

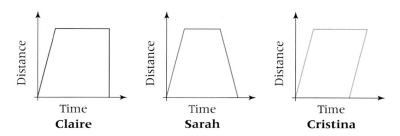

3 Construct a distance–time graph to show each journey.

  a A car travels between Bristol and London. On the outward journey, it travels the 120 miles to London in $2\frac{1}{4}$ hours. The driver remains in London for $1\frac{1}{2}$ hours. The car travels half way back to Bristol at 40 miles per hour, as the motorway is busy, then the remaining distance at 80 miles per hour.

  b Two brothers both went to see each other on the same day. Henry left his home at 2 p.m. to go and see Leo, who lives 5 miles away. Henry walked at an average speed of 4 miles per hour, but he stopped half way for a 15 minute rest. At 2:30 p.m., Leo set out on his bicycle from his home in order to go and visit Henry. He cycled straight there in $\frac{1}{2}$ hour.

  Draw the two journeys on one graph.

4 The following graph represents the journey of a car. Construct a graph of speed (km/h) against time (h) for this journey.

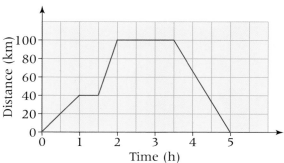

Other real-life graphs

This spread will show you how to:
- Draw and interpret graphs modelling real-life situations

Keyword
Model

You can use graphs to **model** the depth of water flowing in or out of a container at a constant rate.

Imagine water filling this container.

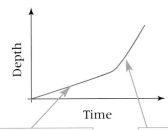

The bottom part of the container has the same diameter, so fills at a steady rate

The top part starts wide then narrows. As the container gets narrower, it fills faster.

- In a real-life graph involving time, time is usually represented by the *x*-axis.

Example

Sketch a graph to show what happens as

1. Sam fills a bath with both taps running
2. He realises it is too hot so turns off the hot tap
3. He turns off the cold tap
4. He gets in
5. Has a long soak
6. Gets out
7. Pulls out the plug.

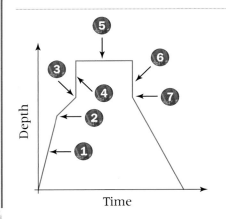

Label the axes with the quantities they represent. A scale is not needed for a sketch graph.

1 Match the four sketch graphs with the containers.

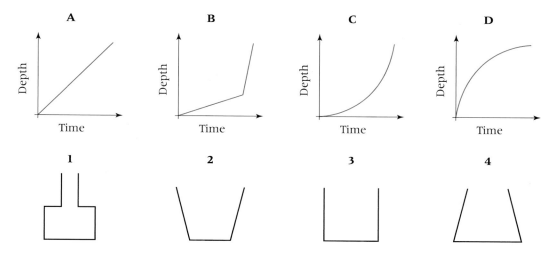

2 The sketch graph shows Andrew's height from 3 to 23 years of age. Explain what the graph shows at each stage and explain why this might be.

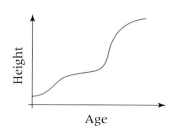

3 Sketch a graph of depth against time as this container is filled.

4 Construct sketch graphs to represent these situations.

a A woman is pregnant and puts on weight. When her baby arrives, she finds it difficult to lose any of the weight for six months, but then joins an exercise class and eats healthily. She is back to her natural weight a year and a half after becoming pregnant. (Graph is weight against time.)

b A frozen chicken is taken out of the freezer and left to defrost. Two hours later it is put in the microwave briefly to speed up and finish off the process. The chicken is then put in the oven to roast for Sunday lunch. (Graph is temperature against time.)

5 A skateboarder at a skate park uses a ramp, as shown. For one go on the ramp, construct a sketch graph of

a speed against time

b acceleration against time.

Direct proportion

This spread will show you how to:

- Understand direct and inverse proportion
- Set up and use equations to solve problems involving direct proportion

Keywords
Proportional to ($\propto$)
Constant of
    proportionality
Direct proportion
Varies

- Two quantities are in **direct proportion** if one quantity increases at the same rate as the other.

For example, pay is directly proportional to hours worked and the cost of tickets is directly proportional to number of people on the trip.

- If P and h are directly proportional (P **varies** with h) write $P \propto h$
- If $P \propto h$ then $P = kh$ for some constant k. You can find k if you know some corresponding values of P and h.

For example, if you work 5 hours for £20, $P = 20$ and $h = 5$, so $k = 4$, the hourly rate of pay.

$\propto$ means 'is proportional to'

k is the **constant of proportionality**
$P = kh$ is a linear relationship.

Example

a Given that y is directly proportional to x and that y is 15 when x is 6, find a formula for y in terms of x.
b Use your formula find y when x is 10.

a $y \propto x$, so $y = kx$
   When y is 15, x is 6: $15 = 6k$
                                $2.5 = k$
   Hence $y = 2.5x$
b When x is 10, $y = 2.5 \times 10 = 25$

Use the values given to find the constant of proportionality.

A formula for y in terms of x means $y = ...$

Example

The time, t minutes, I spend doing my homework varies directly with the number of pieces, n, that I have to complete.
If 3 pieces take 2 hours, how long will 7 pieces take?

$t \propto n$ so $t = kn$
When $n = 3$, $t = 120$ min: $120 = 3k$
                                $k = 40$
When $n = 7$, $t = 40 \times 7$
                 $t = 280$ min or 4 hours 40 min

'Varies directly' is another way of saying 'is directly proportional to.

You could use the unitary method:
3 pieces take 120 min,
so 1 piece takes 40 min,
so 7 pieces take 280 min.

Example

On a map the distance between junctions 9 and 10 on the M4 is 46 cm. The real distance is 11.5 km. Wraysbury Reservoir is $2\frac{1}{4}$ km long. How long would it appear to be on the map?

Map distance $m \propto$ real distance r so $m = kr$
When $m = 46$, $r = 11.5$ $46 = 11.5k$

Hence, $m = 4r$ $k = \frac{46}{11.5} = 4$
When $r = 2.25$
        $m = 4 \times 2.25 = 9$ cm

1 Paul has a 'Pay as you go' mobile phone. For every hour of calls he pays £5.

a Copy and complete the table to show the cost, C (£) for different numbers of hours of calls, h.

h	1	2	3	4	5	6
c						

b Explain what $C \propto h$ means and why it is true for Paul's phone bill.

c Write an algebraic relationship between C and h and use it to find

  i the cost for $17\frac{1}{2}$ hours of calls

  ii the hours spent on calls if the cost is £51.25.

d Which graph best represents the relationship between C and h?

  i 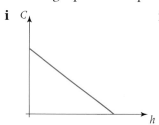   **ii** 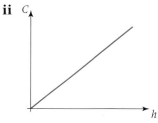   **iii**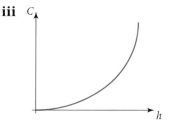

2 A varies directly with b.

a Given that b is 10 when A is 35, find a formula for A in terms of b.

b Use your formula to find

  i the value of A when b is 13   **ii** the value of b when A is 70.

3 It is known that q is directly proportional to t.

a Given that t is 12 when q is 17, find a formula for q in terms of t.

b Use your formula to find

  i the value of q when t is 3.6   **ii** the value of t when q is 13.

4 The distance (d) in miles that a car travels at constant speed is directly proportional to the length of time (t) it travels.

a Given that the car covers 25 miles in $\frac{1}{2}$ hour, find a formula connecting d and t.

b Use your formula to find the distance travelled in 4 hours.

c Use your formula to find the time taken to travel 100 miles.

d Sketch a graph of d against t.

5 In a circuit, the voltage V across a component varies directly with the resistance R of the component. If the voltage is 2.38 volts when the resistance is 3.4 ohms, find a formula connecting V and R and use it to find the voltage when the resistance is 10.2 ohms.

More direct proportion

This spread will show you how to:

- Set up and use equations to solve problems involving direct and inverse proportion
- Relate algebraic solutions to graphical representations of the equations of the original problem

Keywords
Proportional to ($\propto$)
Proportional
Varies

- One quantity can be **proportional** to the square of another quantity.

For example, consider a cuboid of length 5 m and square cross-section which **varies** in size:

 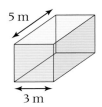 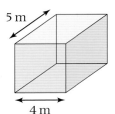

Volume of a cuboid = Area of cross-section × Length
So the volume is proportional to the area of the cross-section.

$$V \propto A \quad \text{or} \quad V = kA \text{ ... in this case, } V = 5A$$

Also, the area of the cross-section is the square of its side length.

$$V \propto l^2 \quad \text{or} \quad V = kl^2 \text{ ... in this case, } V = 5l^2$$

- Similarly, a quantity can be proportional to the cube of another quantity, $y \propto x^3$, or to the square root of another quantity, $y \propto \sqrt{x}$.

Example

Given that y varies directly with the square of x and that y is 75 when x is 5, find y when x is 15.

$y \propto x^2$ so $y = kx^2$
When $y = 75$, $x = 5$ $75 = k \times 5^2$
                              $75 = 25k$
                              $k = 3$
Hence, $y = 3x^2$
When x is 15, $y = 3 \times 15^2 = 675$

Use the values given to find the constant of proportionality, k.

Exam Question

In a factory, chemical reactions are carried out in spherical containers. The time, T minutes, the chemical reaction takes is directly proportional to the square of the radius, R cm, of the spherical container.
When $R = 120$, $T = 32$.
Find the value of T when $R = 150$.

$T \propto R^2$ so $T = kR^2$
When $R = 120$, $T = 32$, so $32 = k \times 120^2$
                              $k = 0.002\ 222\ 2 \ldots$
Hence, $T = 0.002\ 22 \ldots \times 150^2$
         $T = 50$

(Edexcel Ltd., 2004)

1 Norman thinks of a number, squares it and then multiplies the answer by 4.

   a Copy and complete the table to show Norman's starting number, x and his final answer, y.

x	1	2	3	4	5	6
x^2						
y						

   b Explain what $y \propto x^2$ means and why it is true for Norman's numbers.

   c Write an algebraic relationship between y and x and use it to find Norman's starting number when his final answer is 784.

   d Which graph best represents the relationship between y and x?

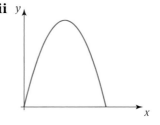

2 R varies with the square of s. If R is 144 when s is 1.2, find

   a a formula for R in terms of s

   b the value of R when s is 0.8

   c the value of s when R is 200.

3 For a circle, explain how you know that the area is directly proportional to the square of its radius. State the value of the constant of proportionality in this case. What other formulae do you know that show direct proportion?

4 The surface area of a sphere varies with the square of its radius. If the surface area of a sphere with radius 3 cm is 113 cm^2, find the surface area of a sphere with radius 7 cm.

5 Given that y varies with the cube of x and that y is 16 when x is 2, find

   a a formula connecting y and x

   b the value of x when y is 128.

6 Given that P is directly proportional to the square root of k and that P is 20 when k is 16, find the value of P when k is 100.

Inverse proportion

This spread will show you how to:

- Set up and use equations to solve problems involving direct and inverse proportion
- Relate algebraic solutions to graphical representations of the equations of the original problem

---

- Two quantities are in **inverse proportion** if one quantity decreases at the same rate as the other increases.

A classic example is 'men digging a trench'. The more men you have, the shorter the time taken to dig the trench.
If it takes 1 man 10 days to dig a trench, it takes 2 men 5 days, 4 men 2.5 days, 10 men 1 day, etc.

- If D and m are inversely proportional (D varies inversely with m) you can write $D \propto \frac{1}{m}$.

- If $D \propto \frac{1}{m}$ then $D = \frac{k}{m}$ for some constant k. You can find k if you know some corresponding values of D and m.

$D = \frac{k}{m}$ is a reciprocal relationship.

Example

T is indirectly proportional to r and when T is 5, r is 4. Form a relationship between T and r and use it to find T when r is 40.

If it takes 2 men 5 days, $5 = k \times \frac{1}{2}$, so $k = 10$.

$T \propto \frac{1}{r}$ so $T = \frac{k}{r}$.
When T is 5, r is 4 $5 = \frac{k}{4}$
                             $k = 20$

The relationship is $T = \frac{20}{r}$.
When $r = 40$, $T = 20 \div 40$ so $T = \frac{1}{2}$.

'Indirectly proportional' means the same as 'inversely proportional'.

Example

The wavelength, w metres, of a radio wave is inversely proportional to its frequency, f kiloHertz.
The frequency of a wave of length 10 m is 30 000 kHz.
Find   **a**   the frequency for a wavelength of 100 m
          b   the wavelength when the frequency is 50 kHz.

a $w \propto \frac{1}{f}$ so $w = \frac{k}{f}$

     When $w = 10$, $f = 30\,000$ $10 = \frac{k}{30\,000}$
                                           $k = 300\,000$

     When $w = 100$, $100 = \frac{300\,000}{f}$

     $f = 3000$ kHz

b   When $f = 50$, $w = \frac{300\,000}{50}$

     $w = 6000$ m

1 On a journey, the speed at which a car travels is inversely proportional to the time taken.

   a Explain what this means using symbols and an example.

   b Copy and complete this table of information for a journey from Birmingham to Manchester undertaken at varying speeds.

Time t (hours)	1	2	4	5	8
Speed S (mph)		50		20	

   c Transform $S \propto \frac{1}{t}$ into an algebraic formula connecting S and t. Use your formula to find the speed at which it takes $3\frac{1}{2}$ hours to travel to Manchester.

   d Which graph best represents the relationship between t and S?

   i   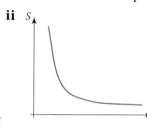   ii   iii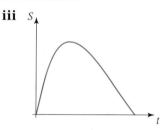

2 A slab of chocolate is being eaten by a group of girls. If just one of the girls eats the chocolate on her own, it takes 4 hours.

   a Draw up a table to show how many hours it takes 1, 2, 3, 4, 5 or 6 girls to eat the chocolate.

   b Plot a graph of time t against the number of girls n.

   c Explain why $t \propto \frac{1}{n}$.

3 Given that y is inversely proportional to x, and that y is 3 when x is 10

   a find a formula for y in terms of x

   b use your formula to find

      i   the value of y when x is 5

      ii the value of x when y is 60.

4 A varies indirectly with c. If A is 50 when c is 2, find

   a an algebraic relationship between A and c

   b the value of

      i A when c is 50

      ii c when A is 400.

More inverse proportion

This spread will show you how to:

- Set up and use equations to solve problems involving direct and inverse proportion
- Relate algebraic solutions to graphical representations of the equations of the original problem

Keywords

Inverse
  proportion
Proportional
Square
Varies

- One quantity can be **inversely proportional** to the **square** of another quantity.

For example, the number of tiles needed to cover a wall is related to the area of the tile.
As the size of the tile gets larger, the number required gets smaller.
The number of tiles is inversely proportional to the area of each tile.
Also, the area of a tile is **proportional** to the square of its length, l^2.

So the number of tiles, N is inversely proportional to l^2: $N \propto \dfrac{1}{l^2}$ or $N = \dfrac{k}{l^2}$.

- Similarly, a quantity can be inversely proportional to the cube of another quantity $y \propto \frac{1}{x^3}$, or to the square root of another quantity, $y \propto \dfrac{1}{\sqrt{x}}$.

Example

Given that y is inversely proportional to the square of x and that y is 4 when x is 5, write a formula for y in terms of x.

$y \propto \dfrac{1}{x^2}$ so $y = \dfrac{k}{x^2}$

When x is 5, y is 4 $4 = \dfrac{k}{5^2}$

$$k = 4 \times 5^2 = 100$$

Hence, $y = \dfrac{100}{x^2}$

Example

The force of attraction, F, between two magnets is inversely proportional to the square of the distance, d, between them.
Two magnets are 0.5 cm apart and the force of attraction between them is 50 newtons.
When the magnets are 3 cm apart what will be the force of attraction between them?

$F \propto \frac{1}{d^2}$, so $F = \frac{k}{d^2}$

When F is 50, d is 0.5 $50 = \frac{k}{0.5^2}$

$$k = 50 \times 0.5^2$$
$$= 12.5$$

So $F = \frac{12.5}{d^2}$

When $d = 3$ $F = \frac{12.5}{3^2}$
$$= 1.4 \text{ newtons}$$

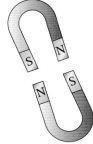

1 Imogen carries out a scientific experiment. She sets two magnets at varying distances apart and notices that the force between them decreases as the square of the distance increases. She performs the experiment three times. These are her results.

Distance d (cm)	1	2	3	4	5
d^2	1	4	9	16	25
Force F (newtons)	100	25	?	?	4

    a Study Imogen's results. Write a formula for F in terms of d and use it to complete the rest of the table.

    b Plot a graph of F against d and join the points to form a smooth curve.

2 Match the four sketch graphs with the four proportion statements.

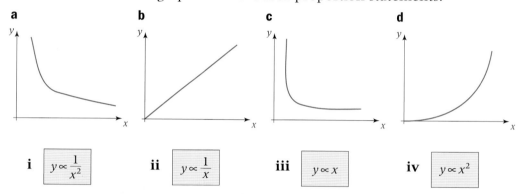

  a   **b**   **c**   **d**

  i $y \propto \dfrac{1}{x^2}$ **ii** $y \propto \dfrac{1}{x}$ **iii** $y \propto x$ **iv** $y \propto x^2$

3 A carpet layer is laying square floor tiles in two identical apartments. The time t (mins) that it takes to lay the tiles is inversely proportional to the square of the length of each tile.
Given that tiles with a side length of 20 cm take 4 hours and 10 minutes to lay, how long will the second apartment take using tiles with side length 40 cm?

4 Given that y is indirectly proportional to the square root of x and that y is 5 when x is 4, find

    a a formula for y in terms of x

    b the value of

      i y when x is 100   **ii** x when y is 12.

5 Given that P varies inversely with the cube of t and that P is 4 when t is 2, find

    a a formula for P in terms of t

    b without a calculator

      i P when t is 3   **ii** t when P is $\frac{1}{2}$.

This spread will show you how to:

- Recognise the characteristic shape of quadratic, cubic, exponential and reciprocal functions
- Discuss and interpret graphs modelling real situations

You can recognise some functions from the shapes of their graphs.

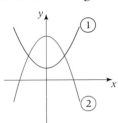 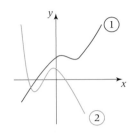 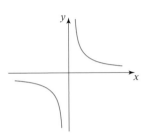 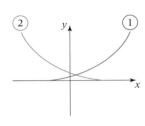

Quadratic functions-
parabola or U shape
①　$x^2 > 0$
②　$x^2 < 0$

Cubic functions-
S shape
①　$x^3 > 0$
②　$x^3 < 0$

Reciprocal functions
- hyperbola
For $y = \frac{1}{x}$, the axes
are asymptotes.

Exponential functions
$f(x) = a^x$
The x-axis is an asymptote.
①　$f(x) = a^x$
②　$f(x) = a^{-x}$

- You can use curves to model some real-life situations.

For example, an exponential curve is often a suitable model for population growth or the half-life of a radioactive isotope as it decays.

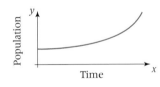 

Example

The graph shows the population P of a village as it grows over time t. After a year, the population of the village is 120 people and after 2 years it is 144 people. Given that $P = ab^t$, find the values of a and b and the time when the village's population will exceed 500.

$P = ab^t$
At (1, 120)　$120 = a \times b^1$ or $ab = 120$　①
At (2, 144)　$144 = a \times b^2$ or $ab^2 = 144$　②
From ①, $a = \frac{120}{b}$. Substituting in ②: $\frac{120}{b} \times b^2 = 144$
$120b = 144$
$b = 1.2$
Hence $a = 120 \div 1.2 = 100$, so $P = 100 \times 1.2^t$.
When $t = 9$, $P = 100 \times 1.2^9 = 515.978$.
The population will exceed 500 after 9 years.

When $t = 1$, $P = 120$.

When $t = 2$, $P = 144$.

Use trial and improvement to find the value of t such that $P > 500$.

1 Match each of the functions with its graph.

i $y = \frac{1}{x-2}$

ii $f(x) = x(1-x^2)$

iii $y = 4^x$

iv $f(x) = x^2 + x - 2$

v $y = x^3 + x^2 - 6x$

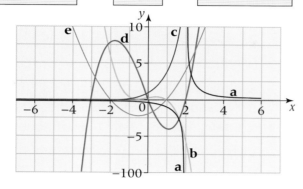

2 Bacteria reproduce by splitting in two. If you begin with one bacterium and this splits, after 1 minute, to make two bacteria and these split, after a further minute, to make four bacteria and so on, sketch a graph of the number of bacteria (N) against time (t). Sketch a graph for a bacterium that splits into 3 to reproduce.

DID YOU KNOW?

Bacteria with an average generation (doubling) time of 20 minutes can produce 1 billion new cells in just 10 hours.

3 The population P, over time t, of a herd of elephants living in a Thai jungle is modelled using the formula $P = 5 \times 1.5^t$

a What is the population of the herd initially (when $t = 0$)?

b What is the population of the herd after   **i** 1 year   **ii** 2 years?

c Find, by trial and error, the time at which the population will exceed 20 elephants.

d Sketch a graph of P against t.

4 The graph shows the profits P of a company over time t. The profit is modelled using the equation $P = mn^t$.

a Use the information in the graph to find the values of m and n.

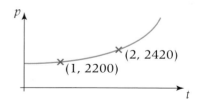

b When will the profits reach £5000?

5 Here is a function: $y = 2^{-x}$

a Draw a table of values for $-2 \leqslant x \leqslant 2$ and sketch a graph of $y = 2^{-x}$.

b Use your calculator to explore what happens to y for very large positive and negative values of x.

Use the power key on your calculator. This is often $\boxed{x^y}$ or $\boxed{\wedge}$.

c Describe what would happen to your graph if it were extended from $x = -100$ to $x = 100$.

d How would your results differ if the original function was $y = 2^x$?

A9

Exam review

Key objectives

- Set up and use equations to solve word and other problems involving direct proportion or inverse proportion and relate algebraic solutions to graphical representation of the equations

- Plot graphs of simple cubic functions, the reciprocal function, the exponential function, and the circular functions, and recognise the characteristic shapes of all these functions

- Discuss and interpret graphs modelling real situations

1 James is filling spherical water balloons with water.
The time taken, T, to fill a balloon in seconds is directly proportional to the cube of its resulting radius, r cm.
A water balloon of radius 2 cm takes 6 seconds to fill.
How long would it take to fill a water balloon of radius 5 cm?　(4)

2 a b 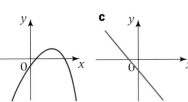 c d 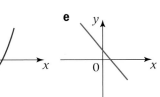 e

f g 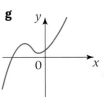 h 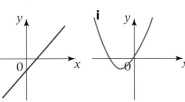 i

Write the letter of the graph which could have the equation　(3)

i $y = 3x - 2$　　**ii** $y = 2x^2 + 5x - 3$　　**iii** $y = \frac{3}{x}$

(Edexcel Ltd., 2004)

This unit will show you how to

- Understand and use vector notation and the associated vocabulary
- Describe a translation by a vector
- Represent, add and subtract vectors graphically
- Work with column vectors
- Understand and use the commutative properties of vector addition
- Calculate a scalar multiple of a vector and represent it graphically
- Understand that vectors represented by parallel lines are multiples of each other
- Understand how the sign of a vector relates to its direction
- Solve simple geometric problems in 2-D using vector methods

Before you start ...

You should be able to answer these questions.

1 Copy and complete this table to show the properties of quadrilaterals. (One has been completed for you.)

Review

Unit S4

	Square	Rhombus	Rectangle	Parallelogram	Trapezium	Kite
1 pair opposite sides parallel				✓		
2 pair opposite sides parallel				✓		
Opposite sides equal				✓		
All sides equal						
All angles equal						
Opposite angles equal				✓		
Diagonals equal						
Diagonals perpendicular						
Diagonals bisect each other				✓		
Diagonals bisect the angle						

2 Copy these regular polygons and draw all lines of symmetry.

Unit S4

a 　b 　c

This spread will show you how to:

- Understand and use vector notation and the associated vocabulary
- Describe a translation by a vector

Keywords
Displacement
Parallel
Pythagoras's
  theorem
Vector

You describe a translation by a **vector**.

You specify the distance moved left or right, then up or down.

The vector $\begin{pmatrix} 4 \\ 3 \end{pmatrix}$ takes ABC to A'B'C'.

The vector $\begin{pmatrix} -4 \\ -3 \end{pmatrix}$ takes A'B'C' to ABC.

You can represent the translation $\begin{pmatrix} 4 \\ 3 \end{pmatrix}$ by an arrowed line parallel to AA'.

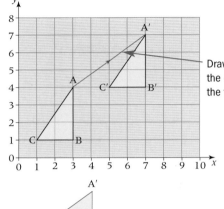

Draw an arrowed line to show the direction and distance of the translation.

Lines joining B and B' or C and C', are **parallel** to AA' and the same length.

You can find the length of the vector taking A to A' using Pythagoras's theorem.
length AA' = $\sqrt{3^2 + 4^2}$ = 5

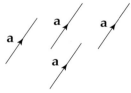

- A vector is a **displacement** that has a fixed length in a fixed direction.

You can draw a vector as an arrowed line. Its position gives the direction of movement, its length gives the distance.

These lines are all parallel and the same length.

They all represent the same vector **a**.

You can tie a vector to a starting point.

The line PQ represents the vector $\overrightarrow{PQ}$ = **p**.

Note that $\overrightarrow{QP}$ = −**p**

A vector does not need to be tied to a fixed starting point.

The line ST represents the vector $\overrightarrow{ST}$ = **s**.

Note that $\overrightarrow{TS}$ = −**s**

The vector −**a** is parallel, the same length, in the opposite direction to **a**.

- Vectors can be described using the notation $\overrightarrow{AB}$ or bold type, **a**.

In handwriting, vectors can be shown with an underline, a.

The arrow shows the direction is from A to B.

1 Draw these vectors on squared paper.

a $\begin{pmatrix} 4 \\ 3 \end{pmatrix}$ b $\begin{pmatrix} 2 \\ 5 \end{pmatrix}$ c $\begin{pmatrix} -1 \\ 4 \end{pmatrix}$ d $\begin{pmatrix} -3 \\ -3 \end{pmatrix}$ e $\begin{pmatrix} 0 \\ 2 \end{pmatrix}$ f $\begin{pmatrix} -4 \\ 0 \end{pmatrix}$

2 a On squared paper draw the vectors $\begin{pmatrix} 1 \\ -2 \end{pmatrix}$ and $\begin{pmatrix} -1 \\ 2 \end{pmatrix}$.

b Write what you notice about these two vectors.

3 Use $\overrightarrow{AB}$ notation to identify equal vectors in this diagram.

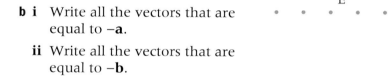

For example,
$\overrightarrow{FG} = \overrightarrow{IJ}$ because
they are parallel, in
the same direction
(left to right) and
the same length.

4 ABCDEF is a regular hexagon.
X is the centre of the hexagon.
$\overrightarrow{XA} = \mathbf{a}$ and $\overrightarrow{AB} = \mathbf{b}$.

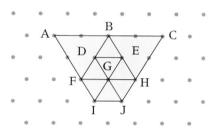

a i Write all the vectors that are
equal to **a**.

ii Write all the vectors that are
equal to **b**.

b i Write all the vectors that are
equal to −**a**.

ii Write all the vectors that are
equal to −**b**.

5 JKLMNOPQ is a regular octagon.
$\overrightarrow{OJ} = \mathbf{j}$ $\overrightarrow{OM} = \mathbf{m}$ $\overrightarrow{OP} = \mathbf{p}$

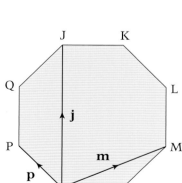

a Write all the vectors that are

i equal to **j**

ii equal to **m**

iii equal to **p**.

b Write all the vectors that are

i equal to −**j**

ii equal to −**m**

iii equal to −**p**.

Working with column vectors

This spread will show you how to:

- Find the magnitude of a column vector
- Work with column vectors

Keywords
Magnitude
Column vector
Position vector

Any vector of the form $\begin{pmatrix} a \\ b \end{pmatrix}$ is called a column vector with a the horizontal displacement and b the vertical displacement.

If $\overrightarrow{KL} = \begin{pmatrix} a \\ b \end{pmatrix}$, you can think of $\begin{pmatrix} a \\ b \end{pmatrix}$ as the instruction that will take you from point K to point L. In this case $\overrightarrow{OK}$ is the position vector of K from the origin O and $\overrightarrow{OL}$ is the position vector of L from the origin O.

$$\overrightarrow{OK} + \overrightarrow{KL} = \overrightarrow{OL}$$

The magnitude of $\overrightarrow{KL} = \begin{pmatrix} a \\ b \end{pmatrix}$ is $|\overrightarrow{KL}| = \sqrt{a^2 + b^2}$.

You can add or subtract column vectors by adding or subtracting the corresponding components.

$$\begin{pmatrix} m \\ n \end{pmatrix} + \begin{pmatrix} p \\ q \end{pmatrix} = \begin{pmatrix} m+p \\ n+q \end{pmatrix}$$

Multiplication by a constant follows the rules of multiplication of brackets.

A vector parallel to $\begin{pmatrix} a \\ b \end{pmatrix}$ but k times its size is $k\begin{pmatrix} a \\ b \end{pmatrix} = \begin{pmatrix} ka \\ kb \end{pmatrix}$.

$\begin{pmatrix} k \\ 0 \end{pmatrix}$ is a vector parallel to the x-axis. $\begin{pmatrix} 0 \\ k \end{pmatrix}$ is a vector parallel to the y-axis.

Example

If A is the point $(2, 3)$, B $(7, -4)$ and $\overrightarrow{AC} = \begin{pmatrix} 5 \\ 4 \end{pmatrix}$ find $\overrightarrow{AB}$, the coordinates of point C and write down the position vector of the midpoint of AB.

$$\overrightarrow{AB} = \begin{pmatrix} 7 & -2 \\ -4 & -3 \end{pmatrix} = \begin{pmatrix} 5 \\ -7 \end{pmatrix}$$

$$\overrightarrow{OC} = \overrightarrow{OA} + \overrightarrow{AC} = \begin{pmatrix} 2 \\ 3 \end{pmatrix} + \begin{pmatrix} 5 \\ 4 \end{pmatrix} = \begin{pmatrix} 7 \\ 7 \end{pmatrix} \Rightarrow C(7, 7)$$

The midpoint of AB has coordinates $\left(\dfrac{2+7}{2}, \dfrac{3+(-4)}{2} \right) = (4.5, -0.5)$

The position vector of the midpoint of AB is $\begin{pmatrix} 4.5 \\ -0.5 \end{pmatrix}$

Example

Find x, y if $\begin{pmatrix} 2 \\ 3 \end{pmatrix} + \begin{pmatrix} x \\ 10 \end{pmatrix} = \begin{pmatrix} -2 \\ y \end{pmatrix}$

$2 + x = -2 \Rightarrow x = -4$ equate horizontal components and solve

$3 + 10 = y \Rightarrow y = 13$ equate vertical components and solve

1 Find the magnitude of these column vectors

a $\begin{pmatrix} 12 \\ 5 \end{pmatrix}$ **b** $\begin{pmatrix} -12 \\ -5 \end{pmatrix}$

2 A is the point $(4, 1)$, B is the point $(-4, -1)$ and C is a point such that $\overrightarrow{AC} = \begin{pmatrix} 6 \\ 2 \end{pmatrix}$.

a Write down $\overrightarrow{AB}$

b Write down $\overrightarrow{BA}$

c Write down the position vector of the midpoint of AB

d Find the coordinates of point C

3 $\mathbf{a} = \begin{pmatrix} 3 \\ 4 \end{pmatrix}$ $\mathbf{b} = \begin{pmatrix} -3 \\ -4 \end{pmatrix}$ $\mathbf{c} = \begin{pmatrix} 7 \\ -7 \end{pmatrix}$

Work out as column vectors

a $\mathbf{a} + \mathbf{b}$ **b** $\mathbf{a} - \mathbf{b}$
c $10\mathbf{a}$ **d** $-10\mathbf{c}$
e $\mathbf{a} + 2\mathbf{b}$ **f** $\mathbf{a} - \mathbf{b} + \mathbf{c}$

4 Find p, q when

a $\begin{pmatrix} 2 \\ 5 \end{pmatrix} - \begin{pmatrix} p \\ -3 \end{pmatrix} = \begin{pmatrix} 10 \\ q \end{pmatrix}$ **b** $\begin{pmatrix} p \\ -3 \end{pmatrix} + 2\begin{pmatrix} 3 \\ q \end{pmatrix} = \begin{pmatrix} 10 \\ 10 \end{pmatrix}$

5 $\mathbf{b} = \begin{pmatrix} -3 \\ -4 \end{pmatrix}$ $\mathbf{c} = \begin{pmatrix} 7 \\ -7 \end{pmatrix}$

a Write down a column vector parallel to $\mathbf{b}$ but twice its length
b Write down a column vector parallel to $\mathbf{c}$ but k times its length

6 $\mathbf{a} = \begin{pmatrix} 3 \\ 4 \end{pmatrix}$ $\mathbf{b} = \begin{pmatrix} -3 \\ -4 \end{pmatrix}$ $\mathbf{c} = \begin{pmatrix} 7 \\ -7 \end{pmatrix}$

Calculate $\mathbf{x}$ when
a $\mathbf{x} + \mathbf{b} = \mathbf{c}$ **b** $\mathbf{a} + 2\mathbf{x} = \mathbf{c}$
c Find the value of p so that the vector $\mathbf{a} - \mathbf{b} + p\mathbf{c}$ is parallel to the y-axis

Combining vectors

This spread will show you how to:

● Represent, add and subtract vectors graphically

Keywords

Resultant vector
Vector

You can add or subtract **vectors** graphically by arranging them 'nose to tail'.

This is the same as combining two translations.

● The **resultant vector** completes the triangle of vectors.

You show resultant vectors with a double arrow.

A resultant vector can be tied to a starting point.

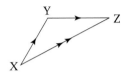

$$\overrightarrow{XY} + \overrightarrow{YZ} = \overrightarrow{XZ}$$

Note the sequence of the letters.

You can add or subtract any number of vectors in this way.

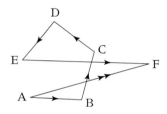

$$\overrightarrow{AB} + \overrightarrow{BC} + \overrightarrow{CD} + \overrightarrow{DE} + \overrightarrow{EF} = \overrightarrow{AF}$$

The pairs of letters in the sequence fit together.

Example

OABC is a parallelogram.
$\overrightarrow{OA} = \mathbf{a}$ $\overrightarrow{OC} = \mathbf{c}$
Write the vector that represents the diagonal

a $\overrightarrow{OB}$ **b** $\overrightarrow{CA}$.

a $\overrightarrow{OB} = \overrightarrow{OA} + \overrightarrow{AB}$
$\quad = \mathbf{a} + \mathbf{c}$

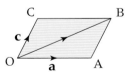

$\overrightarrow{AB} = \mathbf{c}$ because OC is equal and parallel to AB.

b $\overrightarrow{CA} = \overrightarrow{CO} + \overrightarrow{OA}$
$\quad = -\overrightarrow{OC} + \overrightarrow{OA}$
$\quad = -\mathbf{c} + \mathbf{a}$
$\quad = \mathbf{a} - \mathbf{c}$

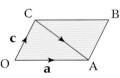

1 The diagram shows vectors **s** and **t**.
On squared paper draw the vectors that represent

a s + s   **b** s + t   **c** t + s

d s − t   **e** t − s   **f** t + t − s

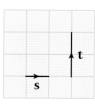

2 The diagram shows vectors **g** and **h**.
On isometric paper draw the vectors
that represent

a g + g   **b** g + h

c h − g   **d** −g + h

e g − h − h

3 OPQR is a rectangle
$\overrightarrow{OP} = \mathbf{p}$ and $\overrightarrow{OR} = \mathbf{r}$.

Work out the vector, in terms of **p** and **r**,
that represents

a $\overrightarrow{PQ}$ **b** $\overrightarrow{OQ}$

c $\overrightarrow{QO}$ **d** $\overrightarrow{RP}$

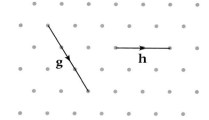

4 OABC is a square.
$\overrightarrow{OA} = \mathbf{a}$ and $\overrightarrow{OB} = \mathbf{b}$.

Work out the vector, in terms of **a** and **b**,
that represents

a $\overrightarrow{BA}$ **b** $\overrightarrow{AB}$

c $\overrightarrow{BC}$ **d** $\overrightarrow{OC}$

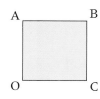

5 OJKL is a rhombus.
$\overrightarrow{OJ} = \mathbf{j}$ and $\overrightarrow{OL} = \mathbf{l}$.

Work out the vector, in terms of **j** and **l**,
that represents

a $\overrightarrow{JK}$ **b** $\overrightarrow{JL}$

c $\overrightarrow{KO}$ **d** $\overrightarrow{KL}$

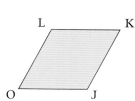

Parallel vectors

This spread will show you how to:

- Understand and use the commutative properties of vector addition
- Calculate a scalar multiple of a vector and represent it graphically
- Understand that vectors represented by parallel lines are multiples of each other

Keywords
Commutative
Vector
Multiple
Parallel
Scalar

You can add **vectors** in any order.

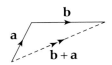 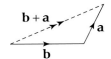

- Vector addition is **commutative**: $\mathbf{a} + \mathbf{b} = \mathbf{b} + \mathbf{a}$

You can extend addition to more than two vectors.

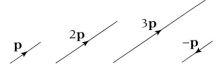

You can write this vector as 4**b**.

$$\mathbf{b} + \mathbf{b} + \mathbf{b} + \mathbf{b} = 4\mathbf{b}$$

In the vector 4**b**, 4 is a **scalar**.
A scalar has magnitude (size) but no direction.

The vector 4**b** is **parallel** to the vector **b**, and four times as long as **b**.

Speed is a scalar; velocity is a vector.

- You can multiply a vector by a scalar.

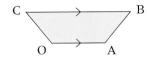

$-1 \times \mathbf{p} = -\mathbf{p}$

The vector 2**p** is **parallel** to the vector **p** and twice the length.
The vector 3**p** is parallel to the vector **p** and three times the length.
The vector −**p** is parallel to the vector **p** and the same length, but in the opposite direction.

- Vectors represented by parallel lines are **multiples** of each other.

Lines representing multiples of the same vector are parallel.

Example

OABC is a trapezium.
The parallel sides CB and OA are such that CB = 3OA.

$\overrightarrow{OA} = \mathbf{a}$

Write, in terms of **a**, the vector

a $\overrightarrow{CB}$ **b** $\overrightarrow{BC}$

---

a $\overrightarrow{CB} = 3\mathbf{a}$ **b** $\overrightarrow{BC} = -3\mathbf{a}$

1 $\overrightarrow{OP} = \mathbf{p}$ $\overrightarrow{OQ} = \mathbf{q}$

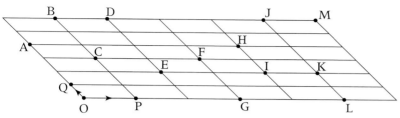

Write, and simplify, the vectors

a $\overrightarrow{OG}$	**b** $\overrightarrow{OL}$	**c** $\overrightarrow{OK}$	**d** $\overrightarrow{OJ}$
e $\overrightarrow{OA}$	**f** $\overrightarrow{OC}$	**g** $\overrightarrow{OB}$	**h** $\overrightarrow{OF}$
i $\overrightarrow{PE}$	**j** $\overrightarrow{PD}$	**k** $\overrightarrow{EF}$	**l** $\overrightarrow{CA}$
m $\overrightarrow{JK}$	**n** $\overrightarrow{JI}$	**o** $\overrightarrow{JE}$	**p** $\overrightarrow{FD}$
q $\overrightarrow{DF}$	**r** $\overrightarrow{DC}$	**s** $\overrightarrow{ME}$	**t** $\overrightarrow{HG}$
u $\overrightarrow{KD}$			

2 The diagram shows vectors **x** and **y**.

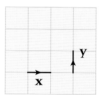

On squared paper draw the vectors

a $2\mathbf{x}$	**b** $3\mathbf{y}$	**c** $2\mathbf{x} + 3\mathbf{y}$
d $3\mathbf{x} - \mathbf{y}$	**e** $\mathbf{y} - 2\mathbf{x}$	**f** $1\frac{1}{2}\mathbf{x} + 1\frac{1}{2}\mathbf{y}$
g $2(\mathbf{x} + \mathbf{y})$	**h** $\frac{1}{2}(2\mathbf{x} - 3\mathbf{y})$	**i** $3\mathbf{x} + 4\mathbf{y}$

3 On squared paper draw a trapezium ABCD with

– parallel sides AB and DC and AB = 4DC

– angle CDA = 90°

– angle DAB = 90°.

If $\overrightarrow{AD} = \mathbf{a}$ and $\overrightarrow{DC} = \mathbf{d}$, write in terms of **a** and **d**, the vectors

a $\overrightarrow{AC}$ **b** $\overrightarrow{DB}$

4 OJKL is a trapezium.
The parallel sides OJ and LK are such that LK = $\frac{1}{4}$OJ.
OJ = 6**j**
Write, in terms of **j**, the vectors that represent

a $\overrightarrow{LK}$ **b** $\overrightarrow{KL}$

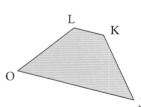

This spread will show you how to:

- Understand how the sign of a vector relates to its direction
- Solve simple geometric problems in 2-D using vector methods

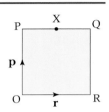

Keywords

Parallel
Resultant
  vector
Symmetry
Vector

- **Vectors** represented by **parallel** lines are multiples of each other.
- An equal vector in the opposite direction is negative.

Example

OPQR is a square. X is the midpoint of the side PQ.

$\overrightarrow{OP} = \mathbf{p}$ $\overrightarrow{OR} = \mathbf{r}$

Find these vectors in terms of **p** and **r**.

a $\overrightarrow{OQ}$ **b** $\overrightarrow{PX}$ **c** $\overrightarrow{OX}$

a, 2**a**, −3**a**, $\frac{1}{2}$**a** are parallel vectors. −3**a** is in the opposite direction.

a $\overrightarrow{OQ} = \overrightarrow{OP} + \overrightarrow{PQ} = \mathbf{p} + \mathbf{r}$
  or $\overrightarrow{OQ} = \overrightarrow{OR} + \overrightarrow{RQ} = \mathbf{r} + \mathbf{p}$

b $\overrightarrow{PX} = \frac{1}{2}\overrightarrow{PQ} = \frac{1}{2}\mathbf{r}$

c $\overrightarrow{OX} = \overrightarrow{OP} + \overrightarrow{PX} = \mathbf{p} + \frac{1}{2}\mathbf{r}$

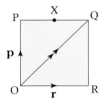

$\overrightarrow{PQ} = \mathbf{r}$
$\overrightarrow{RQ} = \mathbf{p}$

Vector addition is commutative, so **r** + **p** = **p** + **r**.

$\overrightarrow{PX}$ is parallel to $\overrightarrow{PQ}$ and half the length.

You can use a polygon's **symmetry** properties to help you write vectors.

Example

OABCDE is a regular hexagon.
P is a point on AD such that the ratio AP : PD = 1 : 2.

$\overrightarrow{OA} = \mathbf{a}$ and $\overrightarrow{OE} = \mathbf{e}$.
Find these vectors in terms of **a** and **e**.

a $\overrightarrow{AE}$ **b** $\overrightarrow{AD}$ **c** $\overrightarrow{AP}$ **d** $\overrightarrow{AB}$ **e** $\overrightarrow{OB}$

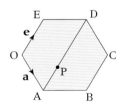

a $\overrightarrow{AE} = \overrightarrow{AO} + \overrightarrow{OE}$
     $= -\mathbf{a} + \mathbf{e}$
     $= \mathbf{e} - \mathbf{a}$

b $\overrightarrow{AD} = 2\overrightarrow{OE}$ (symmetry of regular hexagon)
     $= 2\mathbf{e}$

c $\overrightarrow{AP} = \frac{1}{3}\overrightarrow{AD}$
     $= \frac{1}{3} \times 2\mathbf{e} = \frac{2}{3}\mathbf{e}$

d $\overrightarrow{AB} = \overrightarrow{AD} + \overrightarrow{DC} + \overrightarrow{CB}$
     $= \overrightarrow{AD} + \overrightarrow{OA} - \overrightarrow{OE}$
     $= 2\mathbf{e} + \mathbf{a} - \mathbf{e}$
     $= \mathbf{a} + \mathbf{e}$

e $\overrightarrow{OB} = \overrightarrow{OA} + \overrightarrow{AB}$
     $= \mathbf{a} + \mathbf{a} + \mathbf{e}$
     $= 2\mathbf{a} + \mathbf{e}$

$\overrightarrow{AO} = -\overrightarrow{OA}$

Ratio 1 : 2 means that P is $\frac{1}{3}$ of the way along AD.

Using symmetry
$\overrightarrow{DC} = \overrightarrow{OA}$
$\overrightarrow{CB} = \overrightarrow{EO} = -\overrightarrow{OE}$

1 WXYZ is a square.
The diagonals WY and XZ intersect at M.
$\overrightarrow{WX} = \mathbf{w}$ and $\overrightarrow{WZ} = \mathbf{z}$.
Write these vectors in terms of $\mathbf{w}$ and $\mathbf{z}$.

   a $\overrightarrow{XY}$ **b** $\overrightarrow{WY}$ **c** $\overrightarrow{WM}$

   d $\overrightarrow{MY}$ **e** $\overrightarrow{MX}$ **f** $\overrightarrow{XM}$

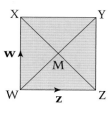

2 ABCDEFGH is a regular octagon.
$\overrightarrow{AB} = \mathbf{a}$ and $\overrightarrow{DE} = \mathbf{d}$.
Work out the vector, in terms of $\mathbf{a}$ and $\mathbf{d}$,
that represents

   a $\overrightarrow{AH}$ **b** $\overrightarrow{FE}$

   c $\overrightarrow{FD}$ **d** $\overrightarrow{HB}$

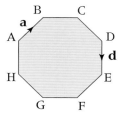

3 ABCDEF is a regular hexagon.
O is the centre of the hexagon.
$\overrightarrow{OB} = \mathbf{b}$ and $\overrightarrow{OC} = \mathbf{c}$.
Work out the vector, in terms of $\mathbf{b}$ and $\mathbf{c}$,
that represents

   a $\overrightarrow{FC}$ **b** $\overrightarrow{BE}$

   c $\overrightarrow{BC}$ **d** $\overrightarrow{AD}$

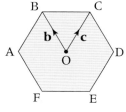

4 ORST is a rhombus.
$\overrightarrow{OR} = \mathbf{r}$ and $\overrightarrow{OT} = \mathbf{t}$.
P lies on OS such that OP : PS = 3 : 1.
Work out the vector, in terms of $\mathbf{r}$ and $\mathbf{t}$,
that represents

   a $\overrightarrow{RS}$ **b** $\overrightarrow{OS}$ **c** $\overrightarrow{OP}$

   d $\overrightarrow{SP}$ **e** $\overrightarrow{TP}$

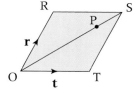

5 OAB is a triangle.
X is the point on AB for which
AX : XB = 1 : 4.
$\overrightarrow{OA} = \mathbf{a}$ and $\overrightarrow{OB} = \mathbf{b}$.

   a Write, in terms of $\mathbf{a}$ and $\mathbf{b}$, an
      expression for $\overrightarrow{AB}$.

   b Express $\overrightarrow{OX}$ in terms of $\mathbf{a}$ and $\mathbf{b}$.
      Give your answer in its simplest form.

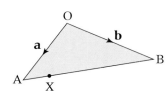

This spread will show you how to:

● Solve simple geometric problems in 2-D using vector methods

Keywords
Collinear
Parallel
Vector

You can use **vectors** in geometric proofs.

● To prove that lines are **parallel**, you show that the vectors they represent are multiples of each other.

Proof

OAB is a triangle.
M is the midpoint of OA.
N is the midpoint of OB.
$\overrightarrow{OA} = \mathbf{a}$ and $\overrightarrow{OB} = \mathbf{b}$.
Show that MN is parallel to AB.

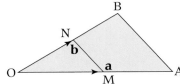

$$\overrightarrow{AB} = \overrightarrow{AO} + \overrightarrow{OB}$$
$$= -\mathbf{a} + \mathbf{b}$$
$$\overrightarrow{MN} = \overrightarrow{MO} + \overrightarrow{ON}$$
$$= -\tfrac{1}{2}\mathbf{a} + \tfrac{1}{2}\mathbf{b}$$
$$= \tfrac{1}{2}(-\mathbf{a} + \mathbf{b})$$
$$-\mathbf{a} + \mathbf{b} = 2 \times \tfrac{1}{2}(-\mathbf{a} + \mathbf{b})$$

So MN is parallel to AB.

Find the vectors $\overrightarrow{AB}$ and $\overrightarrow{MN}$.

$\overrightarrow{AB}$ is a multiple of $\overrightarrow{MN}$.

Always end your proof with a clear statement.

● To prove that points are **collinear**, you show that
  – the vectors joining pairs of the points are parallel
  – the vectors share a common point.

'Collinear' means 'lie on the same straight line'.

Proof

PQRS is an isosceles trapezium.
PQ and SR are parallel sides with PQ = 2 × SR.

$$\overrightarrow{PQ} = 2\mathbf{p} \qquad \overrightarrow{QR} = \mathbf{q}$$

X lies on QR such that QX : XR = 1 : 2.
Y lies on PQ extended such that PQ : QY = 1 : 1.

Prove that S, X and Y are collinear.

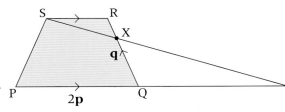

$$PQ = 2SR, \text{ so } \overrightarrow{SR} = \mathbf{p}$$
$$\overrightarrow{RX} = \tfrac{1}{3}\overrightarrow{RQ} = -\tfrac{1}{3}\mathbf{q}$$

$$\overrightarrow{XY} = \overrightarrow{XQ} + \overrightarrow{QY}$$
$$= \tfrac{2}{3}\overrightarrow{RQ} + \overrightarrow{QY}$$
$$= -\tfrac{2}{3}\mathbf{q} + 2\mathbf{p}$$
$$= 2(\mathbf{p} - \tfrac{1}{3}\mathbf{q})$$
$$= 2\overrightarrow{SX}$$

$$\overrightarrow{SX} = \overrightarrow{SR} + \overrightarrow{RX}$$
$$= \mathbf{p} - \tfrac{1}{3}\mathbf{q}$$

So $\overrightarrow{SX}$ and $\overrightarrow{XY}$ are parallel.
SX and XY have the point X in common.

First show that SX and XY are parallel.

QY is PQ extended so it is parallel to PQ.
PQ : QY = 1 : 1, so PQ are QY are the same length.
Therefore $\overrightarrow{QY} = \overrightarrow{PQ} = 2\mathbf{p}$.

So S, X and Y are collinear.

1 In the diagram, $\overrightarrow{OA} = 2\mathbf{a}$, $\overrightarrow{OB} = 2\mathbf{b}$, $\overrightarrow{OC} = 3\mathbf{a}$ and $\overrightarrow{BD} = \mathbf{b}$.

Prove that AB is parallel to CD.

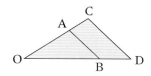

2 RSTU is a rectangle.
M is the midpoint of the side RS.
N is the midpoint of the side ST.
$\overrightarrow{RS} = 2\mathbf{r}$ $\overrightarrow{ST} = 2\mathbf{a}$

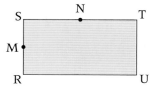

a Work out the vector, in terms of **r** and **s**, that represents

i $\overrightarrow{RT}$ **ii** $\overrightarrow{RM}$ **iii** $\overrightarrow{SN}$ **iv** $\overrightarrow{MN}$.

b Show that MN is parallel to RT.

3 OPQR is a trapezium.
OP is parallel to RQ and $OP = \frac{1}{3}RQ$.
$\overrightarrow{OP} = \mathbf{p}$ and $\overrightarrow{OR} = \mathbf{r}$.

a Work out the vector, in terms of **p** and **r**, that represents

i $\overrightarrow{RQ}$ **ii** $\overrightarrow{OQ}$ **iii** $\overrightarrow{RP}$ **iv** $\overrightarrow{PQ}$

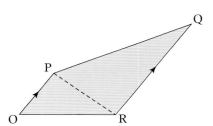

b The point X lies on PR such that
PX : XR = 1 : 3.
Show that O, X and Q lie on the same straight line.

4 OJKL is a parallelogram.
$\overrightarrow{OJ} = \mathbf{j}$ $\overrightarrow{JK} = \mathbf{k}$

a Express, in terms of **j** and **k** these vectors.

i $\overrightarrow{OK}$ **ii** $\overrightarrow{OL}$ **iii** $\overrightarrow{JL}$

b M is the point on KL extended such that KL : LM = 1 : 2.
X is a point on OL such that $OX = \frac{1}{3}OL$.

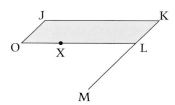

Show that J, X and M lie on the same straight line.

Key objectives

- Understand and use vector notation for translations. Calculate and represent graphically the sum of two vectors, the difference of two vectors and a scalar multiple of a vector
- Calculate the resultant of two vectors
- Understand and use the commutative and associative properties of vector addition
- Work with column vectors
- Solve simple geometrical problems in 2-D using vector methods

1 **v** and **t** are vectors.

$$\mathbf{v} = \begin{pmatrix} 1 \\ 3 \end{pmatrix} \qquad \mathbf{t} = \begin{pmatrix} 2 \\ -4 \end{pmatrix}$$

Calculate these.
Express your answers as column vectors.

a $\mathbf{v} + \mathbf{t}$ **b** $\mathbf{v} - \mathbf{t}$ **c** $\mathbf{t} - 2\mathbf{v}$ (3)

2 OPQ is a triangle.
R is the midpoint of OP.
S is the midpoint of PQ.
$\overrightarrow{OP} = \mathbf{p}$ and $\overrightarrow{OQ} = \mathbf{q}$

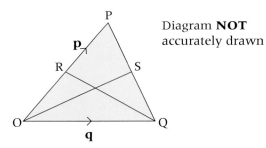

Diagram **NOT** accurately drawn

a Find $\overrightarrow{OS}$ in terms of **p** and **q**.

b Show that RS is parallel to OQ.   (5)

(Edexcel Ltd., 2004)

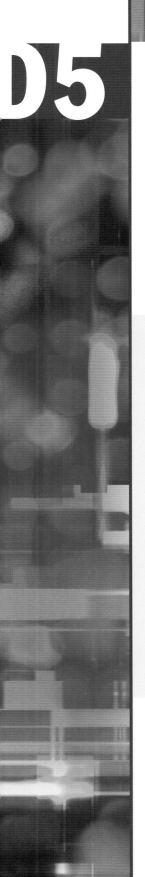

D5

This unit will show you how to

- Show the possible outcomes of two or more events on a tree diagram
- Recognise when two events are independent
- Use tree diagrams to calculate probabilities of combinations of independent events
- Use tree diagrams to calculate probabilities in 'without replacement' sampling
- Understand conditional probability
- Calculate probabilities based on the given conditions

Before you start ...

You should be able to answer these questions.

Review

1 Work out each of these.

Unit N2

   a $1 - 0.45$ **b** $1 - 0.96$

   c $1 - 0.28$ **d** $1 - 0.375$

   e $0.2 + 0.4$ **f** $0.3 + 0.04$

   g $0.65 + 0.25$ **h** 0.5×0.36

   i 0.25×0.68 **j** 0.64×0.3

2 Work out each of these.

Unit N2

   a $1 - \frac{5}{6}$ **b** $1 - \frac{1}{5}$

   c $1 - \frac{7}{9}$ **d** $\frac{1}{5} + \frac{2}{3}$

   e $\frac{3}{4} + \frac{1}{6}$ **f** $\frac{2}{3} \times \frac{5}{6}$

   g $\frac{2}{9} \times \frac{4}{5}$

This spread will show you how to:

Keywords
Event
Independent
Outcome
Tree diagram

● Show the possible outcomes of two or more events on a tree diagram
● Use tree diagrams to calculate probabilities of combinations of independent events

● You can show the possible **outcomes** of two or more **events** on a **tree diagram**.

Example

A box contains 4 red and 6 blue counters.
A counter is chosen at random from the box and its colour noted.
It is then replaced in the box.
The box is shaken and then a second counter is chosen at random.
Draw a tree diagram to show all the possible outcomes.

The counter is replaced in the box each time and the box is shaken, so the two events are **independent**.

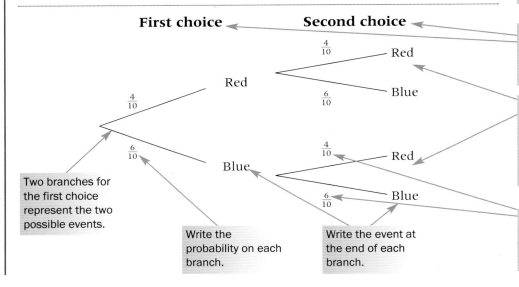

Label the sets of branches 'First choice', 'Second choice'.

For each branch for the first choice, there are two branches/possible events for the second choice.

Two branches for the first choice represent the two possible events.

Write the probability on each branch.

Write the event at the end of each branch.

For each pair of branches, the probabilities should sum to 1.

Example

Records show that there are 10 wet days in every August. The other days are dry. The weather on any day is independent of the weather on the previous day.
Draw a tree diagram to show the possible types of weather on the weekend of 19 and 20 August one year.

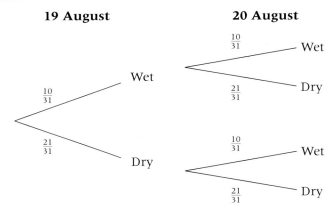

1 A bag contains 8 purple cubes and 9 orange cubes.
A cube is chosen at random from the bag and its colour noted.
It is then replaced in the bag.
The bag is shaken and then a second cube is chosen at random.
Copy and complete the tree diagram to show all the possible
outcomes.

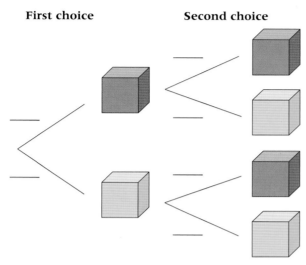

First choice   **Second choice**

2 A set of cards contains 6 number cards and 5 picture cards.
A card is chosen at random from the set.
It is then replaced and the cards are shuffled.
Then a second card is chosen at random.
Draw a tree diagram to show all the possible outcomes.

3 A 50 pence coin and a 20 pence coin are thrown.
Draw a tree diagram to show all the possible outcomes.

4 A bag contains 7 red and 2 blue marbles.
A marble is chosen from the bag and a fair coin is thrown.
Draw a tree diagram to show all the possible outcomes.

5 Kaz has 16 DVDs. Three of them are Star Wars films.
She chooses one of the DVDs at random, notes whether it is a
Star Wars film, and then replaces it.
She then chooses another DVD at random.
Copy and complete this tree diagram for the outcomes.

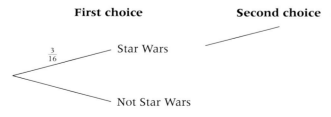

First choice   **Second choice**

$\frac{3}{16}$ — Star Wars

Not Star Wars

Using tree diagrams to find probabilities

This spread will show you how to:

- Use tree diagrams to calculate probabilities of combinations of independent events
- Calculate probabilities based on the given conditions

Keywords
Event
Independent
Outcome
Tree diagram

You can use a **tree diagram** to calculate probabilities of combinations of **independent events**.

- To calculate the probability of independent events, multiply along the branches.

Example

Shireen makes two chocolate rabbits by pouring melted chocolate into moulds and leaving it to set. She then removes the moulds.
The probability that a chocolate rabbit is cracked is 0.1.

a Draw a tree diagram to show all the **outcomes** of the two chocolate rabbits being cracked or not cracked.
b Calculate the probability that both chocolate rabbits will be cracked.
c Work out the probability that one of the chocolate rabbits will be cracked.

a **First chocolate rabbit**   **Second chocolate rabbit**

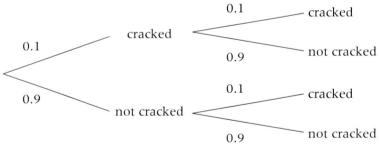

The two events are independent.

'Cracked' and 'not cracked' are mutually exclusive events, so
P(not cracked) = 1 − P(cracked)

b P(cracked, cracked) = 0.1 × 0.1 = 0.01
c Two outcomes result in one cracked rabbit:
(not cracked, cracked) or (cracked, not cracked).
P(one rabbit cracked)
   = P(not cracked, cracked) + P(cracked, not cracked)
   = 0.9 × 0.1   + 0.1 × 0.9
   = 0.09 + 0.09
   = 0.18

The four possible outcomes are:
(cracked, cracked), (cracked, not cracked), (not cracked, cracked) and (not cracked, not cracked).

When A and B are independent P(A or B) = P(A) + P(B).

- If you can take more than one route through the tree diagram:
  - Multiply the probabilities along each route.
  - Add the resultant probabilities for each route.

1 Two torches, one red and one black, are fitted with new batteries.
The probability that a battery lasts for more than 20 hours is 0.7.

  a Draw a tree diagram to show the probabilities of the torch batteries lasting more or less than 20 hours.

  b Find the probability that the battery lasts more than 20 hours

    i in both torches   **ii** in only one torch   **iii** in at least one torch.

2 Rajen has five 10p and three 2p coins in his pocket.
He takes a coin at random from his pocket, notes what it is, and replaces it.
He then picks a second coin at random.

  a Draw a tree diagram to show all the possible outcomes for picking two coins.

  b Find the probability that he picks

    i two 10p coins   **ii** one coin of each type   **iii** at least one 10p coin.

3 A bag contains 7 yellow and 3 green marbles.
A marble is chosen at random from the bag, its colour is noted and it is replaced.
A second marble is chosen at random.

  a Draw a tree diagram to show all the possible outcomes for choosing two marbles.

  b Find the probability of choosing

    i   one marble of each colour

    ii two marbles the same colour

    iii no green marbles.

4 Two vases are fired in a kiln.
The probability that a vase breaks in the kiln is $\frac{1}{50}$.

  a Draw a tree diagram to show all the possible outcomes.

  b Calculate the probability that in the kiln

    i neither vase breaks   **ii** at least one vase breaks.

5 A spinner has 10 equal sectors, 6 green and 4 white.
The spinner is spun twice.

  a Draw a tree diagram to show all the possible outcomes of two spins on the spinner.

  b Find the probability that on two spins the spinner lands on

    i   green both times

    ii green at least one time

    iii one of each colour.

Sampling without replacement

This spread will show you how to:

Keywords
Independent
Tree diagram

● Use tree diagrams to calculate probabilities in 'without replacement' sampling

When an item is chosen and *not* replaced, the probability of choosing a second item changes.

A bag contains 4 white and 3 black counters.

First choice
A counter is chosen at random.
$P(\text{black}) = \frac{3}{7}$
$P(\text{white}) = \frac{4}{7}$

A black counter is chosen and not replaced.
Bag now contains 4 white and 2 black counters.

Second choice
A counter is chosen at random.
$P(\text{black}) = \frac{2}{6}$
$P(\text{white}) = \frac{4}{6}$
The probability of choosing a white counter has increased.

● You can use a **tree diagram** to calculate the probability changes for each choice for 'without replacement' sampling.

Example

A bag contains 7 yellow and 3 blue marbles.
A marble is chosen at random from the bag and its colour noted.
It is *not* replaced in the bag.
The bag is shaken and then a second marble is chosen at random.

a Draw a tree diagram to show all the possible outcomes.
b Find the probability that
   i both marbles are blue
   ii one marble of each colour is chosen.

a   **First choice**   **Second choice**

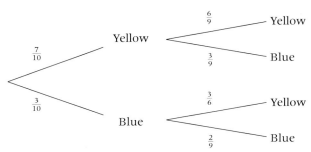

For the second choice there is one less marble in the bag, so the denominator decreases by 1. The numerator depends upon the first choice.

For each pair of branches, the probabilities add to 1.

b i $P(B, B) = \frac{3}{10} \times \frac{2}{9} = \frac{6}{90}$

  ii $P(Y, B) + P(B, Y) = \left(\frac{7}{10} \times \frac{3}{9}\right) + \left(\frac{3}{10} \times \frac{7}{9}\right)$

                          $= \frac{21}{90} + \frac{21}{90} = \frac{42}{90}$

1 A bag contains 3 yellow and 4 blue marbles.
A marble is chosen at random from the bag and its colour noted.
It is *not* replaced in the bag.
The bag is shaken and then a second marble is chosen at random.
The tree diagram shows all the possible outcomes.

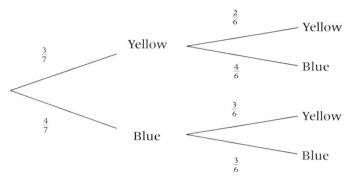

First choice   **Second choice**

Use the tree diagram to find the probability of choosing

a two yellow marbles

b two blue marbles

c one marble of each colour.

2 A bag contains 3 white counters and 8 black counters.
A counter is chosen at random from the bag and its
colour noted. It is *not* replaced in the bag.
The bag is shaken and then a second counter is chosen
at random.

a Draw a tree diagram to show all the outcomes and
their probabilities.

b Find the probability of choosing

   i two white counters

   ii one counter of each colour

   iii at least one white counter.

3 A bag contains 12 lemon and 4 orange sweets.
Reuben chooses a sweet at random and eats it.
He then chooses a second sweet and eats it.

a Draw a tree diagram to show all the outcomes
and probabilities.

b Find the probability of choosing

   i two orange sweets

   ii one sweet of each flavour

   iii at least one orange sweet.

Tree diagrams – conditional probability

This spread will show you how to:

Keywords
Conditional
Event
Outcome

- Use tree diagrams to calculate probabilities in 'without replacement' sampling
- Understand conditional probability

- In **conditional** probability, the probability of subsequent **events** depends on previous events occuring.

Example

Gareth travels through two sets of traffic lights on his way to work.
The probability that the first set of traffic lights is on red is 0.6.

If the first set of lights is on red, then the probability that the second set of lights will be on red is 0.9.
If the first set of lights is not on red then the probability that the second set of lights is on red is 0.25.

The conditional statement 'If ... then ...' describes a conditional probability.

a Draw a tree diagram to show the different possibilities for the traffic lights.
b Work out the probability that on Gareth's way to work
  i both sets of lights will be on red
  ii only one set of traffic lights will be on red.

a **First set of lights**   **Second set of lights**

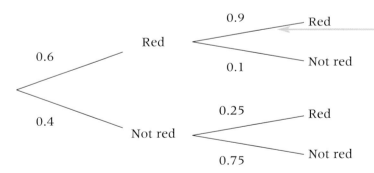

If the first set is red, P(second set red) = 0.9.

Each pair of branches has different probabilities.

On each pair of branches the probabilities add to 1.

b i P(red, red) = 0.6 × 0.9 = 0.54
  ii P(only one on red) = P(red, not red) + P(not red, red)
             = (0.6 × 0.1) + (0.4 × 0.25)
             = 0.06 + 0.1
             = 0.16

1 Richard either drives or cycles to work.
The probability that he drives to work is 0.25.
If Richard drives, the probability that he is late is 0.2.
If Richard cycles, the probability that he is late is 0.4.

  a Draw a tree diagram to show the probability of Richard being late for work.

  b Work out the probability that on any one day Richard will *not* be late for work.

2 Julie's college course offers lectures and accompanying tutorials.
The probability that Julie attends any one lecture is 0.8.
If she attends the lecture the probability that she attends the accompanying tutorial is 0.9.
If she does not attend the lecture the probability that she attends the accompanying tutorial is 0.6.

  a Draw a tree diagram to show the probabilities that Julie attends lectures and tutorials.

  b Work out the probability that Julie attends

   i only one tutorial   **ii** at least one tutorial.

3 40 girls and 60 boys completed a questionnaire.
The probability that a girl completed the questionnaire truthfully was 0.8.
The probability that a boy completed the questionnaire truthfully was 0.3.

> You could rephrase this as:
> If a girl completes the questionnaire, then the probability that she completes it truthfully is 0.8.

  a Copy and complete the tree diagram.

Gender   **Completed**

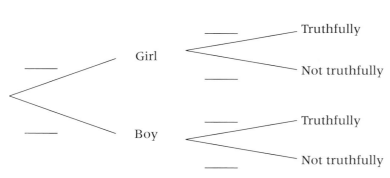

  b One questionnaire is chosen at random.
Work out the probability that

   i the questionnaire has not been completed truthfully

   ii the questionnaire has been completed truthfully by a girl.

This spread will show you how to:

- Understand conditional probability
- Calculate probabilities based on the given conditions

When solving probability problems, look out for **conditional probabilities**.

- You need to calculate probabilities based on the conditions you are given.

Example

The table shows the number of boys and the number of girls in Years 12 and 13 at a school.

	Year 12	Year 13	Total
Boys	48	96	144
Girls	72	54	126
Total	120	150	270

a One student is to be chosen at random.
What is the probability that it is a boy?
b One of the boys is chosen at random.
What is the probability that he is in Year 13?
c One of the Year 13 students is chosen at random.
What is the probability that the student is a boy?

a 144 boys and 270 students in total
$$P(\text{boy}) = \frac{144}{270}$$
b Total 144 boys, 96 boys in Year 13
$$\text{Probability} = \frac{96}{144}$$
c Total 150 in Year 13, 96 are boys
$$\text{Probability} = \frac{96}{150}$$

Conditional probability
If a boy is chosen, what is the probability that he is in Year 13?

If a Year 13 student is chosen, what is the probability that the student is a boy?

Example

At a party, the menu choice is meat or fish.
The table shows the choices made.

	Meat	Fish	Total
Man	8	2	10
Woman	4	11	15
Total	12	13	25

a One of the women is chosen at random.
What is the probability that she ate fish?
b One person who ate fish is chosen at random.
What is the probability that it was a man?

a There are 15 women of whom 11 ate fish.   Probability $= \frac{11}{15}$
b 13 people chose fish, of whom 2 are men.   Probability $= \frac{2}{13}$

1 The table shows the numbers of boys and girls in Years 10 and 11 at a school.

	Year 10	Year 11	Total
Boys	57	60	117
Girls	51	52	103
Total	108	112	220

a One student is chosen at random. What is the probability that the student is a girl?

b One of the girls is chosen at random. What is the probability that she is in Year 10?

c One of the Year 10 students is chosen at random. What is the probability that the student is a girl?

2 The table shows the ages of men and women enrolling for a course at an adult education centre.

	Aged under 40	Age 40 and over	Total
Men	24	30	54
Women	42	54	96
Total	66	84	150

a One person is chosen at random. What is the probability that the person is a man?

b One of the men is chosen at random. What is the probability that he is under 40?

c One person under 40 is chosen at random. What is the probability that this person is male?

3 After the main course, people chose to have tea or coffee with their dessert.
The table shows their choices.

	Tea	Coffee	Total
Man	11	1	12
Woman	4	16	20
Total	15	17	32

a One of the men is chosen at random. What is the probability that he drank coffee?

b One of the coffee drinkers is chosen at random. What is the probability that the person is a man?

4 A spinner has 24 equal sides. Each side is coloured red or black and has a circle or triangle on it.
The table shows the colour and the shape drawn on each side.

	Circle	Triangle	Total
Red	4	12	16
Black	6	2	8
Total	10	14	24

The spinner is spun once. What is the probability that the spinner lands on

a red

b red, given that it shows triangle

c a triangle, given that it shows red?

Key objectives

● Use tree diagrams to represent outcomes of compound events, recognising when events are independent

1 Gavin and Alan take part in weekly doubles darts matches.
If they are both late then they must forfeit the match and their opposition automatically wins.
The probability that Gavin is late for the start of the match is $\frac{3}{10}$.
The probability that Alan is late for the start of the match is $\frac{13}{20}$.
The next match that Gavin and Alan will take part in is tomorrow.

a Draw a tree diagram to show the probabilities of Gavin and Alan being late or on time for this match.
Label the branches of the probability tree clearly. (5)

b Use your tree diagram to calculate the probability that tomorrow's match will be forfeited. (2)

2 Amy is going to play one game of snooker and one game of billiards.
The probability that she will win the game of snooker is $\frac{3}{4}$.
The probability that she will win the game of billiards is $\frac{1}{3}$.

a Copy and complete the probability tree diagram.

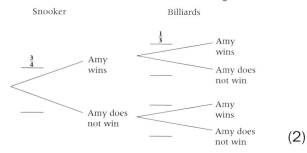

(2)

b Work out the probability that Amy will win exactly one game. (3)

Amy played one game of snooker and one game of billiards on a number of Fridays.
She won at both snooker and billiards on 21 Fridays.

c Work out an estimate for the number of Fridays on which Amy did not win either game. (3)

(Edexcel Ltd., 2005)

This unit will show you how to

- Draw and describe the graphs of trigonometric functions for angles of any size
- Use the graphs of trigonometric functions to solve equations for angles between 0° and 360°
- Use a calculator to find one solution to equations involving trigonometric functions
- Use the symmetry of the graphs of trigonometric functions to find further solutions to equations

Before you start ...

You should be able to answer these questions.

1 a Use a calculator to find
　　i sin 30　**ii** cos 60.
b Use a calculator to find
　　i sin 40　**ii** cos 50.
c Draw a right-angled triangle.
　　Use this to explore other trig ratios
　　with the same value.

Review

Unit S5

This spread will show you how to:

- Draw, sketch and describe the graph of the sine function for angles of any size
- Use the graph of the sine function to solve equations

In right-angled triangles, two angles are always **acute**.

In obtuse-angled triangles, one angle is always **obtuse**.

Your calculator will give a value for the **sine** of any angle.

You can draw a graph of the sine ratio from this table of values.

x	0	45°	90°	135°	180°	225°	270°	315°	360°
sin x	0	0.707	1	0.707	0	−0.707	−1	−0.707	0

These four triangles are congruent:

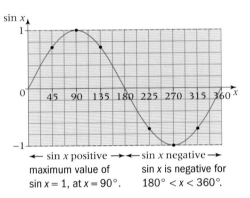

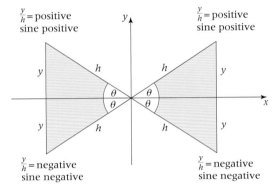

$\frac{y}{h}$ = positive — sine positive

$\frac{y}{h}$ = positive — sine positive

$\frac{y}{h}$ = negative — sine negative

$\frac{y}{h}$ = negative — sine negative

← sin x positive → ← sin x negative →
maximum value of   sin x is negative for
sin x = 1, at x = 90°.   180° < x < 360°.

- You can use the graph of $y = \sin x$ to solve equations for angles between 0° and 360°. The graph is symmetrical.

Most values of sin x give two solutions.

Example

Solve these equations for angles between 0° and 360°.

a sin x = 0.6   **b** sin x = −0.4   **c** sin x = 1

a x = 37° and 143°

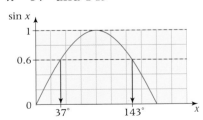

b x = 204° and 336°

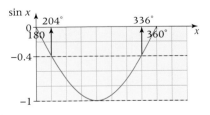

The symmetry of the graph shows that the two answers should add to give a multiple of 180°.

c sin x = 1 is the maximum point of the graph. x = 90°.

1 Use the graph of $y = \sin x$ to solve these equations for angles between 0° and 360°.
Use a calculator to check the accuracy of your answers.

   a $\sin x = 0.7$ **b** $\sin x = 0.8$

   c $\sin x = -0.8$ **d** $\sin x = -0.5$

   e $\sin x = -0.2$ **f** $\sin x = 0.4$

   g $\sin x = -0.6$ **h** $\sin x = 0.3$

   i $\sin x = -0.3$ **j** $\sin x = -1$

   k $\sin x = 0$ **l** $\sin x = 0.5$

2 Use the graph of $y = \sin x$ to find the sine of these angles.
Use a calculator to check the accuracy of your answers.

   a $\sin 30°$ **b** $\sin 150°$

   c $\sin 45°$ **d** $\sin 225°$

   e $\sin 315°$ **f** $\sin 90°$

   g $\sin 270°$ **h** $\sin 180°$

   i $\sin 75°$ **j** $\sin 240°$

   k $\sin 100°$ **l** $\sin 300°$

3 Use the graph of $y = \sin x$ to find the other angle between 0° and 360° with sine the same value as

   a $\sin 50°$ **b** $\sin 68°$

   c $\sin 140°$ **d** $\sin 217°$

   e $\sin 262°$ **f** $\sin 98°$

   g $\sin 330°$ **h** $\sin 243°$

   i $\sin 12°$ **j** $\sin 309°$

   k $\sin 172°$ **l** $\sin 287°$

4 **i** Use the sine rule to find the angle marked θ in each triangle.

   ii Use the graph of $y = \sin x$ to work out a second value for θ, if a second value is possible.

a
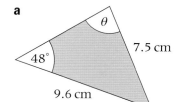

b
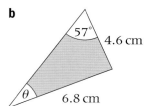

c
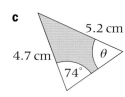

Cosine graph

This spread will show you how to:

- Draw, sketch and describe the graph of the cosine function for angles of any size
- Use the graph of the cosine function to solve equations

Your calculator will give a value for the **cosine** of any angle.

You can draw a graph of the cosine ratio from this table of values.

x	0	45°	90°	135°	180°	225°	270°	315°	360°
cos x	1	0.707	0	−0.707	−1	−0.707	0	0.707	1

These four triangles are congruent:

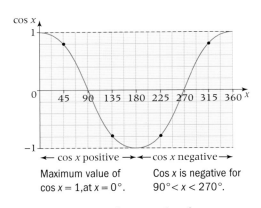

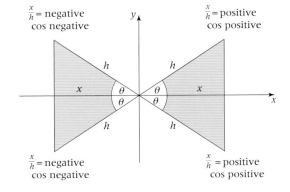

$\frac{x}{h}$ = negative
cos negative

$\frac{x}{h}$ = positive
cos positive

$\frac{x}{h}$ = negative
cos negative

$\frac{x}{h}$ = positive
cos positive

Maximum value of cos x = 1, at $x = 0°$.

Cos x is negative for $90° < x < 270°$.

- You can use the graph of $y = \cos x$ to solve equations for angles between 0° and 360°. The graph is symmetrical.

Most values of cos x give two solutions.

Solve these equations for angles between 0° and 360°.

a $\cos x = 0.6$ **b** $\cos x = -0.4$ **c** $\cos x = -1$

a $x = 53°$ and 307°
b $x = 114°$ and 246°
c This is the minimum value of the graph.
$x = 180°$

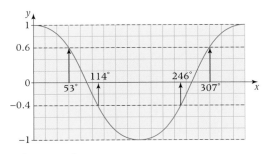

The symmetry of the graph shows that the two answers should add to 360°.

The graph of $y = \cos x$ is the same as the graph of $y = \sin x$ translated 90° along the x-axis.

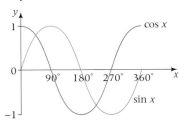

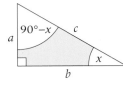

$\sin x = \frac{a}{c} = \cos (90° - x)$

$\sin (90° - x) = \frac{b}{c} = \cos x$

1 Use the graph of $y = \cos x$ to solve these equations for angles between 0° and 360°.
Use a calculator to check the accuracy of your answers.

a $\cos x = 0.7$ **b** $\cos x = 0.8$

c $\cos x = -0.8$ **d** $\cos x = -0.5$

e $\cos x = -0.2$ **f** $\cos x = 0.4$

g $\cos x = -0.6$ **h** $\cos x = 0.3$

i $\cos x = -0.3$ **j** $\cos x = 1$

k $\cos x = 0$ **l** $\cos x = 0.5$

2 Use the graph of $y = \cos x$ to work out the cosine of each angle.
Use a calculator to check the accuracy of your answers.

a $\cos 60°$ **b** $\cos 140°$

c $\cos 45°$ **d** $\cos 225°$

e $\cos 315°$ **f** $\cos 90°$

g $\cos 270°$ **h** $\cos 180°$

i $\cos 55°$ **j** $\cos 250°$

k $\cos 80°$ **l** $\cos 330°$

3 Use the graph of $y = \cos x$ to find the other angle between 0° and 360° with cosine the same value as

a $\cos 30°$ **b** $\cos 66°$

c $\cos 150°$ **d** $\cos 237°$

e $\cos 164°$ **f** $\cos 78°$

g $\cos 125°$ **h** $\cos 234°$

i $\cos 18°$ **j** $\cos 199°$

k $\cos 302°$ **l** $\cos 268°$

4 a Use your calculator to check that
$\cos 60° = \sin 30°$, $\cos 52° = \sin 38°$ and $\cos 17° = \sin 73°$.

b Use the right-angled triangle, ABC, to explain the equalities in part **a**.

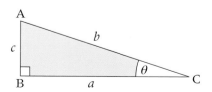

Tangent graph

This spread will show you how to:

- Draw, sketch and describe the graph of the tangent function for angles of any size
- Use the graph of the tangent function to solve equations

Keywords

Acute
Obtuse
Reflex
Tangent ratio

You can draw the graph for the **tangent** ratio.

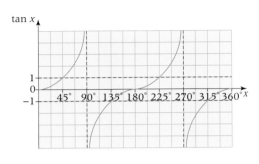

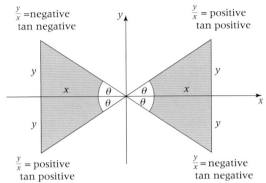

Tan x is negative for $90° < x < 180°$ and $270° < x < 360°$.
As x approaches $90°$ or $270°$, tan x approaches infinity.
$x = 90°$ and $x = 270°$ are asymptotes to the graph of $y = \tan x$.

If you try to find tan $90°$ using a calculator you get an error message.
tan $85° = 11.43 ...$
tan $89° = 57.28 ...$

- You can use the graph of $y = \tan x$ to solve equations for angles between $0°$ and $360°$.

Example

Solve these equations for angles between $0°$ and $360°$.

a $\tan x = 0.6$ **b** $\tan x = -2.5$ **c** $\tan x = -1$

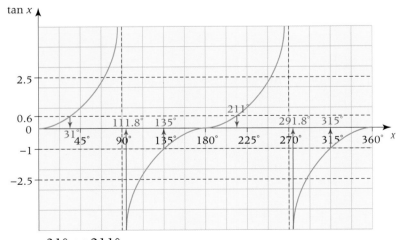

The symmetry of the graph shows the two answers should differ by $180°$.

a $x = 31°$ or $211°$
b $x = 111.8°$ or $291.8°$
c $x = 135°$ or $315°$

1 Use the graph of $y = \tan x$ to solve these equations for angles between 0° and 360°.
Use a calculator to check the accuracy of your answers.

a $\tan x = 0.7$ **b** $\tan x = 0.8$

c $\tan x = -0.8$ **d** $\tan x = -1.5$

e $\tan x = 1.9$ **f** $\tan x = 2.4$

g $\tan x = -0.6$ **h** $\tan x = 1.1$

i $\tan x = -1.1$ **j** $\tan x = -2.2$

k $\tan x = 0$ **l** $\tan x = 1$

2 Use the graph of $y = \tan x$ to work out the tangent of each angle.
Use a calculator to check the accuracy of your answers.

a $\tan 30°$ **b** $\tan 60°$

c $\tan 45°$ **d** $\tan 225°$

e $\tan 315°$ **f** $\tan 150°$

g $\tan 240°$ **h** $\tan 180°$

i $\tan 75°$ **j** $\tan 20°$

k $\tan 135°$ **l** $\tan 300°$

3 Use the graph of $y = \tan x$ to find the other angle between 0° and 360° that has the same value as

a $\tan 70°$ **b** $\tan 57°$

c $\tan 190°$ **d** $\tan 248°$

e $\tan 332°$ **f** $\tan 47°$

g $\tan 312°$ **h** $\tan 133°$

i $\tan 16°$ **j** $\tan 29°$

k $\tan 111°$ **l** $\tan 287°$

4 JKL is an equilateral triangle of side 2 cm.

a Show that $\tan 30° = \dfrac{\sqrt{3}}{3}$.

b Work out $\tan 60°$ from the triangle. Leave your answer in surd form.

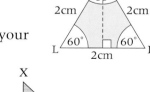

c XYZ is an isosceles right-angled triangle.
ZY = XZ = 2 cm and angle XZY = 90°.

  i   Work out $\tan 45°$.

  ii Work out $\sin 45°$, leaving your answer in surd form.

  iii Write the value of $\cos 45°$.

Use the relationship between cos and sin.

Solving trigonometric equations

Keywords
Cosine
Sine
Tangent

This spread will show you how to:

● Use a calculator to find one solution to equations involving trigonometric functions

● Use the symmetry of the graphs of trigonometric functions to find further solutions to equations

You can use your calculator to solve equations involving trigonometric ratios. It will only give one solution.

You use the symmetry of the trigonometric graphs to find other solutions.

Sine graph

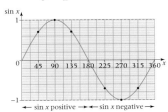

The two solutions add to a multiple of 180°.

You use x and $180° - x$.

Cosine graph

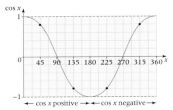

The two solutions add to 360°.

You use x and $360° - x$.

Tangent graph

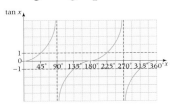

The two solutions differ by 180°.

You use x and $180° + x$.

Example

Use your calculator to solve these equations for angles 0° to 360°.
Give your answers to 1 decimal place.

a $\sin x = 0.904$ **b** $\cos x = -0.849$ **c** $\tan x = 0.427$ **d** $\sin x = -0.176$

a $x = 64.7°$ From calculator
   or $x = 180° - 64.7° = 115.3°$ From graph

b $x = 148.1°$ From calculator
   or $x = 360° - 148.1° = 211.9°$ From graph

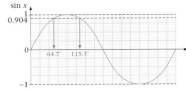

c $x = 23.1°$ From calculator
   or $x = 180° + 23.1° = 203.1°$ From graph

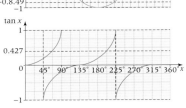

d $x = -10.1°$ From calculator
   The positive solutions are:
   $x = 180° + 10.1° = 190.1°$ and
   $360° - 10.1° = 349.9°$
   in the range 0° to 360°   Extend the graph to −90°, using symmetry.

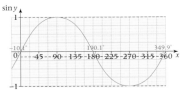

Use your calculator to solve these equations for angles 0° to 360°.

Use the graphs of $y = \sin x$, $y = \cos x$ and $y = \tan x$ to find other solutions in the range 0° to 360°.

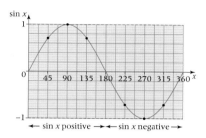

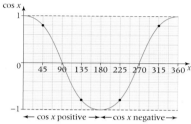

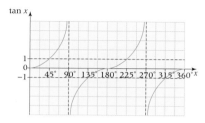

Give all your answers to 1 decimal place.

1 $\sin x = 0.534$ **2** $\sin x = 0.934$

3 $\sin x = -0.997$ **4** $\cos x = 0.521$

5 $\cos x = 0.124$ **6** $\sin x = 0.124$

7 $\tan x = 1.25$ **8** $\tan x = 2.61$

9 $\tan x = -1.67$ **10** $\sin x = 0.62$

11 $\cos x = -0.508$ **12** $\tan x = 1.889$

13 $\tan x = -1.889$ **14** $\sin x = 0.866$

15 $\cos x = -0.866$ **16** $\sin x = 0.732$

17 $\cos x = -0.430$ **18** $\tan x = 4.5$

19 $\tan x = -4.5$ **20** $\tan x = 0.807$

21 $\cos x = -0.725$ **22** $\sin x = -0.467$

23 $\sin x = 0.321$ **24** $\cos x = -0.5$

25 $\tan x = 1$ **26** $\sin x = 0.5$

27 $\tan x = 8.68$ **28** $\tan x = 5.923$

29 $\cos x = 0.402$ **30** $\cos x = -0.695$

Use symmetry to help you sketch the graph for negative values of x to −90°.

Exam review

Key objectives

- Draw, sketch and describe the graphs of trigonometric functions for angles of any size
- Use an extended range of function keys, including trigonometrical and statistical functions relevant across this programme of study

1 a Using a calculator, work out the value of $y = \cos 65°$. Give your solution to 2 significant figures. (2)

b Sketch the graph of $y = \cos x$ for $0 \leqslant x \leqslant 360°$ and use this to find another angle with the same cosine as $65°$. Show your working on your graph and give your solution to 2 significant figures. (3)

2 Given that $y = 5\cos(2x + 30)° - 1$, write down

a the maximum value of y (2)

b the minimum value of y. (2)

(Edexcel Ltd., 2004)

This unit will show you how to

- Work with set notation
- Use a Venn diagram
- Solve problems using set methods and Venn diagrams

Before you start ...

You should be able to answer these questions.

Review

1 Solve

   a $2x + 3 > 5$ **b** $3x + 3 \leqslant x + 7$

   c $2x + 3 \geqslant 5x - 4$ **d** $2 < x - 3 < 5$

   e $4 < 3x - 8 < 10$

Unit A6

2 List the set of integers n for which

   a $1 < 3n - 8 < 5$ **b** $n^2 < 4$

Unit A6

Set language and notation

This spread will show you how to:

- Work with set notation

Keywords

Set
Element
Is a member of
Universal set
Empty set
Union
Intersection
Subset

A **set** is a collection of objects, numbers, etc. which are called the **members** or **elements** of the set.

A set can be defined

- by listing the elements e.g. {2,3,5,7,11}

- in words e.g. {*prime* numbers less than 12}

- algebraically e.g. $A = \{x : 11 < x \leqslant 19 \text{ and } x \text{ is prime}\}$ which gives $A = \{13,17,19\}$

In any given situation $\mathscr{E}$ is the universal set which includes all the elements under consideration.

$\in$	means "**is a member of**" or "is an element of"	$a \in \{a, b, c\}$
$\cap$	stands for the "**intersection**", the elements which are common	$\{a, b, c, d\} \cap \{a, d, e, f\} = \{a, d\}$
$\cup$	stands for the "**union**", all the elements contained	$\{a, b, c, d\} \cup \{d, e, f\} = \{a, b, c, d, e, f\}$
$\varnothing$	stands for the "**empty set**", the set with no elements	$\{a, b, c\} \cap \{d, e, f\} = \varnothing$
$\subset$	stands for "is a **subset** of" or "is part of"	$\{2, 3, 5\} \subset \{2, 3, 5, 7, 11, 13, 17\}$
$n(A)$	stands for "the number of elements in set A"	$A = \{a, b, c, d, e\}$ $n(A) = 5$
A'	stands for "the **complement** of A", the elements of $\mathscr{E}$ which are not in A	$\mathscr{E} = \{a, b, c, d, e\}$ and $A = \{a, b\}$ $A' = \{c, d, e\}$

Two sets are equal if they have exactly the same elements. It is not enough to simply have the same number of elements.

Example

$\mathscr{E} = \{$whole numbers from 10 to 24$\}$
$A = \{$even numbers$\}$
$B = \{$numbers divisible by 5$\}$
Write down the elements of $A \cap B$

$A \cap B = \{10, 20\}$ $A = \{10, 12, 14, 16, 18, 20, 22, 24\}$ $B = \{10, 15, 20\}$

1   Complete with the correct symbol

   a $\{7,11\}$......$\{5,7,9,11,13\}$ **b** 7.......$\{1,2,3,4,5,6,7\}$

   c $\{1,2,3,4\}$......$\{5,6,7\} = \varnothing$ **d** $\{\ \}$.......$\{a,b\} = \{a,b\}$

2   If $A = \{x : 13 < x \leqslant 21$ and x is prime$\}$, write down the elements of A and find $n(A)$.

3   If $\mathscr{E} = \{$integers$\}$ and $Q = \{x : -2 \leqslant 2x - 4 < 7\}$ list the elements of Q and find $n(Q)$.

4   If $A = \{2,3,5,7,11\}$ and $B = \{1,2,3,4,5,6\}$ find

   a $A \cap B$ **b** $A \cup B$ **c** $n(A \cup B)$

5   If $\mathscr{E} = \{1,2,3,4,5,6,7,8,9,10\}$, $A = \{2,4,6,8,10\}$ and $B = P\{1,4,9\}$ find

   a A' **b** $A' \cap B'$ **c** $(A \cup B)'$

   d $A' \cap B$ **e** $n(A')$ **f** $n(A' \cap B')$

6   If $R = \{$rectangles$\}$ and $S = \{$squares$\}$ complete the statement $R \cap S$.

7   Sets are defined as follows

   $\mathscr{E} = \{$cars in an underground car park$\}$ $D = \{$cars with diesel engines$\}$
   $J = \{$cars not made in Japan$\}$ $H = \{$hatchbacks$\}$

   a Write sentences, not in set language, to express
    i $J' \cap H = \varnothing$ **ii** $D \cap H = H$

   b Write in set language "all the cars not made in Japan have diesel engines".

8   The students at The Grammar School must study either Mathematics and Physics, or Physics and Computing or Computing and Biology. Let

    $\mathscr{E} = \{$all the students at The Grammar School$\}$
    $M = \{$students studying Mathematics$\}$
    $P = \{$students studying Physics$\}$
    $C = \{$students studying Computing$\}$
    $B = \{$students studying Biology$\}$

   Complete the following statements

   a $M' =$ **b** $M \cap C =$ **c** $M \cup C =$

9   If $n(A) = n(B)$, does that necessarily imply that sets A and B are equal?

10 If $P \cup Q = P$ what can you deduce about sets P and Q.

This spread will show you how to:

● use a Venn diagram

Keywords

Venn diagram
Intersection
Union
Complement

A Venn diagram can be used to represent sets.

The universal set is represented by a rectangle.

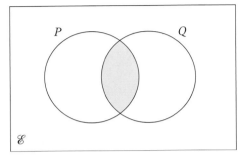

$P \cap Q$

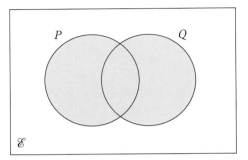

$P \cup Q$

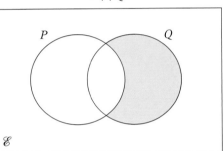

$P' \cap Q$

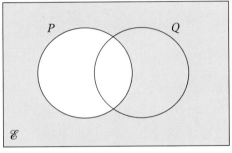

P'

Example

Using the Venn diagram complete the following

a $P \cap Q = \{c, e, d\}$ **b** $P' = \{f, g, h, p, q, w\}$

c $(P \cup Q)' = \{p, q\}$ **d** $n(P) = 8$

e $n(P \cup Q) = 12$ **f** $n(Q') = 7$

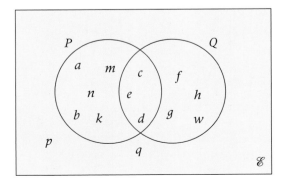

1 Describe using set language the shaded region

a
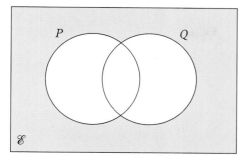

b
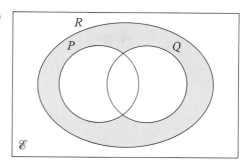

2 On the diagram shade the region $(P \cup Q) \cap (R \cup Q')$

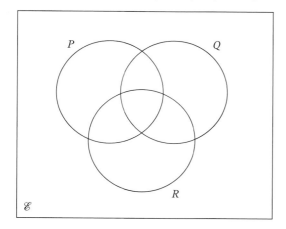

3 Draw separate Venn diagrams to represent the following information:

a $P \cap Q = P$ **b** $P \cap Q = \varnothing$

c $P \cup Q = P$ **d** $R \subset (P \cap Q)$

4 Draw a Venn diagram to show the following information:

$\mathscr{E} = \{1, 2, 3, 4, 5, 6, 7, 8, 9, 10, 11, 12, 13, 14, 15, 16, 17\}$
$A = \{2, 3, 5, 7, 11, 13, 17\}$
$B = \{2, 4, 6, 8, 10, 12, 14, 16\}$
$C = \{3, 6, 9, 12, 15\}$

Use the Venn diagram to find:

a $A \cap B \cap C$ **b** $A \cup B \cup C$ **c** $B \cup C$

d $B \cap C$ **e** $A \cup B'$ **f** $A \cap B'$

g $n(A \cap B \cap C)$ **h** $n(B \cup C)$ **i** $n(\mathscr{E})$

This spread will show you how to:

- solve problems using set methods and Venn diagrams

Keywords

Number of
  elements
Venn diagram

In general for any two sets $n(A \cup B) = n(A) + n(B) - n(A \cap B)$

You can persuade yourself that the above is true by looking at the Venn diagram of any two sets which intersect.

For three sets

$$n(A \cup B \cup C) = n(A) + n(B) + n(C) - (n(A \cap B) + n(A \cap C) + n(B \cap C)) + n(A \cap B \cap C)$$

Example

If $n(P) = 30$, $n(Q) = 22$ and $n(P \cup Q) = 34$, what is $n(P \cap Q)$?

$n(P \cup Q) = n(P) + n(Q) - n(P \cap Q)$

$\qquad 34 = 30 + 22 - n(P \cap Q)$

$n(P \cap Q) = 30 + 22 - 34 = 18$

Example

Disks made on a production line are subject to three kinds of defect, A, B or C. A sample of 1000 articles was inspected and the following results were obtained.

  31 had a type A defect   37 had a type B defect

  42 had a type C defect   11 had both type A and type B defects

   13 had both type B and type C defects   10 had both type A and type C defects

920 had no defects

Find how many disks had all three types of defect.

Draw a Venn diagram to represent the data.
Let x = disks which had all three types of defect
The numbers in the other regions can be found in terms of x.

For example 13 had both type B and type C
defects, so $13 - x$ had only
type B and type C defects.

$1000 = 920 + (10 + x) + (13 + x) + (19 + x)$

$\qquad + (11 - x) + (10 - x) + (13 - x) + x$

$1000 = 920 + 76 + x$

$\qquad x = 4$

The number disks which had all three types
of defect is 4.

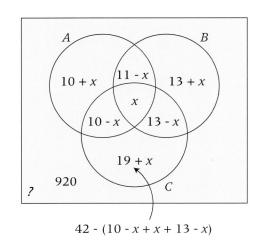

$42 - (10 - x + x + 13 - x)$

1 If $n(P) = 15$, $n(Q) = 7$ and $n(P \cup Q) = 19$, what is $n(P \cap Q)$?

2 If $n(P) = 10$, $n(Q) = 5$ and $n(P \cap Q) = 2$, what is $n(P \cup Q)$?

3 If $n(P) = 10$ and $n(\mathscr{E}) = 25$, what is $n(P')$?

4 If $n(P) = 28$, $n(Q) = 25$, $n(P \cap Q) = 17$ and $n(P \cup Q)' = 0$ find:

    a $n(P \cup Q)$ **b** $n(\mathscr{E})$ **c** $n(Q')$

5 In form 5C there are 27 students of which 17 play tennis, 11 play basketball and 2 play neither game. How many play both tennis and basketball?

6 At The Grammar School there are 100 students in Class 6. They study one or more subjects as follows.

    36 study Physics   42 study Chemistry
    47 study Mathematics   16 study both Physics and Chemistry
    13 study both Chemistry and Mathematics   9 study both Physics and Mathematics
     6 study Physics, Chemistry and Mathematics

Draw a Venn diagram to represent these data.

7 At The Grammar School there are 98 students in Class 5. Of these,

    49 play football   44 play tennis
    35 play basketball   11 play both tennis and football
    14 play both basketball and football.   13 play tennis and basketball

Three students play none of the games and x students play all three games.

    a Draw a Venn diagram to represent these data, showing clearly the number in each separate region.
    b Find the value of x.

8 Three intersecting sets are defined by

    $n(A \cap B \cap C) = 5$ $n(\mathscr{E}) = 68$

          $n(A) = 18$ $n(B) = 15$ $n(C) = 30$

      $n(A \cap B) = 9$ $n(B \cap C) = 10$ $n(A \cap C) = 6$

    a Draw a Venn diagram to represent these data, showing clearly the number in each separate region.
    b Find the value of $n(A \cap (B \cup C)')$.

Exam review

Key objectives

- Work with set notation
- Use a Venn diagram
- Solve problems using set methods and Venn diagrams

1 $\mathscr{E} = \{1, 2, 3, 4, 5, 6, 7, 8\}$
$P = \{2, 3, 5, 7\}$

a List the members of P'

(1)

The set Q satisfies both the conditions $Q \subset P$ **and** $n(Q) = 3$

b List the members of **one** set Q which satisfies both these conditions.

(2)

(IGCSE May 2006 3H Question 15)

2 In a class, all the students study Mathematics (M) or Computing (C).
A number of the students study both of these subjects.
22 students study mathematics.
12 students study computing.
8 study both subjects.

By representing this information in a Venn diagram, or otherwise, find the number of students in the class.

(3)

3 There are 35 students in a group.
18 students play hockey.
12 students play both hockey and tennis.
15 students play neither hockey nor tennis.

Find the number of students who play tennis.

(3)

(IGCSE May 2007 4H Question 15)

This unit will show you how to

- Work with function notation
- Find the range of a function
- Identify values that need to be excluded from the domain
- Find composite and inverse functions
- Find the domain and range of the inverse function

Before you start ...

You should be able to answer these questions.

1 For $y = x^2 - 5x$ complete the table.

x	-5	0	$\frac{1}{2}$	$\frac{2}{3}$
y				

2 Solve

a $\dfrac{3}{x-1} = 5$ **b** $x^2 = 36$

c $\sqrt{\dfrac{2x-5}{3}} = 5$ **d** $x^2 - x = 42$

Review

Unit A1

Unit A3

Function notation

This spread will show you how to:

● Work with function notation

Keywords

Function
Image
Solve

Consider the rule $f(x) = 2x - 1$

f is the name of the function

$f(x)$ tell us what happens to x under f, it is the image of x.

Alternatively, you can write $f: x \mapsto 2x - 1$.

You read this as "f is a function which maps x to $2x - 1$"

Example

For each function write down $f(-1)$

a $f(x) = 2x - 1$ **b** $f(x) = x^3 + 1$

---

a $f(-1) = 2(-1) - 1 = -2 - 1 = -3$ **b** $f(-1) = (-1)^3 + 1 = -1 + 1 = 0$

Example

If $h(x) = 2x^2 + 4$ calculate the values of x for which $h(x) = 9x$

$$\left. \begin{array}{l} h(x) = 2x^2 + 4 \\ h(x) = 9x \end{array} \right\} \, 2x^2 + 4 = 9x$$

$2x^2 - 9x + 4 = 0$ Rearrange to get a standard quadratic $ax^2 + bx + c = 0$

$(2x - 1)(x - 4) = 0$ Factorise

Either $2x - 1 = 0$ or $x - 4 = 0$

$\qquad\qquad x = \dfrac{1}{2} \qquad\qquad x = 4$

Example

For the functions $f(x) = x + 3$ and $g(x) = x^4 - 2$ find the exact values of a given that $f(4) = g(a)$.

$f(4) = 4 + 3 = 7$ Start by finding $f(4)$

$g(a) = a^4 - 2$ To find $g(a)$ replace x by a in $g(x)$

$7 = a^4 - 2$ Since $f(4) = g(a)$

$0 = a^4 - 9$ Make the left hand side of the equation equal to zero

$0 = (a^2 - 3)(a^2 + 3)$ Factorise, using $m^2 - n^2 = (m + n)(m - n)$

Either $a^2 - 3 = 0$ or $a^2 + 3 = 0$ $a^2 + 3 = 0$ is not possible

$\qquad a^2 = 3$

$\qquad a = \pm\sqrt{3}$ Exact values so leave your answer in surd form

1 For each function write down $f(0)$

    a $f(x) = 15 - 3x$ **b** $f(x) = \dfrac{x}{23}$ **c** $f(x) = x^2 - 5$

2 If $f(a) = a - a^3$, find

    a $f(1)$ **b** $f(-1)$ **c** $f\left(\dfrac{1}{2}\right)$

3 If $h:x \mapsto 2x$

    a What is the value of $h(x)$ if $x = 4$?
    b What is the value of $h(-x)$ if $x = 4$?

4 A function is defined by $f(x) = 4x - 2$. Given that $f(a) = a$, find a.

5 If $f(x) = 6x - 5$
    a Evaluate $f(2)$.
    b Calculate the values of x for which $f(x) = x^2$.

6 If $h(x) = \dfrac{3x - 1}{2}$
    a Find $h(0)$ **b** Find x if $h(x) = 7$

7 A cubic polynomial function, f, is defined by $f(x) = ax^3 + bx^2 + cx + d$ where a, b, c and d are constants. For each of the following statements write down the corresponding linear equation.

    a $f(0) = -5$ **b** $f(-1) = -5$ **c** $f(2) = 13$

8 The function g is defined by $g:x \mapsto \dfrac{7}{x}$. Given that $g(m) = n$,

    show that $g(n) = m$.

9 For the functions $f(x) = 2x - 12$ and $g(x) = x^2 + 6$ find the exact values of a given that $f(15) = g(a)$.

10 If $f(x) = x^2$, what is the value of $f(x + 1)$?

Domain and range

This spread will show you how to:

- Find the range of a function
- Identify values that need to be excluded from the domain

Keywords
Function
Domain
Range

Domain is the set of values that you can put into a function.
Range is the set of values produced, the image set.

If you draw a graph for the function $y = f(x)$ then:

- The domain of $f(x)$ is the set of x values.
- The range of $f(x)$ is the set of y values.

For a relation to be a **function** each value in the domain has exactly one image in the range, i.e. *each x value maps to one y value.*

Example

Find the range of each of these functions.

a $f(x) = x^2$ with domain $\{2, 4, 6, 8\}$

b $f(x) = x^2$ for $-3 \leqslant x \leqslant 3$

---

a $f(x) = \{2^2, 4^2, 6^2, 8^2\} = \{4, 16, 36, 64\}$

b Consider a sketch of $y = x^2$ for $-3 \leqslant x \leqslant 3$
The corresponding y values range
from 0 to 9.
The range is $0 \leqslant y \leqslant 9$, or $0 \leqslant f(x) \leqslant 9$.

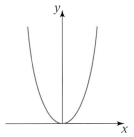

Example

Which value(s) must be excluded from the domain of each function?

a $f(x) = \sqrt{x}$ **b** $h(x) = \dfrac{x+1}{x-1}$

---

a Since $\sqrt{}$ is only defined for positive values, the largest possible domain on which $f(x)$ could be defined is $x \geqslant 0$. All negative values of x must be excluded.

b Since division by zero is not defined, the denominator cannot be zero. The largest possible **domain** on which $h(x)$ could be defined is for all values of x except 1, i.e. $x \neq 1$

1 On two parallel lines, draw a mapping of the function $f : x \mapsto 2x - 1$ for the domain D, where $D = \{-3, -2, -1, 0, 1, 2, 3\}$. Write down the range of f.

2 The diagram shows a sketch graph of $y = \sin x°$

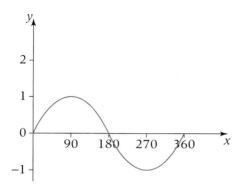

a A function f is given by $f : x \mapsto \sin x°$ for $0° \leqslant x \leqslant 360$. Write down the range of f.

b A function h is given by $h : x \mapsto \sin x°$ for $0° \leqslant x \leqslant 180$. Write down the range of f.

3 A function f if defined by $f(x) = 5x$ for $-2 \leqslant x \leqslant 2$. Write down the range of f.

4 A function f is defined by $f(x) = x^2 - 10x + 24$ for $4 \leqslant x \leqslant 6$.
   a Draw a sketch graph of $y = f(x)$ showing clearly where the graph cuts the axes.

   b Using symmetry show that the coordinates of the lowest point are $(5, -1)$.

   c Write down the range of f.

5 Find the largest possible domain on which each of the following functions could be defined.

   a $f(x) = 1 + \sqrt{x}$ **b** $g(x) = \dfrac{7}{x}$ **c** $h(x) = \dfrac{1}{x + 2}$

6 Which value(s) must be excluded from the domain of each of each of the following functions?

   a $f(x) = \dfrac{x - 5}{x + 5}$ **b** $g(x) = \dfrac{2x + 5}{2x - 5}$

   c $m(x) = \dfrac{8}{8 + x^3}$ **d** $n(x) = \dfrac{1}{\sqrt{x}}$

This spread will show you how to:

- Find composite functions

Keywords
Composite
  function
Order

$fg(x)$ is called the **composite function** of f on g, it is the result of using g first and then f.

$gf(x)$ is called the **composite function** of g on f, it is the result of using f first and then g.

You will usually get different results for fg and gf, so **order** is important.

Example

The functions f, g and h are defined by

$$f : x \mapsto 2x + 1 \qquad g : x \mapsto x^2 - 3 \qquad h : x \mapsto 5x$$

Find

a $fg(2)$ **b** $hgf(1)$

a $fg(2) = f(2^2 - 3) = f(1) = 2 \times 1 + 1 = 3$ **b** $hgf(1) = hg(3) = h(6) = 30$

Example

The functions f and g are defined by

$$f : x \mapsto 2x + 1 \qquad g : x \mapsto x^2 - 3.$$

Find $fg(x)$ and $gf(x)$.

$fg(x) = f(x^2 - 3) = 2(x^2 - 3) + 1 = 2x^2 - 5$ replace x with $x^2 - 3$ in f

$gf(x) = g(2x + 1) = (2x + 1)^2 - 3$ replace x with $2x + 1$ in g

---

Example

The functions f and g are defined by

$$f = x + 3 \qquad g(x) = x^2.$$

Solve the equation $gf(x) = 16$.

$\left. \begin{array}{l} gf(x) = g(x + 3) = (x + 3)^2 \\ gf(x) = 16 \end{array} \right\} \quad (x + 3)^2 = 16$

$x + 3 = \pm 4$ if $a^2 = b \Rightarrow a = \pm\sqrt{b}$

Either $x + 3 = 4$ or $x + 3 = -4$

           $x = 1$ $x = -7$

1 Given that $f:x \mapsto x+5$, $g:x \mapsto x^2+1$ and $h:x \mapsto x-2$ find

   a $fg(2)$ **b** $gf(2)$ **c** $fgh(2)$ **d** $fff(2)$

2 Given that $f(x) = 2x - 5$ and $g(x) = \dfrac{x+5}{2}$ find $fg(x)$ and $gf(x)$.

3 Given that
$$f(x) = x^2 - 4 \text{ and } g(x) = \frac{1}{x-1}, \ x \neq 1$$
find $fg(x)$.

4 Given that
$$f(x) = 3x + 4, \ g(x) = \frac{1}{x}, \ x \neq 0 \text{ and } h(x) = \frac{3+4x}{x}, \ x \neq 0$$
show that $fg(x) = h(x)$.

5 If $f(x) = 2x + 5$ and $g(x) = 4x + a$, find the value of a such that $fg(x) = gf(x)$.

6 The functions f and g are defined by
$$f:x \mapsto x+2 \quad \text{and} \quad g:x \mapsto ax^2 + b$$
where a and b are integers.

Given that $gf:x \mapsto 3x^2 + 12x + 2$ find the values of a and b.

7 Given that $f(x) = x$, $g(x) = x - 2$ and $h(x) = x^2$, solve the equation $fgh(x) = hgf(x) + 2$.

8 The functions f and g are defined by
$$f:x \mapsto \frac{x+2}{x-2}, \ x \neq 2 \quad \text{and} \quad g:x \mapsto \frac{x}{x+3}, \ x \neq -3$$
find

   a $fg(x)$ **b** $gf(x)$ **c** $ff(x)$

9 The diagram shows the graph of $y = f(x)$
Use the graph to write down

   a $f(2)$ **b** $ff(2)$

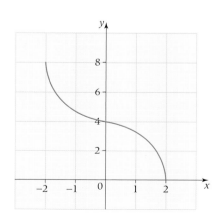

This spread will show you how to:

- Find the inverse function
- Find the domain and range of the inverse function

Keywords

Inverse function
Domain
Range

A function is one–to–one, if each value in the range is the image of only one value in the domain. In this case you can find the **inverse function**.

If $f:x \mapsto \ldots$ the inverse function is $f^{-1}:x \mapsto \ldots$ with the key property $f^{-1}f(x) = ff^{-1}(x) = x$, i.e. $f^{-1}(x)$ cancels the effect of $f(x)$

To find the inverse function $f^{-1}(x)$

- Step 1: Let $f(x) = y$ and rearrange to make x the subject
- Step 2: Replace y with x and call the new function $f^{-1}(x)$

The domain of the inverse function $f^{-1}(x)$ is the range of $f(x)$.
The range of $f^{-1}(x)$ is the domain of $f(x)$

Example

Given that $f(x) = 4x$ and $g(x) = x + 4$ write down f^{-1} and $f^{-1}g$.

$y = 4x$	To find $f^{-1}(x)$, let $f(x) = y$
$x = \dfrac{y}{4}$	Rearrange to make x the subject
$f^{-1}(x) = \dfrac{x}{4}$	Replace y with x and call the new function $f^{-1}(x)$
$f^{-1}g = f^{-1}(x + 4) = \dfrac{x + 4}{4}$	

Example

Find the inverse function of $f(x) = \dfrac{1}{\sqrt{x + 5}}$, $x > 5$ and state the range of the inverse function.

$y = \dfrac{1}{\sqrt{x + 5}}$	
$y^2 = \dfrac{1}{x + 5}$	Square both sides
$x + 5 = \dfrac{1}{y^2} \rightarrow x = \dfrac{1}{y^2} - 5$	Rearrange to make x the subject
$f^{-1}(x) = \dfrac{1}{x^2} - 5$ with range $f^{-1}(x) > 5$	

1 Find the inverse function

a $f(x) = \dfrac{3x - 2}{4}$

b $g(x) = \dfrac{x - 1}{x}, \; x \neq 0$

c $h(x) = \dfrac{x + 5}{x - 5}, \; x \neq 5$

d $p(x) = \sqrt{x - 4}, \; x \geqslant 4$

2 For the function $f(x) = \dfrac{5(2x - 3)}{2}$ find $f^{-1}(15)$.

3 The function f is defined by $f : x \mapsto (x - 3)^2, \; x \geqslant 3$

a State the range of the inverse function $f^{-1}(x)$.

b Find $f^{-1}(x)$.

4 If $f(x) = 2x + 3$ and $f(g(x)) = x$ find the function $g(x)$.

5 Given that $f(x) = \dfrac{5x + 2}{4}$ and $g(x) = \dfrac{4x - 2}{5}$

a Find $fg(x)$

b Describe clearly the relationship between the functions f and g.

c Write down the exact value of $fg\left(\sqrt{5}\right)$.

6 The function f is defined by

$$f : x \mapsto \dfrac{3}{2x + 1}, \; x \neq -0.5$$

Solve the equation $f^{-1}(x) = f(x)$.

7 Given that $f(x) = \dfrac{2}{2x + 1} + \dfrac{3}{(x - 1)(2x + 1)}, \; x > 1.$

a Prove that $f(x) = \dfrac{1}{x - 1}$.

b Find $f^{-1}(x)$.

c Find the range of $f^{-1}(x)$.

8 Given that

$$f(x) = 3x + 4 \text{ and } g(x) = \dfrac{1}{x}, \; x \neq 0$$

show that $g^{-1}f^{-1} = (fg)^{-1}$.

Exam review

Key objectives

- Work with function notation
- Find the range of a function
- Identify values that need to be excluded from the domain
- Find composite and inverse functions
- Know and use the fact that $ff^{-1}(x) = f^{-1}f(x) = x$

1 Let $f(x) = x^2 + 2x - 1$

   a Find the value of $f(0)$. (1)

   b Find a simplified expression for $f(x - 1)$. (3)

   c Solve the equation $f^{-1}(x) = 4$. (3)

2 Let $f(x) = 2x + 1$ and $g(x) = 2x^2 - 1$.

   Find $gf(x)$. You must simplify your answer. (4)

3 Let $f(x) = \dfrac{1}{x + 2}$

   a Find $f(0)$. (1)

   b Find $ff(-1)$. (2)

   c Find $f^{-1}(x)$. (2)

   d When value of x must be excluded from the domain of $f(x)$? (1)

   e Solve the equation $ff(x) = 1$. (4)

4 Write down the maximum possible domain of $f(x) = \sqrt{x} + \dfrac{1}{2x - 3}$ (2)

5 If $f(x) = 2x + 3$ and $g(f(x)) = x$, find $g(x)$. (3)

6 Sketch the curve of $y = \sin x$ for x from $0°$ to $180°$.

   If $f(x) = \sin x$ for $0° \leqslant x \leqslant 180°$ write down the range of $f(x)$. (3)

This unit will show you how to

- Find and apply rates of change in simple cases
- Differentiate functions of the form ax^n
- Use differentiation to find the gradient of a curve at a given point
- Use differentiation to find point(s) on a curve with a given gradient
- Find the equation of a tangent to a curve at a given point
- Use differentiation to find the turning points of a curve
- Distinguish between maxima and minima in simple cases
- Find the equation of the line of symmetry of a quadratic graph
- Use differentiation to solve optimisation problems
- Use differentiation to find the speed and acceleration of a moving body

Before you start ...

You should be able to answer these questions.

1 For the straight line $y = mx + c$
  write down
  a the gradient   **b** the y-intercept

2 For the straight line $8x - 3y = 9$
  write down
  a the gradient   **b** the y-intercept

3 For the straight line $ax + by + c = 0$
  write down
  a the gradient   **b** the y-intercept

Review

Unit A4

Unit A4

Unit A4

This spread will show you how to:

● Find and apply rates of change in simple cases

Keywords
Gradient
Rate of change
dy by dx

The gradient of the straight line $y = mx + c$ is m.

The **gradient** tells you how y changes with respect to x.

It is the **rate of change** of y with respect to x.

You can think of the gradient as the $\dfrac{\text{difference in } y}{\text{difference in } x} = \dfrac{dy}{dx}$

You read $\dfrac{dy}{dx}$ as "dy by dx" or "the rate of change of y with respect to x"

For any straight line of the form $y = mx + c$, $\dfrac{dy}{dx} = m$.

Example

Consider the straight line $9y - x = 9$

a What is the gradient of the line?

b What is $\dfrac{dy}{dx}$?

c What is the rate of change of y with respect to x?

---

a $y = \dfrac{1}{9}x + 1$ Write in the form $y = mx + c$

   gradient $= \dfrac{1}{9}$

b $\dfrac{dy}{dx} = \dfrac{1}{9}$

c the rate of change of y with respect to x is $\dfrac{1}{9}$

Example

The **radius** of a circular oil slick is $t + 2$ metres at time t minutes.

a Express the **circumference** C metres at time t minutes in terms of t.

b Find the **rate**, i.e. $\left(\dfrac{dC}{dt}\right)$, at which the circumference is increasing

---

a $C = 2\pi(t + 2)$ for a circle $C = 2\pi r$

b $C = 2\pi t + 4\pi$ compare to $y = mx + c$ and $\dfrac{dy}{dx} = m$

   $\dfrac{dC}{dt} = 2\pi$

1 A straight line has equation $y = -2x + 5$.

   a What is the gradient of the line?

   b What is $\dfrac{dy}{dx}$?

   c What is the rate of change of y with respect to x?

2 A straight line has equation $2y + 3x - 4 = 0$.

   a What is the gradient of the line?

   b What is $\dfrac{dy}{dx}$?

   c What is the rate of change of y with respect to x?

3 Variables s and t are related though the formula $s = \dfrac{t}{3} + 6$.

   a What is $\dfrac{ds}{dt}$?

   b What is the rate of change of s with respect to t?

4 What can be deduced about the equation of a line which has $\dfrac{dy}{dx} = 1$?

5 A firm charges a basic fee of £13 plus £7 per hour for each engineer sent out on repair work. If one engineer is called out for t hours, and C is the charge in £, complete the following:

   a $\dfrac{dC}{dt} =$ **b** $C =$

6 The circumference C of a circular oil slick of radius r is given by $C = 2\pi r$.

   a What is $\dfrac{dC}{dr}$?

   b What is the rate of change of C with respect to r?

7 The **radius** of a circular oil slick is $5t + 4$ metres at time t minutes.

   a Express the **circumference** C metres at time t minutes in terms of t.

   b Find the **rate**, i.e. $\left(\dfrac{dC}{dt}\right)$, at which the circumference is increasing.

This spread will show you how to:

• Differentiate functions of the form ax^n

Keywords

Differentiation
Differentiate
Derivative

Differentiation is just a way to work out $\dfrac{dy}{dx}$, the derivative.

If $y = x^n$ then $\dfrac{dy}{dx} = nx^{n-1}$

If $y = ax^n$ then $\dfrac{dy}{dx} = anx^{n-1}$ where a is any constant.

Example

Differentiate

a $x + 5$ **b** π **c** x^{10}

---

a 1   the derivative of $mx + c$ is m

b 0 π is a constant, it does not change!

c $10x^{10-1} = 10x^9$

Example

Find $\dfrac{dy}{dx}$

a $y = 2x^3 + 5x - 10$ **b** $y = \dfrac{x^3}{3} + \dfrac{x^2}{2} - 123$

---

a $\dfrac{dy}{dx} = 6x^2 + 5$ Differentiate each term separately and then add them together

b $y = x^2 + x$ $\dfrac{x^3}{3}$ is the same as $\dfrac{1}{3}x^3$ and when you

differentiate $3\left(\dfrac{1}{3}x^{3-2}\right) = x^2$

Example

Find $\dfrac{ds}{dt}$

a $s = (2t - 3)^2$ **b** $s = \dfrac{t^2 + t}{t^3}$

---

a $y = 4t^2 - 12t + 9$ get rid of the brackets by expanding and simplifying

$\dfrac{ds}{dt} = 8t - 12$

b $s = \dfrac{t^2}{t^3} + \dfrac{t}{t^3} = t^{-1} + t^{-2}$ get rid of fractions using $\dfrac{x^n}{x^m} = x^{n-m}$

$\dfrac{ds}{dt} = -t^{-2} - 2t^{-3}$ remember $-1 -1 = -2$ and $-2 -1 = -3$

1 Differentiate

  a $2x - 5$
      b $2x^2 + 5x$
      c $x^2 - x + 5$

  d $3x^2 + 5x - 3$
      e $10x^5 + 3x^4 + \dfrac{1}{2}x^2$
  f $\dfrac{1}{4}x^4 - \dfrac{1}{3}x^3 + x^2 - 5$

2 Find $\dfrac{dy}{dx}$

  a $y = 5 - 3x$
      b $y = 5$
      c $y = 6x^3 - 7x^2 + x - 2$

  d $y = x^{-1}$
      e $y = 2x + 4x^{-2}$
      f $y = x^{-1} - x^{-3}$

3 Find $\dfrac{dh}{dt}$

  a $h = t^2$
      b $h = 0.5t^2$
      c $h = 30 + 5t - 4t^2$

4 Find $\dfrac{dp}{dv}$ if $p = 10 - \dfrac{5v}{33} + \dfrac{v^2}{100}$.

5 Find $\dfrac{dA}{dr}$, where π is the well known constant

  a $A = r^2$
      b $A = \pi r^2$

  c $A = 4\pi r^2$
      d $A = 10\pi r + 2\pi r^2$

6 Find $\dfrac{dV}{dr}$, where π is the well known constant

  a $V = 6\pi r^2$
      b $V = \dfrac{4}{3}\pi r^3$

7 Find $\dfrac{dy}{dx}$

  a $y = x(x - 1)$
      b $y = (2x + 3)(x - 1)$
      c $y = x^2(x + 5)^2$

8 Find $\dfrac{dy}{dx}$

  a $y = \dfrac{5x^2 + 3}{2}$
      b $y = \dfrac{x^2 + x}{x}$
      c $y = x + \dfrac{1}{x}$

  d $y = \dfrac{3x + 6x^2 + x^4}{3x^4}$
      e $y = \left(\dfrac{1}{x} + x\right)^2$
      f $y = (2x + 1)\left(\dfrac{1}{2x} + 2\right)$

Gradient of a curve

This spread will show you how to:

Keywords

Gradient function
Gradient

- Use differentiation to find the gradient of a curve at a given point
- Use differentiation to find point(s) on a curve with a given gradient

To find the gradient of a curve at a given point you use differentiation.

- Step 1: Find $\dfrac{dy}{dx}$ (this is the derivative, also known as the gradient function)

- Step 2: Evaluate $\dfrac{dy}{dx}$ at the required point by substituting for x

Example

Find the gradient of the curve $y = x^2 + 3x - 1$ at $x = 1$ and at $x = -2$.

$\dfrac{dy}{dx} = 2x + 3$ first find $\dfrac{dy}{dx}$

When $x = 1$, $\dfrac{dy}{dx} = 2 \times 1 + 3 = 5$ gradient of the curve at $x = 1$ is $\dfrac{dy}{dx}$ at $x = 1$

The gradient of the curve at $x = 1$ is 5

When $x = -2$, $\dfrac{dy}{dx} = 2 \times (-2) + 3 = -1$

The gradient of the curve at $x = -2$ is -1

Example

Find the coordinate of the point on the curve $y = x^2 + 3x - 1$ where the gradient is 5.

$\dfrac{dy}{dx} = 2x + 3$

$2x + 3 = 5$ at the point where the gradient is 5, $\dfrac{dy}{dx} = 5$

$x = 1$, $y = 1^2 + 3 \times 1 - 1 = 3$ to find the y coordinate, substitute

                                     $x = 1$ into the equation of the curve

The gradient of the curve is 5 at the point $(1,3)$.

Example

Consider the curve $y = x^3 - 6x^2 + mx$. The gradient of the curve at $x = 1$ is 2. Find the value of m.

$\dfrac{dy}{dx} = 3x^2 - 12x + m$

$3 - 12 + m = 2$ $\dfrac{dy}{dx} = 2$ when $x = 1$

$m = 11$

1 If $y = x^2$ find the gradient of the curve at the point with $x = 2$.

2 If $y = x^2 + 3x - 1$ find the gradient at $(1,3)$.

3 If $y = x^3 + 2x^2 - 5x + 26$ find the gradient of the curve at the point with $x = \dfrac{1}{2}$.

4 Find the coordinates of the points on $y = x^3 - x^2 - 2x$ where the gradient of the curve is -1.

5 Find the x – coordinate of the point on the curve $y = x^2 - 3x + 5$ at which the gradient is $\dfrac{1}{3}$.

6 Find the gradient of the curve $y = x^3 + \dfrac{1}{x^3}$ at the point with $x = 1$.

7 If $s = 4t^2 - t^3$ find the rate of change of s with respect to t when $t = 1$.

8 The temperature of a liquid at time t seconds is given by T where $T = 2t^2 - 5t + 13$. Find the time when the rate of change of the temperature is $-2°$/second.

9 Find the coordinates of the point on the curve $y = x^2 - 3x$ at which the gradient is the same as the gradient of the line $y - 2x + 7 = 0$

10 For the curve $y = x^3 - 2x^2 - 4x + 5$ find

   a $\dfrac{dy}{dx}$

   b The gradient of the curve at the point with $x = 1$

   c The values of x for which $\dfrac{dy}{dx} = 0$

11 The gradient of the curve $y = px^2 + 8$ at the point $(2, q)$ is 8.
Find the values of the constants p and q.

12 A curve has equation $y = px^2 + 2x + q$, where p and q are constants.
The curve passes though $(0, 5)$.

At the point with $x = 1$ the gradient of the tangent to the curve is 8.
Find the values of the constants p and q.

This spread will show you how to:

- Find the equation of a tangent to a curve at a given point.

Keywords
Tangent
Gradient
Equation
Parallel lines

A **tangent** is a line that touches a curve at one point only.

The gradient of a curve at a given point is the same as the gradient of the tangent at that point.

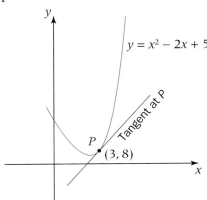

$y = x^2 - 2x + 5$

Tangent at P

P

$(3, 8)$

To find the *equation of a tangent*

- Step 1: Find $\dfrac{dy}{dx}$ and evaluate at the point where you want the tangent.
  This is the gradient of the tangent, *m*.
- Step 2: Use $y = mx + c$. To find *c* you need to use the coordinates of a point on the tangent.

Parallel lines have the same gradient.

Example

Find the equation of the tangent (see diagram above) to the curve $y = x^2 - 2x + 5$ at the point where $x = 3$

The point on the curve is at $x = 3$, $y = 3^2 - 2 \times 3 + 5 = 8$, i.e. (3,8)

$\dfrac{dy}{dx} = 2x - 2$

At the point (3, 8), $\dfrac{dy}{dx} = 2 \times 3 - 2 = 4$ the gradient of the tangent is $\dfrac{dy}{dx}$ evaluated at $x = 3$

The equation of the tangent is $y = 4x + c$ use $y = mx + c$

$8 = 4 \times 3 + c$ so $c = -4$ the point (3, 8) must be on the tangent

The equation of the tangent is $y = 4x - 4$

1 Find the equation of the tangent to the curve $y = x^2 - 2x + 4$ at the point $(1, 3)$.

2 Find the equation of the tangent to the curve $y = x^3 - x^2 + 4x + 2$ at the point $(1, 6)$.

3 Find the equation of the tangent to the curve $y = \dfrac{1}{x^2}$ at the point $(1, 1)$.

4 Find the equation of the tangent to the curve $y = x - 2x^2$ at the point where $x = 2$.

5 Find the equation of the tangent to the curve $y = 3 + 2x - x^3$ at the point where $x = -2$.

6 Find the equation of the tangent to the curve $y = x - \dfrac{4}{x^2}$ at the point where $x = \dfrac{1}{2}$.

7 Find the equation of the tangent to the curve $y = \dfrac{x^3 + 3x^2}{2x}$ at the point where $x = 2$.

8 Find the equation of the tangent to the curve with $\dfrac{dy}{dx} = -3 + 4x$ at the point $(-2, 3)$.

9 The gradient of the curve C is given by $\dfrac{dy}{dx} = (2x - 1)^2$.

   Show that there is no point on C at which the tangent is parallel to the line $y = -2x$.

10 Find the equation of the tangent to the curve $y = 2x^2 + 3x - 4$ at the point where it crosses the y-axis.

11 **a** Find the points of intersection of the curve $y = x^2 - 6x$ and the line $y = -6x + 4$.

   b Show that equation of the tangent to the curve at one of the points of intersection is $y = -2x - 4$.

   c Find the equation of the tangent at the other point of intersection.

12 The curve C has equation $y = \dfrac{1}{3}x^3 - x^2 + x + 5$. The point P has $x = -1$.

   Another point Q also lies on C. The tangent to C at Q is parallel to the tangent to C at P. Find the x-coordinate of Q.

This spread will show you how to:
- Use differentiation to find the turning points of a curve
- Distinguish between maxima and minima in simple cases
- Find the equation of the line of symmetry of a quadratic graph

You can use differentiation to find a turning point.

At a **turning point** the curve "turns" and the gradient must be zero.

You can decide if it's a maximum or a minimum by referring to the shape of the graph.

A quadratic graph is symmetric about its turning point (vertex). If the turning point has coordinates (a, b) then $x = a$ is the equation of the **line of symmetry** of the graph.

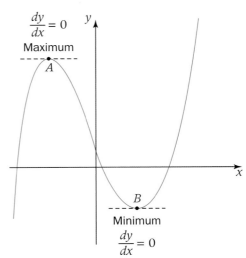

A, B are turning points

$\frac{dy}{dx} = 0$

Maximum

A

B

Minimum

$\frac{dy}{dx} = 0$

Example

a Find the turning point of the curve $y = x^2 - 2x + 5$.

b Is it a maximum or a minimum?

c Write down the equation of the line of symmetry of the graph $y = x^2 - 2x + 5$.

---

a $\frac{dy}{dx} = 2x - 2$

$2x - 2 = 0$ $\frac{dy}{dx} = 0$ at a turning point

$x = 1, y = 4 \rightarrow (1, 4)$

b $(1, 4)$ is a minimum $y = x^2 - 2x + 5$ is $\cup$ shaped

c $x = 1$ the graph is symmetric about the maximum

Example

Find the maximum value of $s = t(5 - 2t)$

$s = 5t - 2t^2$ get rid of the brackets by expanding and simplifying

$\frac{ds}{dt} = 5 - 4t$

$5 - 4t = 0$ $\frac{dy}{dx} = 0$ at a maximum

$t = 1.25$ it is maximum because $s = 5t - 2t^2$ is $\cap$ shaped

maximum $s = 3.125$ use $t = 1.25$ in $s = 5t - 2t^2$

1 Find the turning point(s)

   a $y = 2x^2 + 6x + 7$ **b** $y = -2x^2 + 6x + 7$ **c** $y = x^3 - 27x + 10$

2 Find the minimum value of $y = x^2 - 7x$.
Explain how you know it is a minimum.

3 Find the maximum value of $y = 4x - x^2 - 5$.
Explain how you know it is a maximum.

4 The graph of $y = x^2 + \dfrac{16}{x}$ has a minimum point for $x > 0$.

   Calculate the coordinates of this minimum point.

5 The sketch shows the graph of $y = x^3 - 12x + 5$.

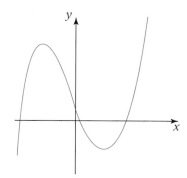

   Use differentiation to find the coordinates of the maximum and
minimum points. You must identify which is the maximum
and which the minimum.

6 For each function find the equation of the line of symmetry of the
corresponding graph.

   a $y = 8x - x^2$ **b** $y = x^2 - 8x$ **c** $y = (x - 2)^2$

7 Find the maximum and minimum values of $(x - 2)^2 (x + 3)$.

8 Find the maximum value of $v = t(4 - t)$.
Explain how you know it is a maximum.

This spread will show you how to:

● Use differentiation to solve practical problems

Keywords
Maximum
Minimum
Optimisation

You can use differentiation to find the *optimal* value that *maximizes* or *minimises* a quantity.

Given a quantity $h(x)$ i.e. h as a function of x, you can find the value of x for maximum or minimum h by solving $\dfrac{dh}{dx} = 0$.

Example

A rectangle has a perimeter of 36 m. Find the maximum possible area.

Let the width of the rectangle be x
The length of the rectangle is $18 - x$

$\dfrac{36 - 2x}{2} = 18 - x$

The area is A where $A = x(18 - x) = 18x - x^2$

$$\frac{dA}{dx} = 18 - 2x$$

$18 - 2x = 0$ for maximum A, $\dfrac{dA}{dx} = 0$

$\qquad x = 9$

Maximum area $= 9(18 - 9) = 81\,\text{m}^2$ it is a maximum because
$\qquad\qquad\qquad\qquad\qquad\qquad\qquad\quad$ $A = 18x - x^2$ is $\cap$ shaped

The maximum possible area is 81 m².

Example

A ball was thrown vertically upward and, after t seconds, its height h metres above the ground was given by $h = 100 + 50t - 5t^2$. Find the greatest height above the ground reached by the ball.

$$\frac{dh}{dt} = 50 - 10t$$

$50 - 10t = 0$ for maximum h, $\dfrac{dh}{dt} = 0$

$\qquad t = 5$

Maximum height

$\qquad = 100 + 50 \times 5 - 5 \times 5^2 = 225\,\text{m}$ use $t = 5$ in $h = 100 + 50t - 5t^2$

1 A rope of length 100 m is used to fence off a rectangular area as shown in the diagram. The width of the rectangle is x m.

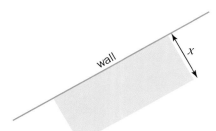

a Write down an expression for the length of the rectangle.
b Write down an expression for the area, A, of the rectangle.
c Find $\dfrac{dA}{dx}$.
d Hence find the maximum value of the area.
Explain how you know it is a maximum.

2 A rectangle has an area of 100 cm². The width of the rectangle is x cm.

a Show that the perimeter of the rectangle is given by P where $P = 2x + \dfrac{200}{x}$.
b Find the minimum value of the perimeter.

3 A solid cylinder has a volume of 8π m³. The height of the cylinder is x and the radius r.

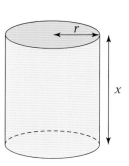

a Show that $x = \dfrac{8}{r^2}$.
b Show that the total surface area of the cylinder is given by A where $A = 2\pi r^2 + \dfrac{16\pi}{r}$.
c Find the minimum possible surface area.

4 The sum of two numbers a and b is 100.
a Show that the product of the two numbers is given by P where $P = 100a - a^2$.
b Find the values of a and b so that P is a maximum.
Explain how you know it is a maximum.

This spread will show you how to:

- Use differentiation to find the speed and acceleration of a moving body

s is the distance displacement for a particle at time t.

The rate of change of displacement with respect to time is
the speed (velocity) of a particle v where $v = \dfrac{ds}{dt}$.

The rate of change of velocity with respect to time is the acceleration of
the particle a where $a = \dfrac{dv}{dt}$.

"initial" usually means when $t = 0$ "instantaneous rest" means $v = 0$

Maximum velocity can be found by finding first the time when $\dfrac{dv}{dt} = 0$
and using this in the expression for v.

Example

Given that the displacement of a particle is given by $s = 100 - 5t - 2t^2 + t^3$
(metres), find:

a an expression for the speed of the particle at time t seconds
b an expression for the acceleration of the particle at time t seconds
c the initial value of the acceleration of the particle.

a $v = \dfrac{ds}{dt} = -5 - 4t + 3t^2$

b $a = \dfrac{dv}{dt} = -4 + 6t$

c $a(t = 0) = -4$ ms^{-2} initially $t = 0$

Example

A lorry accelerates from rest. At time t seconds the velocity of the lorry
is v ms^{-1}.

Given that $v = 3t - \dfrac{t^2}{2}$, for $0 \leqslant t \leqslant 3$ find the acceleration when

the velocity is 4 ms^{-1}.

$3t - \dfrac{t^2}{2} = 4$ find t when $v = 4$ ms^{-1}

$6t - t^2 = 8$ or $0 = t^2 - 6t^2 + 8$

$(t - 2)(t - 4) = 0$

$t = 2$ or $t = 4$ not allowed since $0 \leqslant t \leqslant 3$

$a = \dfrac{dv}{dt} = 3 - t$ and when $t = 2$, $a = 3 - 2 = 1$ ms^{-2}.

1 A particle moves in a straight line. Its displacement t seconds after leaving the fixed point O is x metres, where $x = 100 - 5t^2 + t^3$. Find:

    a an expression for the velocity of the particle at time t

    b an expression for the acceleration of the particle at time t

    c the initial value of the displacement

    d the initial value of the velocity

    e the initial value of the acceleration.

2 Given that the velocity of a particle is given by $v = t^2 - 10t + 30$, find the velocity when the acceleration is zero.

3 Given that the displacement of a particle is given by $x = 30 - \dfrac{20}{t^2}$, find the acceleration when the velocity is 5 m s^{-1}.

4 A flare is launched from a balloon and moves in a vertical line. At time t seconds, the height of the flare is h metres, where $h = 563 - 10t - \dfrac{260}{t}$ for $t \geq 2$

The flare is launched when $t = 2$.

    a Find the height of the flare immediately after it is launched.

    b Find the velocity of the flare immediately after it is launched.

5 A particle moves in a straight line. Its displacement t seconds after leaving the fixed point O is s metres, where $s = t(t - t^2)$. Find:

    a an expression for the velocity of the particle at time t

    b an expression for the acceleration of the particle at time t

    c the maximum value of the velocity.

6 A particle moves along a straight line so that after t seconds its distance x metres from a fixed point O on the line is given by

$$x = 15 - 14t + 6t^3.$$

Find

    a an expression for the velocity of the particle at time t

    b an expression for the acceleration of the particle at time t

    c the distance travelled by the particle in the first second

    d the speed of the particle when $t = 1$

    e the acceleration of the particle when $t = 1$

    f the value of t for which the particle is instantaneously at rest.

    g the minimum value of x.

Key objectives

- Find and apply rates of change in simple cases
- Differentiate functions of the form ax^n
- Use differentiation to find the gradient of a curve at a given point and find point(s) on a curve with a given gradient
- Find the equation of a tangent to a curve at a given point
- Use differentiation to find the turning points of a curve, distinguishing between maxima and minima in simple cases and find the equation of the line of symmetry of a quadratic graph
- Use differentiation to solve problems including finding the speed and acceleration of a moving body

1 Differentiate with respect to x

    a $-x^2 - 4x$ (2)

    b $(x^2 - 3)(1 - x)$ (4)

    c $\dfrac{x^3 + 2x^5}{x^5}$ (3)

    d $\dfrac{7}{x^6}$ (2)

2 The volume of water, V, in cm^3, that remains in a leaking tank after t seconds is given by $V = 60\,000 - 900t + 0.75t^2$

Find the rate of change of the volume with respect to t when $t = 10$ seconds.   (4)

3 The speed, v, in metres per second, of a body after t seconds is given by $v = 15t - 6t^2$

Find the acceleration when $t = 0$.   (4)

4 Show that the equation of the tangent to the curve $y = 2x^2 + x$ at the point $x = a$ is given by $y = (4a + 1)x - 2a^2$ (5)

5 Use differentiation to find the minimum value of y, given that $y = 6x^2 - 12x + 16$

Explain why it is a minimum and not a maximum.   (5)

In your Edexcel IGCSE examination you will be given a formula sheet like the one on this page.

You should use it as an aid to memory. It will be useful to become familiar with the information on this sheet.

Pythagoras' Theorem

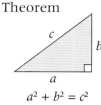

$a^2 + b^2 = c^2$

Volume of cone $= \frac{1}{3}\pi r^2 h$

Curved surface area of cone $= \pi r l$

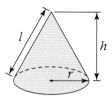

Volume of sphere $= \frac{4}{3}\pi r^3$

Surface area of sphere $= 4\pi r^2$

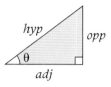

$\text{adj} = \text{hyp} \times \cos\theta$

$\text{opp} = \text{hyp} \times \sin\theta$

$\text{opp} = \text{adj} \times \tan\theta$

or

$\sin\theta = \dfrac{\text{opp}}{\text{hyp}}$

$\cos\theta = \dfrac{\text{adj}}{\text{hyp}}$

$\tan\theta = \dfrac{\text{opp}}{\text{adj}}$

In any triangle ABC

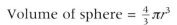

Sine rule $\dfrac{a}{\sin A} = \dfrac{b}{\sin B} = \dfrac{c}{\sin C}$

Cosine rule $a^2 = b^2 + c^2 - 2bc \cos A$

Area of triangle $= \frac{1}{2} ab \sin C$

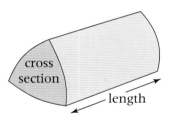

Volume of prism = area of cross section × length

Area of a trapezium $= \frac{1}{2}(a + b)h$

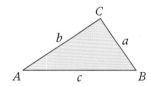

Circumference of circle $= 2\pi r$

Area of circle $= \pi r^2$

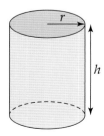

Volume of cylinder $= \pi r^2 h$

Curved surface area of cylinder $= 2\pi rh$

The Quadratic Equation

The solutions of $ax^2 + bx + c = 0$, where $a \neq 0$, are given by

$$x = \frac{-b \pm \sqrt{b^2 - 4ac}}{2a}$$

Answers

N1 Before you start ...

1 a 1, 2, 3, 4, 6, 12

  b 1, 2, 3, 5, 6, 10, 15, 30

  c 1, 2, 3, 4, 5, 6, 8, 10, 12, 15, 20, 24, 30, 40, 60, 120

  d 1, 2, 3, 4, 5, 6, 8, 9, 10, 12, 15, 18, 20, 24, 30, 36, 40, 45, 60, 72, 90, 120, 180, 360

2 2, 3, 5, 7, 11, 13, 17, 19, 23, 29, 31, 37, 41, 43, 47, 53, 59, 61, 67, 71, 73, 79, 83, 89, 97

3 a 6   **b** 60

4 a 10^2 **b** 10^3 **c** 10^4 **d** 10^6

N1.1

1 a 10^2 **b** 10^1 **c** 10^3 **d** 10^0

  e 10^4 **f** 10^6 **g** 10^5 **h** 10^8

2 a 10^{-2} **b** 10^{-1} **c** 10^{-3} **d** 10^{-5}

  e 10^{-4} **f** 10^{-7} **g** 10^{-6} **h** 10^0

3 a 1000   **b** 1 000 000

  c 100 000   **d** 1 000 000 000

  e 10 000   **f** 10

  g 100   **h** 10 000 000

4 a 1   **b** 0.01   **c** 0.000 01 **d** 0.001

  e 0.000 000 1   **f** 0.1

  g 0.0001   **h** 0.000 001

5 a 10^5 **b** 10^9 **c** 10^8 **d** 10^3

  e 10^4 **f** 10^4

6 a 10^{-2} **b** 10^{-4} **c** 10^{-8} **d** 10^{-8}

  e 10^{-8} **f** 10^6

7 a 10^2 **b** 10^{12} **c** 10^{-6} **d** 10^0

8 a 3   **b** 5   **c** 3   **d** 2.5

9 a 2   **b** 5   **c** −1   **d** 1.3

10 a 4.005   **b** 5   **c** 7   **d** −4

N1.2

1 a 1.375×10^3 **b** 2.0554×10^4

  c $2.314\,55 \times 10^5$ **d** 5.8×10^9

2 a 1.4×10^3 **b** 2.1×10^4

  c 2.3×10^5 **d** 5.8×10^9

3 9×10^3, 1.08×10^4, 3.898×10^4, 4.05×10^4, 4.55×10^4, 5×10^4

4 a 63 500   **b** 910 000 000 000 000 000

  c 111   **d** 299 800 000

5 a 2.15×10^4 **b** 7×10^3

  c $1.225\,16 \times 10^{20}$ **d** 1.5×10^7

6 a 6×10^7 **b** 4×10^{12} **c** 1.5×10^9

  d 7×10^0 **e** 1.5×10^{10} **f** 1.2×10^{11}

7 a 5.48×10^{16} **b** 8.41×10^{27}

  c 3.01×10^{15} **d** 1.11×10^{19}

8 ai 80 240 000   **aii** 4 400 000

  aiii 9 928 000   **aiv** 534 650 000

  b 4 400 000, 9 928 000, 80 240 000, 534 650 000

9 a 5.7×10^{17} **b** 2.6×10^2 **c** 4.3×10^4

10 5.1×10^4 km

11 1.09×10^{27} cm^3

N1.3

1 a 3.4×10^{-4} **b** 1.067×10^{-1}

  c 9.1×10^{-6} **d** 3.15×10^{-1}

  e 5.05×10^{-5} **f** 1.82×10^{-2}

  g 8.45×10^{-3} **h** 3.06×10^{-10}

2 a 2.86×10^{-4} **b** 8.20×10^{-6}

  c 2.35×10^{-5} **d** 2.36×10^{-6}

  e 1.54×10^{-3} **f** 1.09×10^2

  g 1.00×10^{-5} **h** 2.36×10^{-7}

3 a 0.0045   **b** 0.000 031 7

  c 0.000 001 09   **d** 0.000 000 979

4 a 0.02   **b** 0.1   **c** 1.5×10^{-8} **d** 2000

  e 0.000 002 5   **f** 31 000

5 a 3.44×10^{-2} **b** 7.19×10^{-13}

  c 2.89×10^{-5} **d** 6.51×10^{-2}

  e 2.29×10^{-6} **f** 2.35×10^3

6 a 3×10^{-2} **b** 6×10^{-13}

  c 3×10^{-5} **d** 6.4×10^2

  e 2.4×10^{-6} **f** 2×10^3

7 3.3×10^{-9} seconds

8 a 1.2×10^{-2} kg   **b** 2.0×10^{-26} kg

9 5.1135×10^{12} days

N1.4

1 a Divisible by 3   **b** Divisible by 5

  c Divisible by 3   **d** Divisible by 3

  e Divisible by 2

2 a 3×7 **b** 2^3 **c** 3×5

  d $2 \times 3^2 \times 5$ **e** $2^2 \times 31$

3 a $2^2 \times 3^2 \times 5^2$ **b** $2 \times 3^2 \times 5 \times 7$

  c $7 \times 11 \times 13$ **d** $3^2 \times 5 \times 7^2$

  e 3×457 **f** $3^4 \times 11$

  g $2^2 \times 17 \times 41$ **h** $3^3 \times 53$

  i 11×307 **j** $2 \times 3^3 \times 5 \times 13 \times 701$

4

Number	Prime? (Yes/No)	Prime factor decomposition
2000	No	$2^4 \times 5^3$
2001	No	$3 \times 23 \times 29$
2002	No	$2 \times 7 \times 11 \times 13$
2003	Yes	
2004	No	$2^2 \times 3 \times 167$
2005	No	5×401
2005	No	$2 \times 17 \times 59$
2007	No	$3^2 \times 223$
2008	No	$2^3 \times 251$
2009	No	$7^2 \times 41$
2010	No	$2 \times 3 \times 5 \times 67$

5 ai $2^6 \times 3^3$ **aii** $3^2 \times 47$

  aiii $2^2 \times 7 \times 29$ **aiv** 23

  bi 3   **bii** 47   **biii** 203   **biv** 23

6 2×42, 3×28, 7×12, 4×21, 6×14, 7×12

7 a $3 \times 5 \times 11^2$

  b $3 \times 5 \times 121$, $3 \times 11 \times 55$, $5 \times 11 \times 33$, $11 \times 11 \times 15$

N1.5

1 **a** 5　　**b** 16　　**c** 3　　**d** 5　　**e** 14
2 **a** 48　　**b** 800　　**c** 66　　**d** 416　　**e** 280
3 **a** 60, 1260　　**b** 7, 8085　　**c** 48, 1680
　d 2, 9216　　**e** 2, 314 706
4 **a** $2^5 \times 3^2$　　**b** $3^3 \times 5^2 \times 7$　　**c** $5^3 \times 7^2 \times 11$
5 **ai** 1, 475　　**aii** 1, 828
　aiii 1, 323　　**aiv** 1, 99
　b The HCF is 1 (unless both numbers are the same)
　c The LCM is the product of the two numbers (unless both numbers are the same)
6 **a** Not co-prime　　**b** Co-prime
　c Co-prime　　**d** Not co-prime
7 **a** HCF = 3, LCM = 15 015
　　 HCF = 1, LCM = 45 045
　　 HCF = 1, LCM = 21 175
　　 HCF = 7, LCM = 45 045
　b HCF = 1, LCM = the product of the two numbers
8 **a** True
9 **a** 6, 1890　　**b** 2, 34 650　　**c** 1, 510 510

N1 Exam review

1 **ai** $2^2 \times 3 \times 7$　　**aii** $3^2 \times 7$
　b 21
2 **a** 4.8×10^7 km　　**b** 230 000 km

S1 Before you start …

1 **i** 78.5 cm^2, 172.0 cm^2　　**ii** 31.4 cm, 46.5 cm
2 **a** 600 cm^3　　**b** 351.9 cm^3　　**c** 45.92 cm^3
3 **a** 500 cm^2　　**b** 276.5 cm^2　　**c** 106.9 cm^2

S1.1

1 **a** 15 cm^2　　**b** 33.48 cm^2　　**c** 45.6 cm^2
　d 31.5 cm^2　　**e** 13.34 cm^2
2 **a** 12 cm^2　　**b** 20 cm^2　　**c** 20.5 cm^2
　d 600 mm^2　　**e** 1250 mm^2
3 205.5 cm^2

S1.2

1 **a** 15.1 cm^2　　**b** 15.7 cm^2　　**c** 101.8 cm^2
　d 15.7 cm^2　　**e** 6.0 cm^2　　**f** 12.3 cm^2
2 **a** 7.5 cm　　**b** 10.5 cm　　**c** 33.9 cm
　d 6.3 cm　　**e** 5.0 cm　　**f** 7.0 cm
3 **a** 45.4 mm　　**b** 14.8 cm　　**c** 29.5 cm
4 23.6 cm^2
5 **a** 11.1 m^2　　**b** 13

S1.3

1 **a** 142 cm^2　　**b** 98 cm^2　　**c** 114 cm^2
　d 65.6 cm^2　　**e** 96 cm^2　　**f** 102 mm^2
2 **a** 44.0 cm^2　　**b** 165 cm^2　　**c** 75.4 cm^2
　d 66.4 cm^2
3 **a** 330 cm^2　　**b** 468 cm^2

S1.4

1 **a** 105 cm^3　　**b** 60 cm^3　　**c** 72 cm^3
　d 36 cm^3　　**e** 64 cm^3　　**f** 28 mm^3
2 **a** 75.4 cm^3　　**b** 628.3 cm^3　　**c** 201.1 cm^3
　d 160.8 cm^3
3 **a** 270 cm^3　　**b** 600 mm^3　　**c** 1080 cm^3

S1.5

1 **a** 27 cm^3　　**b** 343 cm^3
　c 64 cm^3　　**d** 0.125 m^3
2 **a** 155.9 cm^3　　**b** 86.6 cm^3
　c 304.8 cm^3　　**d** 125.7 mm^3
3 **a** 384 cm^2　　**b** 600 cm^2
　c 216 cm^2　　**d** 6 cm^2
4 **a** 1 cm × 1 cm × 80 cm, 2 cm × 2 cm × 20 cm, 4 cm × 4 cm × 5 cm
　bi 112 cm^2　　**bii** 322 cm^2
5 **a** 1 cm × 1 cm × 24 cm, 2 cm × 2 cm × 6 cm; 56 cm^2; 98 cm^2
　b 1 cm × 1 cm × 64 cm, 2 cm × 2 cm × 16 cm, 4 cm × 4 cm × 4 cm; 96 cm^2; 258 cm^2

S1.6

1 **a** 42 cm^3　　**b** 1.77 cm^3　　**c** 69.2 cm^3
　d 32 cm^3　　**e** 263.9 cm^3　　**f** 87.1 cm^3
2 **a** Y　　**b** 134 cm^3
3 3
4 69.2 cm^3

S1.7

1 **a** 443.4 mm^2　　**b** 249.4 cm^2　　**c** 46.8 cm^2
　d 106.0 cm^2　　**e** 138.5 cm^2　　**f** 197.2 cm^2
　g 85.1 cm^2　　**h** 94.8 cm^2
　i 9856.1 mm^2
2 **a** 93.5 mm^2　　**b** 4704.3 mm^2

S1.8

1 **a** 128.8 cm^2　　**b** 56.5 cm^2　　**c** 4241.2 mm^2
　d 23.6 cm^2　　**e** 221.0 cm^2　　**f** 784.6 cm^2
2 **a** 188.5 cm^2　　**b** 82.4 cm^2　　**c** 316.5 cm^2
3 **a** 15.7 cm^2　　**b** 15.7 cm^2
　c 6804.7 mm^2　　**d** 184.4 cm^2
　e 101.8 cm^2　　**f** 40.2 m^2
4 156.6 cm^2

S1.9

1 **a** 1436.8 cm^3　　**b** 57 906 mm^3
　c 1047.4 cm^3　　**d** 24 429 cm^3
　e 70 276 cm^3　　**f** 10 306 000 mm^3
2 **a** 615.8 cm^2　　**b** 7238.2 mm^2
　c 498.8 cm^2　　**d** 4071.5 cm^2
　e 8235.5 cm^2　　**f** 229 022 mm^2
3 7 cm
4 **a** 134.0 cm^3　　**b** 100.5 cm^2
5 785.4 cm^3
6 **a** Cube　　**b** Cube

7 ai 6 cm   **aii** 12 cm
b 452.4 cm^2 **c** 452.4 cm^2
di $4\pi r^2$ **dii** $2r \times 2\pi r$
e They both equal $4\pi r^2$.

S1 Exam review
1 a 52.7 cm^2 **b** 22.0 cm^3
2 156π cm^3

A1 Before you start …
1 a 196   **b** 1   **c** -32 **d** $\frac{8}{27}$
  e 0.000 008 **f** 10 000 000 000 **g** -5 **h** $\frac{2}{5}$
2 $(1 + 2x)^3$, $(\frac{1}{x})^2$, x^2, $\sqrt{(12 - 2x)}$, $(-x)^4$ or $(1 + 2x)^3$,
  $(\frac{1}{x})^2$, $\sqrt{(12 - 2x)}$, x^2, $(-x)^4$
3 a $4n + 7$ **b** $3(n - 6)$ **c** $10 - n^2$
  d $\frac{3n - 6}{2}$ **e** n^5

A1.1
1 a y^{12} **b** k^4 **c** m^{12} **d** g^3 **e** h^{-6}
  f b^{-12} **g** j^{-6} **h** t^{10} **i** n^{-2}
2 a $15h^{13}$ **b** $3p^2$ **c** $4p^{16}$ **d** $60r^{-1}$
  e $27h^{-9}$ **f** $3b^8$ **g** $2m^{-9}$ **h** 2
3 $30w^{-9}$
4 $64p^{12} \div 64p^{14} = p^{-2}$
5 False, it simplifies to 3^{x+y}.
6 a 36   **b** 3   **c** 64   **d** -7 **e** 49
7 The first expression is the odd one out.
8 $x = 1$
9 a $9^x + 3^{x+1} = 3^{2x} + 3 \times 3^x = u^2 + 3u$
  b $3^{3x} + 9 \times 3^x$ or $3^{3x} + 3^{x+2}$
  c $3^{2x} - 3^{-x}$
  d $u^4 - \frac{u^2}{9}$

A1.2
1 a $15x + 27$ **b** $8p^2 - 16p$
  c $15m - 6m^2$ **d** $21y + 17$
  e $10x^2 + 10xy - 45x$ **f** $57 - 2t$
  g $4h + 16$ **h** $3x^3 + x^4$
2 a $x^2 + 13x + 42$ **b** $2x^2 - 2x - 24$
  c $12p^2 + 23p + 10$ **d** $6m^2 - 32m + 42$
  e $10y^2 + 17y - 63$ **f** $9t^2 - 12t + 4$
  g $2x^2 + 4x - 15$ **h** $-25b^2 + 10b - 1$
3 a $(3y + 8)(2y - 1) = 6y^2 + 13y - 8$
  b $\frac{1}{2}(2x - 3)(3x - 2) = 3x^2 - 6.5x + 3$
4 $\sqrt{(3p - 1)^2 + (2p + 2)^2} = \sqrt{13p^2 + 2p + 5}$
5 $x(12 - x) = 12x - x^2$
6 $(2n + 1)^2 - (2n - 1)^2$
  $= 4n^2 + 4n + 1 - 4n^2 + 4n - 1 = 8n$
7 a $x^3 + 6x^2 + 12x + 8$ **b** $2y^3 + 3y^2 - 32y + 15$

A1.3
1 a $3(x + 2y + 3z)$ **b** $5(2p - 3)$
  c $x(5y + 7)$ **d** $3m(2n + 3t)$
  e $4x(4x - 3y)$ **f** $3p(1 + 3q)$

g $7x(y - 8x)$ **h** $3x(x + 4x^2 - 2)$
i $(m + n)(3 + m + n)$ **j** $(p - q)[4 + (p - q)^2]$
k $(a + b)(x + y)$
2 a $(x + 2)(x + 5)$ **b** $(x + 3)(x + 5)$
  c $(x + 2)(x + 6)$ **d** $(x + 5)(x + 7)$
  e $(x - 5)(x + 2)$ **f** $(x - 7)(x + 5)$
  g $(x - 5)(x - 3)$ **h** $(x - 5)(x + 4)$
  i $(x - 20)(x + 12)$ **j** $(x - 9)(x + 12)$
  k $(x - 5)(x + 5)$ **l** $(x - 3)(x + 2)$
3 a $(x + y)^2$ **b** 400
4 a $2(3x + 1)$ **b** $2(x - 6)(x - 4)$
5 $(x + 7)(x + 3) = 0$
6 a $2(x + 6)(x + 2)$ **b** $3(y + 3)(y + 12)$
  c $4(m - 5)(m + 4)$ **d** $x(x + 5)(x + 3)$
  e $x(y - 12)(y + 9)$ **f** $y(x - 4)(x + 4)$
7 a $(p - 4)(p + 3)$ **b** $3p(p + 2)$
  c $10(x + 3)(x + 4)$ **d** $y(x - 9)(x + 7)$
  e $a(a - b)^2$ **f** $3a(m + n + b + c)$

A1.4
1 a $(2x + 3)(x + 1)$ **b** $(3x + 2)(x + 2)$
  c $(2x + 5)(x + 1)$ **d** $(2x + 3)(x + 4)$
  e $(3x + 1)(x + 2)$ **f** $(2x + 1)(x + 3)$
  g $(2x + 7)(x - 3)$ **h** $(3x + 1)(x - 2)$
  i $(4x - 3)(x - 5)$ **j** $(6x - 1)(x - 3)$
  k $(3x + 2)(4x + 5)$ **l** $(2x - 3)(4x + 1)$
  m $3(2x - 5)(x - 2)$ **n** $(2x - 3)(2x + 3)$
  o $(2x + 3)(3x - 1)$ **p** $(6x - 1)(3x + 4)$
2 You cannot find two numbers that multiply to 6
  and add to 4.
3 a $9x(2x - 1)$ **b** $4ab(1 - 2b)(1 + 2b)$
  c $m(3n + 8 - m^2)$ **d** $(x - 9)(x + 2)$
  e $(2x + 5)(x - 3)$ **f** $x(x + 3)(x + 4)$
  g $p(2x + 3)(x + 4)$ **h** $50(x - 5)(x + 4)$
  i $10(4p - 3)(p - 5)$
4 Total $= 8x^2 + 52x + 60$,
  so mean $= 2x^2 + 13x + 15 = (2x + 3)(x + 5)$
5 a $(2y + 3)(1 - 3y)$ **b** $(3 - 2p)(4p + 1)$
  c $(3y - 5)(2 - y)$ **d** $3(2m - 5)(2 - m)$
  e $y(3x + 1)(2 - x)$

A1.5
1 a $(x - 10)(x + 10)$ **b** $(y - 4)(y + 4)$
  c $(m - 12)(m + 12)$ **d** $(p - 8)(p + 8)$
  e $(x - \frac{1}{2})(x + \frac{1}{2})$ **f** $(k - \frac{5}{6})(k + \frac{5}{6})$
  g $(w - 50)(w + 50)$ **h** $(7 - b)(7 + b)$
  i $(2x - 5)(2x + 5)$ **j** $(3y - 11)(3y + 11)$
  k $(4m - \frac{1}{2})(4m + \frac{1}{2})$ **l** $(20p - 13)(20p + 13)$
  m $(x - y)(x + y)$ **n** $(2a - 5b)(2a + 5b)$
  o $(3w - 10v)(3w + 10v)$ **p** $(5c - \frac{1}{2}d)(5c + \frac{1}{2}d)$
  q $x(x - 4)(x + 4)$ **r** $2y(5 - y)(5 + y)$
  s $(\frac{4}{7}x + \frac{8}{9}y)(\frac{4}{7}x - \frac{8}{9}y)$
2 a 400 **b** 240 000 **c** 199 **d** 157 000
3 a $30^2 - 7^2 = (30 - 7)(30 + 7) = 23 \times 37$
  b $100^2 - 3^2 = (100 - 3)(100 + 3) = 97 \times 103$

c $34^2 - 23^2 = (34 - 23)(34 + 23) = 11 \times 57$

d $20^2 - 9^2 = (20 - 9)(20 + 9) = 11 \times 29$

4 $\sqrt{1200}$

5 a $3(2x^2 - 5xy + 3y^2)$　　**b** $(4a + 3b)(4a - 3b)$

c $(x - 4)(x - 7)$　　**d** $(2x - 3)(x + 7)$

e $x(x - 3)(x + 6)$　　**f** $5ab(1 + 2ab)$

g $(x + 5)(2 - x)$　　**h** $10(1 - x)(1 + x)$

i $(y + 9)(y - 7)$　　**j** $2x(x - 8)(x + 8)$

k $(3x - 2)(2x - 3)$　　**l** $(x - y)(x + y)(x^2 + y^2)$

6 $97^2 - 57^2 = (97 - 57)(97 + 57) = 40 \times 154$

= 6160 cm^2

A1 Exam review

1 $(5 - x)(2 + 2x) = 8x - 2x^2 + 10$ cm^2

2 a k^3　　**bi** $7x - 1$　　**bii** $x^2 + 5xy + 6y^2$

c $(p + q)(5 + p + q)$　　**d** m^8　　**e** $6r^3t^6$

N2 Before you start ...

1 a 443　　**b** 373　　**c** 21.3　　**d** 265

e 132.17　　**f** 266.81　　**g** 27.03　　**h** 156.23

2 a 696　　**b** 1104　　**c** 722　　**d** 13

e 15.54　　**f** 2023　　**g** $505\frac{7}{9}$　　**h** $73\frac{6}{11}$

3 a 4.5　　**b** 3.67　　**c** 1.1　　**d** 3.4

e 1.06　　**f** 17.82

N2.1

1 a 310　　**b** 450　　**c** 530　　**d** 2170　　**e** 56 690

2 a 43　　**b** 1　　**c** 23　　**d** 32　　**e** 44

3 ai 1300　　　**aii** 1000

bi 1100　　　**bii** 1000

ci 0　　　　**cii** 0

di 500　　　**dii** 1000

ei 41 500　　**eii** 41 000

4 a 0.3　　**b** 0.74　　**c** 0.151　　**d** 0.68

5 a 0.056　　**b** 3.2　　**c** 15　　**d** 950

6 a 0.52　　**b** 34 600　　**c** 72 700　　**d** 0.0045

7 a 400　　**b** 0.6　　**c** 0.005　　**d** 700 000

8 a 0.42　　**b** 0.86　　**c** 1800　　**d** 4300

9 ai $500 \div 20$　　　　**aii** $20 + 40 \div 4$

aiii $2000 + 30 \times 50$　**aiv** $2 + 3$

bi 25　　**bii** 30　　**biii** 3500　　**biv** 5

ci 25.056　　**cii** 26.5　　**ciii** 3320

civ 5.1186

10 ai e.g. $250 + 350 = 600$

aii e.g. $60 \div 30 = 2$

aiii e.g. $60 \times 200 = 12\ 000$

aiv e.g. $100 - 60 = 40$

bi The example given produces an underestimate.

bii The example given produces an underestimate.

biii The example given produces an underestimate.

biv The example given produces an overestimate.

ci 620　**cii** 2.37　**ciii** 13 279.86　**civ** 31.9

N2.2

1 a 5.85 m, 5.75 m　　**b** 16.55 litres, 16.45 litres

c 0.95 kg, 0.85 kg　　**d** 6.35 N, 6.25 N

e 10.15 s, 10.05 s　　**f** 104.75 cm, 104.65 cm

g 16.05 km, 15.95 km　　**h** 9.35 m/s, 9.25 m/s

2 a 6.75 m, 6.65 m　　**b** 7.745 litres, 7.735 litres

c 0.8135 kg, 0.8125 kg　　**d** 6.5 N, 5.5 N

e 0.0015 s, 0.0005 s　　**f** 2.545 cm, 2.535 cm

g 1.1625 km, 1.1615 km　　**h** 15.5 m/s, 14.5 m/s

3 a 174 kg, 162 kg　　**b** 28.4 kg, 27.6 kg

4 a 37.5 mm, 32.5 mm　**b** 42.5 mm, 37.5 mm

c 112.5 mm, 107.5 mm　　**d** 4.75 cm, 4.25 cm

5 The number of nails should be assumed to be exact so maximum = $38 \times 12.5 = 475$ g, minimum = $38 \times 11.5 = 437$ g.

6 Sum of upper bounds = 461 kg, which exceeds the limit.

7 526.75 cm^2, 481.75 cm^2

8 65 g, 63 g

9 49 crates

N2.3

1 a Larger　　**b** Smaller　　**c** Smaller

d Smaller　　**e** Larger　　**f** Larger

g Smaller　　**h** Same

2 a $8 \div 2$　　**b** $10 \div 5$　　**c** $12 \div 4$　　**d** $18 \div 10$

3 a $15 \div 5$　　**b** $28 \div 4$　　**c** $10 \div \frac{5}{2}$　　**d** $72 \div 8$

4 a 18×2　　**b** 24×4　　**c** $8 \times \frac{3}{2}$　　**d** 5.5×10

e 5.9×1　　**f** $66 \times \frac{5}{3}$　　**g** $7 \times \frac{10}{7}$　　**h** 8×0.8

5 a 24　　**b** 144　　**c** 65　　**d** 72

e 2　　**f** 9　　**g** 8　　**h** 30

6 a 27　　**b** 60　　**c** 16　　**d** 21

e 7　　**f** 90　　**g** 40　　**h** 6

7 a 16　　**b** 24　　**c** 12　　**d** 16

e 30　　**f** 24　　**g** 36　　**h** 28

8 a 2.5　　**b** 2　　**c** 3.5　　**d** 0

e 1　　**f** −7　　**g** 4.8　　**h** −3.85

9 a False. If you divide a negative number by 2 the answer is bigger.

b True

c False. If you multiply a negative number by 5 the answer is smaller.

d False. Multiplying zero by 10 gives zero.

10 a Repeatedly multiplying a positive number by 0.9 will make it smaller and smaller, tending to zero. Repeatedly multiplying a negative number by 0.9 will make it bigger and bigger, tending to zero.

b The answer will alternate between being positive and negative, and tend to zero.

N2.4

1 a $\frac{1}{2}$　　**b** $\frac{3}{10}$　　**c** $\frac{1}{4}$　　**d** $\frac{2}{5}$

e $\frac{3}{5}$　　**f** $\frac{1}{20}$　　**g** $\frac{9}{20}$　　**h** $\frac{3}{8}$

2 a 0.75   **b** 0.4   **c** 0.625   **d** 0.15
   e 0.04   **f** 0.16   **g** 0.35   **h** 0.06

3 a $\frac{3}{4}$ **b** $\frac{3}{10}$ **c** $\frac{1}{2}$ **d** $\frac{9}{20}$
   e $\frac{5}{8}$ **f** $\frac{5}{12}$ **g** $\frac{8}{15}$ **h** $\frac{5}{8}$

4 a $\frac{1}{4}$ **b** $\frac{3}{8}$ **c** $\frac{7}{10}$ **d** $\frac{7}{12}$
   e $\frac{1}{9}$ **f** $\frac{1}{6}$ **g** $\frac{5}{24}$ **h** $\frac{19}{56}$

5 a $\frac{2}{15}$ **b** $\frac{3}{8}$ **c** $\frac{1}{6}$ **d** $\frac{1}{2}$
   e $\frac{1}{6}$ **f** $\frac{1}{8}$ **g** $\frac{3}{7}$ **h** $\frac{1}{16}$

6 a 4   **b** 4   **c** 4   **d** $\frac{8}{9}$
   e $\frac{5}{2}$ **f** $\frac{7}{5}$ **g** $\frac{2}{3}$ **h** $\frac{7}{5}$

7 a 0.7   **b** 1.2   **c** 2.1   **d** 0.2
   e 0.9   **f** 6.5   **g** 0.72   **h** 0.33
   i 1.85   **j** 16.23   **k** 10.214   **l** 7.883

8 a 0.96   **b** 31   **c** 101.6   **d** 1.29
   e 1.12   **f** 7   **g** 0.45   **h** 15

9 a 48   **b** 51   **c** 152   **d** 246
   e 24   **f** 392.2   **g** 58.5   **h** 3160

10 a £392   **b** £735   **c** £58.50   **d** £471.90
   e £741   **f** £208.25

N2.5

1 a 126.741   **b** 17.29   **c** 58.32
   d 62.04   **e** 62.641   **f** 50.73
   g 282.2   **h** 9.768   **i** 67.31

3 a $\frac{31}{35}$ **b** $\frac{20}{27}$ **c** $\frac{5}{16}$
   d $\frac{25}{56}$ **e** $\frac{37}{30}$ **f** $\frac{5}{18}$
   g $\frac{1}{6}$ **h** $\frac{17}{40}$ **i** $\frac{4}{3}$

5 a 43.7   **b** 15.7   **c** 103
   d 78.7   **e** 66.7   **f** 181
   g 263   **h** 15.0   **i** 113

7 a 457.2   **b** 801.85   **c** 477.375   **d** 39.4

9 a 6.25%   **b** 24%   **c** 18.2%
   d 1.3%   **e** 54.2%   **f** 14.9%

N2 Exam review

1 ai $54 + 26 = 80$ **aii** $3 \div 0.5 = 6$
   bi 80.48   **bii** 6.875
   ci 23   **cii** 1.7

2 The pen could be as long as 10.5 cm, and the pen case as short as 10.05 cm.

A2 Before you start …

1 a 8   **b** −6   **c** 121   **d** −7   **e** 79

2 a $x + 16y$ **b** $9x^2 + 3x$ **c** $7ab$
   d $7p - 9$ **e** $14x^3$ **f** $2x + 3y - x^2$

3 a $24x - 27$ **b** $5y + 11$
   c $3x^2 - 2xy$ **d** $x^2 + 2x - 63$
   e $6w^2 - 32w + 32$ **f** $p^2 + q^2 - 2pq$

4 a $\frac{2}{5}$ **b** $\frac{19}{45}$ **c** $\frac{1}{2}$ **d** $5\frac{17}{30}$ **e** $\frac{7}{12}$ **f** $3\frac{1}{8}$

A2.1

1 a $x = 4\frac{1}{5}$ **b** $x = 4\frac{5}{6}$ **c** $x = 9$
   d $x = -\frac{1}{4}$ **e** $x = -4\frac{1}{3}$ **f** $y = -1\frac{1}{5}$
   g $x = 9$ **h** $x = -3$ **i** $x = 3$

2 a $x = 3\frac{4}{5}$ **b** $x = 2\frac{2}{7}$ **c** $y = 1\frac{2}{5}$
   d $y = 5$ **e** $w = -3\frac{2}{5}$ **f** $x = -\frac{1}{3}$
   g $y = -\frac{1}{3}$ **h** $z = 1\frac{3}{16}$ **i** $p = 2\frac{4}{5}$
   j $q = 2\frac{3}{8}$ **k** $r = \frac{7}{9}$

3 a Length = 5   **b** $p = -5$

4 a Square length = 7.2   **b** 32

A2.2

1 a $x = 47$ **b** $y = 1\frac{13}{17}$ **c** $z = 3$
   d $p = 9\frac{1}{3}$ **e** $q = 9\frac{9}{14}$ **f** $m = \frac{4}{23}$

2 False

3 A $9\frac{4}{9}$ mm², B $15\frac{1}{9}$ mm²

4 $1\frac{4}{11}$

5 35

6 3 cm, 4 cm; 1 cm, 4 cm

A2.3

1 a $3w$ **b** $\frac{b}{3}$ **c** $2c$ **d** $\frac{4b}{d}$
   e $4bd^2$ **f** $x + 3$ **g** $x + 1$ **h** $\frac{y-2}{3}$
   i $x + 2$ **j** $x - 7$ **k** $\frac{1}{x-7}$ **l** $x - 2$
   m $2y - 5$ **n** $\frac{1}{x+9}$ **o** $2x - 5$ **p** $\frac{3x+4}{x-2}$
   q $x - 4$

2 The first fraction cannot be simplified as the numerator and denominator have no common factors.

3 $a(ab - 1)$

4 a 3   **b** $\frac{12a}{7}$ **c** $\frac{2}{5}m$ **d** $\frac{15}{g^2}$
   e $2\frac{2}{3}$ **f** $\frac{f}{2p^2}$ **g** $\frac{5}{y}$ **h** $3(x+7)$
   i $\frac{2(x-2)(x-9)}{x^2 - 17x + 18}$ **j** $\frac{2(x-5)}{x^2(x-6)}$

5 $\frac{15}{16}$ m by $\frac{4}{15}$ m

6 −3

A2.4

1 a $\frac{4p}{5}$ **b** $\frac{4y}{7}$ **c** $\frac{3}{p}$ **d** $\frac{11y}{8}$
   e $\frac{p}{15}$ **f** $\frac{6y-7x}{xy}$ **g** $\frac{4x+2}{x^2}$

2 $\frac{x}{3} - \frac{x}{4} = \frac{x}{12}$, $\frac{5x}{12} - \frac{3x}{12} = \frac{x}{6}$, $\frac{2}{3}x - \frac{1}{3}x = \frac{4x^2}{12x}$, $\frac{x}{6} + \frac{x}{4} = \frac{5x}{12}$

3 a $\frac{14x+3}{20}$ **b** $\frac{12x+13}{77}$ **c** $\frac{y-1}{15}$ **d** $\frac{41p-112}{35}$
   e $\frac{5x-13}{(x-7)(x+4)}$ **f** $\frac{8x+9}{(x-2)(x+3)}$
   g $\frac{11-y}{(y-2)(y+1)}$ **h** $\frac{-3p-17}{(p+3)(p-1)}$
   i $\frac{12(w-2)}{w(w-8)}$ **j** $\frac{4x-3}{(x-2)^2}$

4 $\dfrac{2(8p-7)}{(p+1)(p-2)}$

A2.5

2 a $x = 16\frac{9}{14}$ **b** $y = -16\frac{3}{4}$

3 The first two are possible, the third is not.

A2 Exam review

1 $x = 4$

2 a $23 - 6x$ **b** $32x^5y^{15}$ **c** $\dfrac{2(n-1)}{n-2}$

D1 Before you start ...

1 a 3   **b** 3   **c** 6   **d** 3.57
2 a 5   **b** 3   **c** 4   **d** 3.4

D1.1

1 ai 7   **aii** 6   **aiii** 5.82   **aiv** 6   **av** 2
  bi 75   **bii** 63   **biii** 60.1   **biv** 63   **bv** 27
  ci 8   **cii** 96   **ciii** 95.6   **civ** 96   **cv** 2
  di 71   **dii** 22, 37   **diii** 40.4   **div** 37   **dv** 38
  ei 26   **eii** 88, 89   **eiii** 84.2   **eiv** 87   **ev** 7
  fi 72   **fii** 27   **fiii** 46.9   **fiv** 34   **fv** 37
  gi 8   **gii** 105   **giii** 105.2   **giv** 105   **gv** 3

2 Range is unduly affected by one extreme value, which IQR ignores.

3 Mode is the lowest value.

4 a 1, 6, 8, 2, 8, 5, 6, 9, 3, 5, 7, 4, 4, 5, 5
  bi 8   **bii** 5   **biii** 5.2   **biv** 5   **bv** 3
  c Range and IQR stay same.
  d Answers are as for Q1 less 100.
  e Adding 100 does not affect spread of values but does affect averages.

5 ai 200, 200   **aii** 100, 100   **aiii** 100, 100
  aiv 100, 100   **av** 2, 2
  b All measures the same although sets of numbers are different.

D1.2

1 45.6

2 ai 6   **aii** 6   **aiii** 5.84
  aiv 4   **av** 3
  bi 5   **bii** 5   **biii** 5
  biv 4   **bv** 2
  ci 4   **cii** 6   **ciii** 6.63
  civ 6   **cv** 5
  di 6   **dii** 6   **diii** 5.68
  div 6   **dv** 3

3 1.5

4 7

D1.3

1 81 minutes

2 7.6 hours

3 26 lessons

4 The mean is higher for city dwellers than for small town dwellers, perhaps because they live closer to more museums.

5 The mean spend is higher for online customers than for in store customers, perhaps because they have to pay delivery costs.

6 $\dfrac{36c + 42d}{78}$

7 $\dfrac{mx + fy}{m + f}$

D1.4

1 ai $10 < t \leqslant 15$ **aii** $10 < t \leqslant 15$
  aiii 15.5 minutes   **aiv** 20 minutes
  bi $10 < t \leqslant 20$ **bii** $20 < t \leqslant 30$
  biii 24.2 minutes   **biv** 40 minutes
  ci $10 < t \leqslant 15$ **cii** $10 < t \leqslant 15$
  ciii 14.6 minutes   **civ** 25 minutes
  di $5 < t \leqslant 15$ **dii** $15 < t \leqslant 25$
  diii 26.5 minutes   **div** 50 minutes

2 a $150 < B \leqslant 200$ **b** £171

3 a £88   **b** $8 < M \leqslant 120$
  c No, the 21st value is in the $8 < M \leqslant 120$ interval.

D1 Exam review

1 41.7 seconds

2 $\dfrac{12n + 15m}{27}$

N3 Before you start ...

1 a $\frac{3}{4}$ **b** $\frac{1}{4}$ **c** $\frac{1}{2}$ **d** $1\frac{1}{6}$

2 a $\frac{1}{3}$ **b** $1\frac{3}{5}$ **c** $\frac{3}{10}$ **d** 8

3 a 60%   **b** $\frac{7}{20}$ **c** $\frac{13}{20}$ **d** 0.35

4 a £12   **b** 10.8 cm
  c 16.5 m   **d** 1 hour 57 minutes

N3.1

1 a 1   **b** $\frac{3}{4}$ **c** $\frac{5}{6}$ **d** $\frac{3}{10}$ **e** $\frac{7}{12}$

2 a $\frac{1}{3}$ **b** $\frac{1}{3}$ **c** $\frac{1}{12}$ **d** $\frac{1}{15}$ **e** $\frac{5}{21}$

3 a $1\frac{3}{4}$ **b** $1\frac{2}{3}$ **c** $3\frac{3}{10}$ **d** $3\frac{1}{8}$

4 a $\frac{3}{2}$ **b** $\frac{4}{5}$ **c** $\frac{7}{16}$ **d** $\frac{5}{16}$

5 a $\frac{2}{9}$ **b** $\frac{5}{2}$ **c** $\frac{1}{9}$ **d** $\frac{9}{4}$

6 a $\frac{3}{5}$ **b** $\frac{3}{8}$ **c** $\frac{6}{35}$ **d** $\frac{8}{15}$

7 a 3   **b** $5\frac{3}{5}$ **c** $8\frac{1}{4}$ **d** $3\frac{3}{32}$

8 a $6\frac{1}{8}$ **b** $\frac{5}{8}$ **c** $10\frac{5}{16}$ **d** $\frac{2}{3}$

9 a $3\frac{1}{3}$ **b** $\frac{35}{64}$ **c** $\frac{15}{38}$ **d** $4\frac{57}{160}$

10 32   **11** 8

N3.2

1 a 0.5   **b** 2.4   **c** 0.15   **d** 0.04
2 a 0.375   **b** 0.8   **c** 0.0625   **d** 0.12
4 a $0.\dot{3}$ **b** $0.\dot{6}$ **c** $0.1\dot{6}$ **d** $0.\dot{1}$
  e $0.8\dot{3}$

5 $0.\dot{1}42\,85\dot{7}$, $0.\dot{2}85\,71\dot{4}$, $0.\dot{4}28\,57\dot{1}$, $0.\dot{5}71\,42\dot{8}$, $0.\dot{7}14\,28\dot{5}$, $0.\dot{8}57\,14\dot{2}$

6 a $\frac{1}{2}$ **b** $\frac{3}{10}$ **c** $\frac{3}{4}$ **d** $\frac{19}{20}$ **e** $\frac{13}{20}$

7 a $\frac{1}{9}$ **b** $\frac{2}{9}$ **c** $\frac{5}{33}$ **d** $\frac{125}{999}$ **e** $\frac{8}{37}$

8 a $\frac{19}{90}$ **b** $\frac{13}{18}$ **c** $\frac{91}{110}$ **d** $\frac{421}{666}$ **e** $\frac{1349}{1650}$

9 a $\frac{8}{11}$ **b** $0.\dot{7}$ **c** $0.0\dot{7}\dot{4}$

10 a Let $x = 0.5757\,...$, then $100x = 57.5757\,...$ and $100x - x = 99x = 57$ so $x = \frac{57}{99}$

  b $\frac{59}{165}$

N3.3

1 a 35% **b** 60.7% **c** 99.5% **d** 100%
  e 215% **f** 0.056% **g** 1700% **h** 10.1%

2 a 55.6% **b** 34.4% **c** 75.8% **d** 51.3%
  e 43.7% **f** 83.9% **g** 83.9% **h** 105.6%

3 a 0.22 **b** 0.185 **c** $0.\dot{5}$ **d** $0.3\dot{5}$
  e $0.06\dot{5}$ **f** $0.61\dot{4}\dot{9}$ **g** $0.54\dot{4}\dot{6}$ **h** $1.5\dot{2}$

4 a 80% **b** 15% **c** 87.5% **d** 75%
  e 28% **f** 18.75% **g** 45% **h** 14%

5 a 66.7% **b** 6.7% **c** 42.9% **d** 83.3%
  e 77.8% **f** 36.4% **g** 8.3% **h** 13.3%

6 a $\frac{1}{6}, \frac{1}{2}, \frac{3}{5}, \frac{2}{3}$ **b** $\frac{1}{12}, \frac{7}{24}, \frac{1}{3}, \frac{3}{8}, \frac{1}{2}$

  c $\frac{1}{20}, \frac{7}{40}, \frac{3}{5}, \frac{5}{8}, \frac{3}{4}$ **d** $\frac{7}{36}, \frac{5}{18}, \frac{4}{9}, \frac{1}{2}, \frac{2}{3}, \frac{3}{4}$

7 a $\frac{2}{5}, \frac{4}{10}, \frac{3}{8}, \frac{1}{4}$ or $\frac{4}{10}, \frac{2}{5}, \frac{3}{8}, \frac{1}{4}$ **b** $\frac{2}{5}, \frac{1}{3}, \frac{3}{10}, \frac{3}{11}, \frac{2}{9}$

  c $\frac{4}{9}, \frac{7}{20}, \frac{1}{3}, \frac{2}{7}, \frac{1}{5}$ **d** $\frac{5}{9}, \frac{3}{8}, \frac{1}{3}, \frac{4}{13}, \frac{2}{7}, \frac{3}{11}$

8 a 0.33, 33.3%, $33\frac{1}{3}\%$

  b $0.\dot{4}$, 44.5%, 0.45, 0.454

  c 22.3%, 0.232, 23.22%, 0.233, $0.2\dot{3}$

  d $0.6\dot{5}$, 0.66, 66.6%, 0.6666, $\frac{2}{3}$

  e 14%, $14.\dot{1}\%$, 0.142, $\frac{1}{7}$, $\frac{51}{350}$

  f $\frac{5}{6}, \frac{6}{7}$, 86%, 0.866, $0.8\dot{6}$

9 Abby is correct. Since $0.\dot{3}$ is exactly equal to $\frac{1}{3}$, $0.\dot{9}$ is exactly equal to $3 \times \frac{1}{3} = 1$.

N3.4

1 a £12 **b** 240 m **c** 52.2 kg
  d 0.18 km **e** €70.40 **f** 18 seconds
  g 86.1 cm **h** 34.2 g

2 a 0.25 **b** 0.4 **c** 0.9 **d** 0.05
  e 1.1 **f** 0.3 **g** 1.05 **h** 0.95

3 a 0.3 **b** 0.32 **c** 0.325
  d 0.0125 **e** 1.12 **f** 0.000 06

4 a 1.15 **b** 1.025 **c** 1.225
  d 1.875 **e** 2.08 **f** 1.000 45

5 a 0.95 **b** 0.9175 **c** 0.7725
  d 0.6175 **e** 0.02 **f** 0

6 a 1.1 **b** 1.21

7 a 1.113 **b** 0.972 **c** 1.014

8 It makes no difference which she does first.

N3.5

1 a 2.5% **b** 3.33% **c** 55.2% **d** 1.25%

2 41.4%

3 16%, 12.7%, 36.6%, 22.8%

4 £8450

5 13 600

6 a £182.25 **b** £385.01

7 a £82.94 **b** £313.42 **c** £140.92

N3 Exam review

1 ai 0.2 **aii** 20%

  bi $\frac{5}{11}$ **bii** 45.5%

2 £6500

S2 Before you start ...

1 a   **b**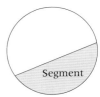

  c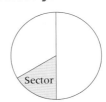

2 $a = 43°$, $b = 107°$, $c = 233°$, $d = 139°$

3 $e = 102°$, $f = 29°$

4 $g = 123°$, $h = 73°$

S2.1

1 ai

  aii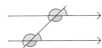

  bi

  bii

ci

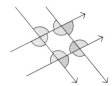

cii

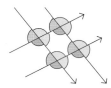

2 a $a = 17°$, angles on a straight line; $b = 17°$, alternate angles; $c = 163°$, corresponding angles

   b $d = 125°$, alternate angles; $e = 105°$, vertically opposite angles

   c $f = 134°$, vertically opposite angles; $g = 134°$, corresponding angles; $h = 24°$, angles in a triangle

3 a $180°$

   b x (bottom left) and z (bottom right)

   c angles in the triangle are x, y, z (part **b**) and add up to $180°$ (part **a**)

4 a $a = 25°$, $b = 155°$, $c = 25°$, $d = 155°$

   b Parallelogram

5 $a = 80°$, $b = 110°$, $c = 70°$, $d = 30°$, $e = 150°$

S2.2

1

Shape	△	□	⬠	⬡	⬢	⬤
Number of sides	3	4	5	6	8	10
Number of triangles the shape splits into	1	2	3	4	6	8
Sum of the interior angles in the shape	180°	360°	540°	720°	1080°	1440°
Size of one interior angle	60°	90°	108°	120°	135°	144°
Size of one exterior angle	120°	90°	72°	60°	45°	36°

2 a $x = 80°$ **b** $x = 77°$, $2x = 154°$

   c $x = 55°$, $3x = 165°$

3 a $x = 125°$, $y = 50°$ **b** $x = 110°$, $y = 65°$

4 ai $120°$ **aii** $108°$ **aiii** $118°$

   b The sum of the two opposite angles

5 Drawing a diagonal splits the quadrilateral into two triangles, so the sum of the interior angles is $2 \times 180° = 360°$

S2.3

1 66°		**2** 43.5°		**3** 126°
4 15°		**5** 97°		**6** 107°
7 131°		**8** 29°		**9** 133°

S2.4

1 90°		**2** 56°		**3** 63°
4 128°		**5** 59°		**6** 32°

7 $g = 152°$, $h = 110°$ **8** $i = 134°$, $j = 67°$

9 $k = 48°$, $l = 132°$ **10** $m = 117°$, $n = 126°$

11 $p = q = 98°$ **12** $30°$

13 $15°$ **14** $t = 86°$, $u = 94°$

15 $122°$ **16** $w = x = 90°$

17 $36°$ **18** $40°$

S2.5

1 $90°$ **2** $b = 44°$, $c = 46°$

3 $d = e = 28°$ **4** $f = g = 20°$

5 $48°$ **6** 5 cm

7 $28°$, 8.1 cm   **8** $m = 12°$, $n = 84°$

9 $p = 46°$, $q = 67°$ **10** $r = 73°$, $s = 53.5°$

11 $t = 112°$, $u = 56°$ **12** $v = 65°$, $w = 32.5°$

13 $x = 90°$, $y = 66°$ **14** $53°$

S2.6

1 43°		**2** 72°		**3** 57°

4 $19°$ **5** $e = 50°$, $f = 40°$

6 $g = 35°$, $h = 55°$ **7** $26°$

8 $42°$ **9** $k = l = 36°$

10 $m = n = 59°$ **11** $67°$

12 $q = 44°$, $r = 92°$ **13** $s = 62°$, $t = 56°$

14 $u = 64°$, $v = 128°$, $w = 26°$

15 $x = 58°$, $y = 32°$ **16** $17°$

S2.7

1 Angles PQR and RSP do not sum to $180°$.

2 Rectangle

3 Since $120^2 + 64^2 = 136^2$ the triangle contains a right angle, hence DF is a diameter.

4 If C was the centre of the circle then the angle at C would be twice the angle at the circumference.

5 $\angle XQP = \angle XPQ$ (isosceles triangle) so $\angle RQP = \angle SPQ$ (angles on straight lines sum to $180°$); $\angle RQP + \angle RSP = 180°$ and $\angle SPQ + \angle QRS = 180°$ (cyclic quadrilateral), so $\angle RSP = \angle QRS$ (as $\angle RQP = \angle SPQ$); therefore RQPS is an isosceles trapezium.

6 Angle PXO $= 100°$, but it would be a right angle if PX were a tangent.

S2 Exam review

1 a $18°$ (angle between tangent and radius is a right angle)

   b $72°$ (angles in a triangle add to $180°$)

2 a 55° (angle in a semicircle is a right angle)
b 35° **c** 110°

A3 Before you start ...

1 a 20, 27 **b** 32, 64 **c** $\frac{1}{25}, \frac{1}{36}$

d 125, 216 **e** 31, 50 **f** 243, −729

2 a 16 **b** 32 **c** 62 **d** 1 **e** 1

3 a $x(x-5)$ **b** $(x+3)(x+7)$

c $(x-5)(x+5)$ **d** $(y-3)^2$

e $(p-10)(p+10)$ **f** $3a(b-3a)$

g $(4m-7)(4m+7)$

4 a −10 **b** 29

A3.1

1 a 29, 34 **b** 65, 58

c 16, 21 **d** 999 999, 9 999 999

e 13, 21 **f** 3.375, 1.6875

2 a 7, 13 **b** 8, 16

c 93, 91, 89 **d** 8, 125

3 a 3, 6, 9, 12, 15 **b** 2, 4, 8, 16, 32

c 2, 3, 5, 7, 11 **d** 121, 144, 169, 196, 225

4 a 6, 10, 15

b 1, 3, 6, 10, 15, 21, 28, 36, 45, 55

c They form a pattern of squares when drawn.

d They form a pattern of cubes when drawn.

5 a 110 **b** $\frac{10}{11}$ **c** 100 **d** 100 000

6 a $6667^2 = 44\ 448\ 889$, $66\ 667^2 = 4\ 444\ 488\ 889$

b 4 444 444 444 488 888 888 889

c 666 666 667

A3.2

1 a 5, 12, 19, 26, 33 **b** −3, 0, 5, 12, 21

c 1, 8, 27, 64, 125 **d** 8, 17, 32, 53, 80

e 4, 9, 16, 25, 36 **f** 7, 4, 1, −2, −5

g 1, 4, 9, 16, 25 **h** 6, 12, 20, 30, 42

2 a 1, 3, 5, 7, 9; odd numbers

b 7, 14, 21, 28, 35; multiples of 7

c 2, 4, 8, 16, 32; powers of 2

d 10, 100, 1000, 10 000, 100 000; powers of 10

e 1, 3, 6, 10, 15; triangular numbers

f 100, 81, 64, 49, 36; square numbers decreasing from 100

g 1, 2, 3, 4, 5; positive integers

h 1, 1, 2, 3, 5; Fibonacci sequence

3 a Convergent, has a limit of 1

b Divergent

c Convergent, has a limit of 0

d Convergent, has a limit of 0

e Oscillating

f Divergent

4 a D **b** A **c** C **d** B

A3.3

1 a $T_n = 3n + 10$ **b** $T_n = 3n - 1$

c $T_n = 5n + 20$ **d** $T_n = n + 3$

e $T_n = 0.2n + 0.8$ **f** $T_n = 53 - 3n$

g $T_n = 14 - 4n$

2 a Each cube has four faces showing, and the two end cubes each have an extra face showing.

b Area of rectangle = Height × Length, and the length is one more than the height.

c There is a blue bead above each red bead, and there is an extra blue bead on the left and one on the right. Each red bead has a link to the left and one above, and the last red bead also has a link to the right.

3 a $E = 5H + 1$ **b** $M = 2L(L+1)$

4 $3 \times 3 \times 3$ cube: 1, 6, 12, 8; $n \times n \times n$ cube: $(n-2)^3$, $6(n-2)^2$, $12(n-2)$, 8;

$m \times n \times p$ cuboid: $(m-2)(n \times 2)(p-2)$,

$2(p-2)(n-2) + 2(p-2)(m-2) + 2(m-2)(n-2)$,

$4(p-2) + 4(m-2) + 4(n-2)$, 8

A3.4

1 a $x = 0$ or 3 **b** $x = 0$ or −8 **c** $x = 0$ or 4.5

d $x = 0$ or 3 **e** $x = 0$ or 5 **f** $x = 0$ or 7

g $x = 0$ or 12 **h** $x = 0$ or 2 **i** $x = 0$ or 6

j $y = 0$ or 3 **k** $w = 0$ or 7

2 a $x = -3$ or −4 **b** $x = -2$ or −6 **c** $x = -5$

d $x = -5$ or 3 **e** $x = -7$ or 2 **f** $x = 5$ or −1

g $x = 2$ or 3 **h** $x = 6$

i $x = -3$ or $-\frac{1}{2}$ **j** $x = -2$ or $-\frac{1}{3}$

k $x = -2$ or $-\frac{1}{2}$ **l** $y = -\frac{1}{2}$ or $-\frac{2}{3}$

m $x = 2$ or 6 **n** $x = 1.5$ or −5

o $x = 2$ or 3 **p** $x = -3$ or −7

3 a $x = 4$ or −4 **b** $x = 8$ or −8 **c** $y = 5$ or −5

d $x = \frac{2}{3}$ or $-\frac{2}{3}$ **e** $y = \frac{1}{2}$ or $-\frac{1}{2}$ **f** $x = 13$ or −13

g $x = 2.5$ or −2.5 **h** $y = 2$ or −2

4 a $x = 0$ or $\frac{1}{3}$ **b** $x = 5$ or −3 **c** $x = 3$ or $\frac{2}{3}$

d $y = 1\frac{1}{3}$ or $-1\frac{1}{3}$ **e** $x = 1\frac{1}{4}$ or $-1\frac{1}{4}$

f $x = 0$ or 1 **g** $x = -\frac{1}{4}$ or $\frac{3}{5}$ **h** $x = 2$ or 6

5 a $x = \frac{1}{2}$ or $-\frac{2}{3}$ **b** $x = 3$ **c** $x = -\frac{1}{2}$ or $\frac{3}{5}$

d $x = 1$ **e** $x = 2, -2, 3$ or −3

6 a $w(w+7)$ **b** $w(w+7) = 60 \Rightarrow w^2 + 7w - 60 = 0$

c 5 cm by 12 cm

A3.5

1 a $x = -0.785$ or −2.55 **b** $x = 1$ or 0.2

c $x = 5$ or −1.5 **d** $x = -0.268$ or −3.73

e $x = 0.158$ or −3.16 **f** $y = 2.21$ or −0.377

g $x = 3$ or 0.333 **h** $x = 0.158$ or −3.16

i $x = 1.67$ or −4

2 a $x = 1.61$ or −5.61 **b** $x = 4.46$ or −2.46

c $x = 1.17$ or −0.284 **d** $x = 2.87$ or −4.87

e $x = 2.32$ or −0.323

3 a $(x-2)(x+7)$

b $(x-2)(x+7) = 20 \Rightarrow x^2 + 5x - 34 = 0$

c 1.84 cm by 10.84 cm

4 a $x = -4$ or -1.5; $x = -4.35$ or -1.15

   b Only the first can be solved by factorization:
$$2x^2 + 11x + 12 = (2x + 3)(1x + 4)$$

   c $b^2 - 4ac$

5 $b^2 - 4ac$ is negative so you cannot take the square root.

A3 Exam review

1 a $(3x + 47)(x - 2)$

   b $x = 2$ or -15.7

2 a $5n$

   bi $5n + 5(n + 1) = 10n + 5$, which is even + odd = odd

   bii $5n \times 5(n + 1) = 25n^2 + 25n = 25n(n + 1)$; one of n or $n + 1$ must be even, so the product is even

D2 Before you start ...

1 a 62   **b** 60   **c** 18.5

   d 15   **e** 7   **f** 30

   g 108   **h** 150   **i** 10.5

2 a £50   **b** £70   **c** 2 days

D2.1

1 a

Height, h cm	$h <$ 155	$h <$ 160	$h <$ 165	$h <$ 170	$h <$ 175
Cumulative frequency	9	36	81	97	100

   b

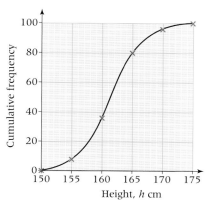

2 a

Age, A	$A <$ 30	$A <$ 40	$A <$ 50	$A <$ 60	$A <$ 70
Cumulative frequency	20	56	107	134	145

   b

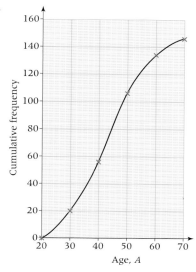

3 a

Time, t minutes	$t <$ 10	$t <$ 20	$t <$ 30	$t <$ 40	$t <$ 50	$t <$ 60
Cumulative frequency	6	24	53	88	109	120

   b

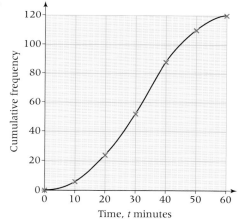

4 a

Weight, w minutes	$w <$ 3000	$w <$ 3500	$w <$ 4000	$w <$ 4500	$w <$ 5000
Cumulative frequency	7	30	64	89	100

   b

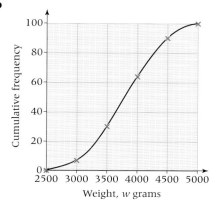

413

5 a

Height, h cm	$h < 60$	$h < 80$	$h < 100$	$h < 120$	$h < 140$	$h < 160$
Cumulative frequency	3	22	51	91	112	120

b

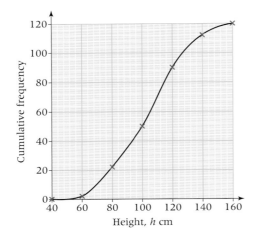

6 a

Test result, t %	$t < 40$	$t < 50$	$t < 60$	$t < 70$	$t < 80$	$t < 90$
Cumulative frequency	5	21	46	79	97	100

b

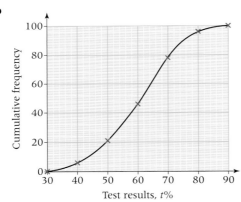

D2.2

1 a 162   **b** 7   **c** 35
2 a 43   **b** 15   **c** 10
3 a 32   **b** 19   **c** 56
4 a 3800   **b** 800   **c** 82
5 a 105   **b** 33   **c** 55
6 a 98   **b** 16
   ci 10   **cii** 15

D2 Exam review

1 ai 58   **aii** 8.75   **aiii** 23
2 a 52   **b** 29   **c** 14

A4 Before you start …

1 a

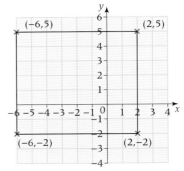

b 56 square units

2 a–d

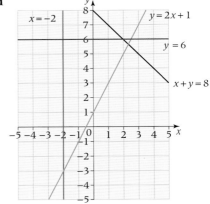

3 ai Two pairs   **aii** Three pairs
   aiii Two pairs   **aiv** None
   av One pair
   bi Adjacent sides are perpendicular
   bii None   **biii** None
   biv One pair   **bv** None
4 a $<$ **b** $>$ **c** $<$ **d** $=$

A4.1

1 a $y = 7$ is horizontal, $x = 9$ and $x = -0.5$ are vertical, $y = 2x - 1$ is diagonal, $y = x^2 + x$ is none of these

2 a $y = 3$ **b** $y = \frac{3}{4}$ **c** $y = -3$
   d $x = -2$ **e** $x = \frac{1}{4}$ **f** $x = 2.5$

3 a Vertical line cutting x-axis at $x = 5$
   b Horizontal line cutting y-axis at $y = 2$
   c Vertical line cutting x-axis at $x = 1.6$
   d Horizontal line cutting y-axis at $y = -3$
   e Horizontal line cutting y-axis at $y = 1$
   f Vertical line cutting x-axis at $x = -1\frac{1}{4}$

4 a $(5, 2)$ **b** $(4, -3)$ **c** $(-2, 9)$ **d** $(-2, -4)$
5 a For example $x = 4$, $x = 7$, $y = 1$, $y = 6$
   b For example $x = 4$, $x = 7$, $y = 3$, $y = 6$
   c For example $x = 4$, $y = 3$, $4y + 3x = 36$

6 a $(1, 2)$

b For example, below $y = 5$, above $y = 3$, right of $x = 2$, on the line $y = x + 1$

A4.2

1 a Gradient 3, y-intercept -2

b Gradient $\frac{1}{2}$, y-intercept 7

c gradient 3, y-intercept -2

d Gradient 2, y-intercept $2\frac{1}{2}$

2 a $y = 6x + 2$ **b** $y = -2x + 5$

c $y = -x + \frac{1}{2}$ **d** $y = -3x - 4$

3 a $y = 2x + c$ for any $c \ne -1$

b $y = -5x + c$ for any $c \ne 2$

c $y = -\frac{1}{4}x + c$ for any $c \ne 2$

d $y = c - 4x$ for any $c \ne 7$

e $y = c + \frac{3}{4}x$ for any $c \ne 6$

f $2y = 9x + c$ for any $c \ne -1$

4 a $y = 3x - 2$, $y = 4 + 3x$ and $2y - 6x = -3$

b $y = x + 3$ and $y = 3 - x$

c $y = 5$

d $y = 3x - 2$

e $y = x + 3$ and $y = 3 - x$

5 a $y = -4x - 2$

b $y = 1\frac{1}{2}x - 2$

6 a True   **b** False

c False   **d** False

7 a $(1, 5)$

b $(3, 4)$ and $(-1.4, -4.8)$

A4.3

1 a 4   **b** 2   **c** $\frac{1}{3}$

2 a – d

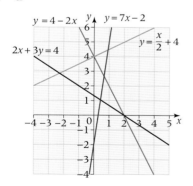

3 a $x = -3$ **b** $y = 4x - 2$ **c** $y = 3 - 2x$

d $y = -\frac{3}{4}x - 1$ **e** $y = \frac{1}{2}x + 1$

4 a $y = 4x + 5$ **b** $y = \frac{1}{2}x + 7$ **c** $y = -\frac{1}{3}x + 2$

5 a $5\frac{1}{3}$ **b** $1.28, -0.78$

6 a $y = 2.5x + 5$ **b** $y = 1 - x$

A4.4

1 a

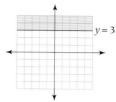

b

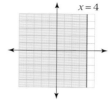

c

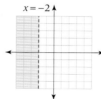

d

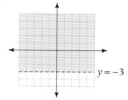

e

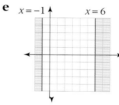

f

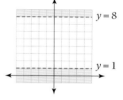

g

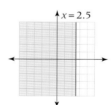

h

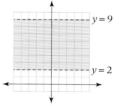

i

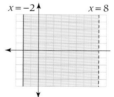

j

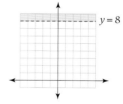

k

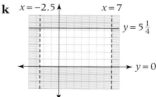

2 a $-2 < y \le 5$ **b** $1 \le x < 7.5$

3 $(-1, 0), (0, 0), (1, 0), (-1, 1), (0, 1), (1, 1)$

4 a e.g. $0 < x < 2, 0 < y < 1$

b e.g. $x > 1$ or $x < 0$ and $y > 1$ or $y < 0$

c e.g. $0 < x < 2, 0 < y < 1$

5 a $y \le 9, y \ge x^2$

b $y < 5, y \ge (x - 3)(x + 2)$

A4.5

1 a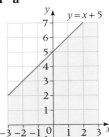

b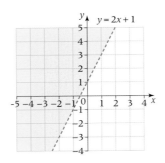

c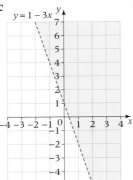

d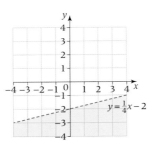

e f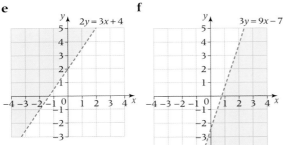

g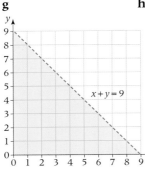

h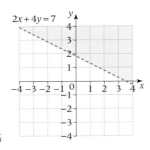

2 a $y < 2x + 4$ **b** $y \geqslant 2\frac{2}{3}x - 1$

3 a $(-1, 1), (-1, 0), (-1, -1), (-1, -2), (-1, -3),$
$(-1, -4), (0, 1), (0, 0), (0, -1)$
  b $(3, 1), (3, 0), (4, 0)$

4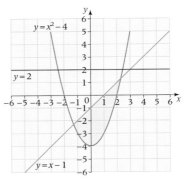

5 $x \geqslant 0$ and $y \geqslant 0$ because the number of each type of coach cannot be negative. The total number of seats is $20x + 48y$, which must be at least enough for 316 people, so $20x + 48y \geqslant 316$ or $5x + 12y \geqslant 79$. To have at least 2 adults per coach, there can be no more than 8 coaches, so $x + y \leqslant 8$.

6 a e.g. $x > 0$, $y > 0$, $x + y < 1$
  b e.g. $y > 1$ or $y < 0$, $y > x$ or $y > 3 - x$

A4 Exam review

1 $y = 3x - 2$

2 a $y = -\frac{5}{6}x + 2\frac{1}{2}$ **b** 20
  ci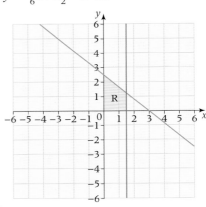

  cii $(1, 1)$

D3 Before you start ...

1 a $\frac{3}{5}$ **b** $\frac{3}{7}$ **c** $\frac{5}{8}$

2 a $\frac{5}{6}$ **b** $\frac{9}{20}$ **c** $\frac{23}{24}$

3 a $\frac{1}{6}$ **b** $\frac{3}{20}$ **c** $\frac{2}{15}$

4 a 0.7 **b** 0.75 **c** 0.375 **d** 0.4 **e** $0.\dot{3}$

D3.1

1 a $\frac{9}{25}$ **b** $\frac{16}{25}$ **c** $\frac{8}{25}$ **d** $\frac{17}{25}$ **e** $\frac{2}{5}$ **f** $\frac{19}{25}$

2 a The Hawks' chance of winning is twice the Jesters' chance of winning.
  b 0.15

3 0.25

4 a $\frac{1}{9}$ **b** $\frac{8}{9}$ **c** $\frac{4}{9}$ **d** $\frac{1}{9}$ **e** $\frac{1}{3}$
  f $\frac{2}{3}$ **g** $\frac{2}{3}$ **h** $\frac{1}{3}$ **i** $\frac{4}{9}$

5 a 30

bi $\frac{2}{5}$ **bii** $\frac{4}{15}$ **biii** $\frac{17}{30}$

biv $\frac{1}{3}$ **bv** $\frac{1}{6}$ **bvi** Can't tell

6 a $\frac{1}{32}$ **b** $\frac{5}{16}$ **c** $\frac{13}{16}$ **d** $\frac{15}{32}$

D3.2

1 120

2 $\frac{1}{5}$ or 0.2

3 ai 0.23   **aii** 0.33   **aiii** 0.44

   aiv 70.72   **bi** 56   **bii** 112

4 a 152   **b** 126   **c** 215

5 a

	Canoeing	Abseiling	Potholing	Total
Male	13	4	5	22
Female	7	9	12	28
Total	20	13	17	50

bi $\frac{2}{5}$ **bii** $\frac{3}{5}$ **biii** $\frac{37}{50}$ **biv** $\frac{3}{5}$

ci 120   **cii** 168

D3.3

1 No, $\frac{124}{360} = 0.34$ which is not close to 0.5.

2 Yes, as the probability of picking a black is close to 0.5.

3 Yes, all frequencies have similar values.

4 No, 3 occurs much more than other outcomes.

5 a 0.2, 0.3, 0.4, 0.1, 0.1, 0.2 0.3, 0.2, 0.1, 0.3

   b 0.22

6 a 0.306, 0.222, 0.222, 0.25

   b Probably not, as the relative frequencies are similar, but more trials are needed to be sure.

D3.4

1 a

	1	2	3	4	5	6
Head	H and 1	H and 2	H and 3	H and 4	H and 5	H and 6
Tail	T and 1	T and 2	T and 3	T and 4	T and 5	T and 6

bi $\frac{1}{12}$ **bii** $\frac{1}{12}$

The probabilities are the same as both coin and dice are fair.

2 a

Blue dice

		1	2	3	4	5	6
	1	1, 1	1, 2	1, 3	1, 4	1, 5	1, 6
	2	2, 1	2, 2	2, 3	2, 4	2, 5	2, 6
Red dice	**3**	3, 1	3, 2	3, 3	3, 4	3, 5	3, 6
	4	4, 1	4, 2	4, 3	4, 4	4, 5	4, 6
	5	5, 1	5, 2	5, 3	5, 4	5, 5	5, 6
	6	6, 1	6, 2	6, 3	6, 4	6, 5	6, 6

bi $\frac{1}{36}$ **bii** $\frac{1}{36}$ **biii** $\frac{1}{12}$ **biv** $\frac{1}{18}$

3 $\frac{1}{6}$

4 a $\frac{4}{25}$ **b** $\frac{1}{25}$ **c** $\frac{3}{50}$ **d** $\frac{1}{50}$ **e** $\frac{3}{100}$

5 a

		2 pence coin	
		Head	**Tail**
10 pence coin	**Head**	Head, Head	Head, Tail
	Tail	Tail, Head	Tail, Tail

b $\frac{1}{4}$

6 Yes, as $0.24 \times 0.15 = 0.036$

D3.5

1 0.99

2 $\frac{31}{45}$

3 $\frac{193}{288}$

4 a $\frac{99}{400}$ **b** $\frac{407}{1875}$

5 a $\frac{6}{23}$ **b** $\frac{65}{136}$

D3 Exam review

1 The dice is probably biased, as you would expect each number to come up about 17 times, but 3 came up 26 times.

2 Fred is wrong, as $0.05 \times 0.06 \neq 0.011$.

N4 Before you start ...

1 a 80   **b** 21   **c** 19.6   **d** 168

2 672 g

3 3 hours

4 a 96 : 64   **b** 14.4 : 9.6   **c** 5 : 20   **d** 20 : 16

N4.1

1 ai $\frac{7}{20}$ **aii** 35%

   bi $\frac{2}{25}$ **bii** 8%

   ci $\frac{3}{10}$ **cii** 30%

   di $\frac{3}{40}$ **dii** 7.5%

   ei $\frac{9}{40}$ **eii** 22.5%

   fi $\frac{3}{20}$ **fii** 15%

2 a $\frac{3}{25}$ **b** 12%

3 a 16.8 g   **b** 8.75 g   **c** 9 g

   d 67.2 g   **e** 18.75 g   **f** 48.72 g

4 a 45 kg   **b** 84 g   **c** 48 cm

   d 285 m   **e** 15.95 cc   **f** 260 mm

5 a 135 kg   **b** 32.085 m

   c £105   **d** 22.5 cm^3

   e 510.15 g   **f** £16.50

6 a £640   **b** €180

   c 240 kg   **d** 304 cc

   e 1640 kg   **f** 23.3 m

7 a £1500

   bi $\frac{5}{12}$ **bii** 41.7%

N4.2

1 16.6 m^2
2 25.83 kg
3 **b**, **d** and **e**
4 Two variables that are in direct proportion have a straight-line graph through (0, 0).
5 **a** $w \propto l$ **b** $w = kl$ **c** 2.48 kg/m **d** 7.192 kg
6 €4.08
7 £17.12
8 Super

N4.3

1 **a** Doubled **b** Halved
   c Multiplied by 6 **d** Divided by 10
   e Multiplied by a factor of 0.7
2 **a** Halved **b** Doubled
   c Divided by 6 **d** Multiplied by 10
   e Divided by 0.7
3 **a** 2 **b** 8 **c** 0.4 **d** 16 **e** 0.16
4 $y = \frac{500}{w}$
5 800
6 **a** $t = \frac{10}{n}$
   bi 10 hours **bii** $2\frac{1}{2}$ hours
   biii $1\frac{1}{4}$ hours **biv** 2 hours
   bv 1 hour 24 minutes
   c Yes
7 **a** 2 amps **b** 4 amps **c** 6 amps
   d 8 amps **e** 10 amps **f** 0.67 amps
   g 0.5 amps **h** 2.4 amps
8 The graph tends to both axes: when R is small, I is big and when I is small, R is big.
9 9
10 **a** If $x = ky$, then $y = \frac{1}{k}x$. Since k is a constant, $\frac{1}{k}$ is a constant, and y is directly proportional to x.
   b $y = \frac{k}{x}$ can be rearranged to $x = \frac{k}{y}$, so x is inversely proportional to y.

N4.4

1 188 mm, 42.75 litres, 229.4 cm^3, £177.01
2 **a** £330 **b** £246.50
   c £805.35 **d** £342.42
3 **a** £182.32 **b** £299 250
4 **a** £167.60 **b** £339 023
5 **a** £213.75
   b No, the compound interest would only be £178.30.
   c No, the compound interest would be £390.07 but the simple interest would be £427.50.
6 6 years
7 Yes, as if it had exactly doubled she would have sold it for £122 500 × 1.035^{21} = £252 280.
8 **a** £2747.76 **b** £1560

N4.5

1 **a** 2 : 1 **b** 5 : 1 **c** 3 : 2 **d** 1 : 2 : 3
   e 5 : 2 **f** 1 : 2 : 6

2 **a** £33 : £22 **b** 75 cm : 45 cm
   c 48 seats : 36 seats : 12 seats
   d 27 tickets : 9 tickets : 6 tickets
   e 96 books : 36 books : 12 books
   f 80 hours : 50 hours : 30 hours
3 £80 : £60, 51 cm : 34 cm, 10 h : 8 h, 7 cc : 42 cc, 21 min : 24 min, €570 : €150
4 Benny £222.22, Amber £177.78
5 Peggy £3200, Grant £3600, Mehmet £2800
6 Steven £6315.79, Will £3789.47, Phil £1894.74
7 $141 \text{ cm}^2 : 47 \text{ cm}^2 : 12 \text{ cm}^2$,
   19 cm : 12.7 cm : 6.3 cm,
   208 m : 104 m : 138 m,
   131 mm : 327 mm : 262 mm,
   $47.50 : $17.81 : $29.69
8 John £24.41, Janine £20.59

N4.6

1 9 mph
2 £2.10 per metre
3 32 mph
4 89.25 mph
5

Speed (km/h)	Distance (km)	Time
105	525	5 hours
48	106	2 hours 12.5 minutes
$37\frac{1}{3}$	84	2 hours 15 minutes
86	215	2 hours 30 minutes
37.1	65	1 hour 45 minutes

6 **a** 5 g/cm^3 **b** 87.88 g
7 **a** 7.59 g/cm^3 to 3 sf **b** 105
8 **a** 9.46 kg to 3 sf **b** 6.15 litres to 3 sf
9 1358 m

N4 Exam review

1 £250
2 28 g cheese, 42 g topping

S3 Before you start …

1
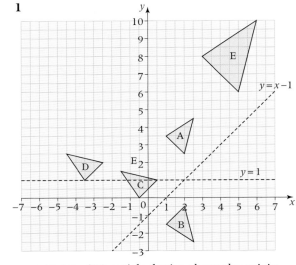

   c Rotate 90° anticlockwise about the origin.
   e Shapes A, B, C and D are congruent.

S3.1

1 **a** Congruent **b** Not congruent **c** Congruent
2 Congruent by RHS
3 WX = 5 cm by Pythagoras, so congruent by RHS
4 **a** Only one side is the same and no angles are the same.
  b Three sides are the same.
5 **a** HFG and FHE, HEG and FGE
  b Opposite sides and angles in a parallelogram are equal, so congruent by SSS or SAS.

4 **a** and **b**

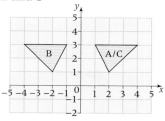

c A and C are the same.

S3.2

1
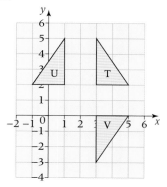

2
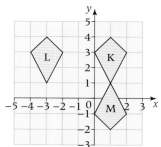

3
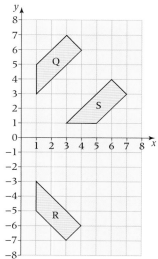

S3.3

1
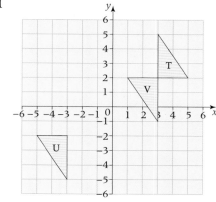

2
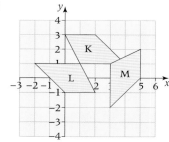

3
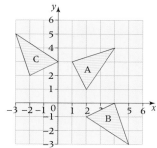

4 **a** and **b**

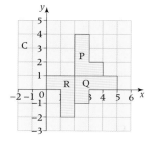

c It maps to R. Two 90° rotations are the same as one 180° rotation.

S3.4

1

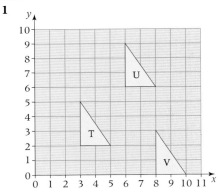

2

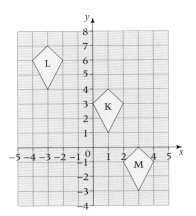

3 a

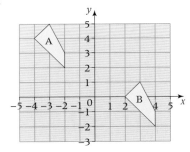

b B translates back to A. The translation has been reversed.

S3.5

1 a Reflection in *y*-axis
b Rotation by 180° about (0, 0)
c Reflection in *x*-axis
d Rotation by 180° about (0, 0)
e Reflection in *y*-axis
f Reflection in *y*-axis

2 a Translation by $\begin{pmatrix} 16 \\ 2 \end{pmatrix}$ **b** Translation by $\begin{pmatrix} 5 \\ 3 \end{pmatrix}$

c Translation by $\begin{pmatrix} 9 \\ 8 \end{pmatrix}$ **d** Translation by $\begin{pmatrix} -4 \\ -5 \end{pmatrix}$

e Translation by $\begin{pmatrix} 7 \\ -6 \end{pmatrix}$ **f** Translation by $\begin{pmatrix} -7 \\ 6 \end{pmatrix}$

3 a Rotation by 90° clockwise about (0, 0)
b Rotation by 180° about (0, 0)
c Rotation by 90° anticlockwise about (0, 0)
d Rotation by 90° clockwise about (0, 0)
e Rotation by 90° clockwise about (0, 0)
f Rotation by 180° about (0, 0)

S3.6

1

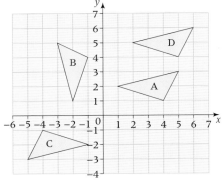

c Rotation 180° about (0, 0)

e Translation by $\begin{pmatrix} 1 \\ 3 \end{pmatrix}$

2

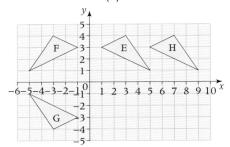

c Rotation 180° about (0, 0)

e Translation by $\begin{pmatrix} 4 \\ 1 \end{pmatrix}$

3

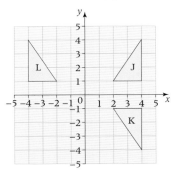

c Reflection in *y*-axis

4 a and **b**

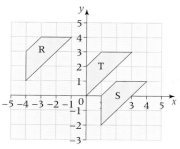

c Translation by $\begin{pmatrix} -4 \\ 1 \end{pmatrix}$

S3.7

1

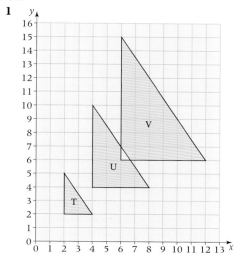

2

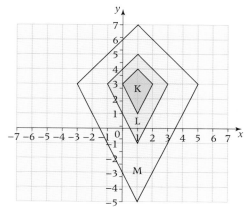

3

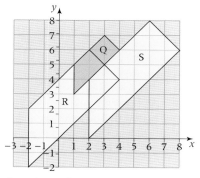

d Twice as large

4

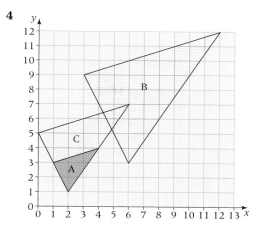

d Three times as large

S3.8

1

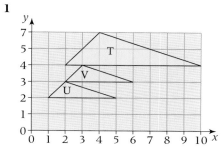

2

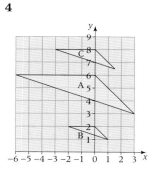

3   **4**

S3.9

1 a Enlargement with centre (0, 0) and scale
    factor 3
  b Enlargement with centre (0, 0) and scale
    factor $\frac{1}{3}$

2 ai Enlargement with centre $(-3, 0)$ and scale factor 2

aii Enlargement with centre $(-3, 0)$ and scale factor $\frac{1}{2}$

b 16 units

3 a Enlargement with centre $(0, 0)$ and scale factor 2

b Enlargement with centre $(0, 0)$ and scale factor $\frac{1}{2}$

4 ai Enlargement with centre $(0, 0)$ and scale factor $\frac{1}{2}$

aii Enlargement with centre $(0, 0)$ and scale factor 2

aiii Enlargement with centre $(0, 0)$ and scale factor 4

b 4 times

S3.10

1 $a = 6.4\ cm$, $b = 4.5\ cm$
2 $c = 9\ cm$, $d = 1.9\ cm$
3 RT = 5 cm, QR = 4 cm, QS = 12 cm
4 a XY = 6.6 cm, VY = 6 cm **b** 15.5 cm
5 12.8 cm, 24.8 cm
6 a $1:3$ **b** $1:3$
   c For any two circles, the ratio of the circumferences is the same as the ratio of the radii because each circumference is just the radius multiplied by a constant (2π).

S3.11

1 a 20 cm **b** 1.72 cm
2 20 cm
3 a 4cm, 1.73 cm
4 a 2.29 cm **b** 11.1 cm

S3.12

1 a $107.5\ cm^2$ **b** $193.75\ cm^3$
2 $244\ cm^2$
3 $125\ cm^3$
4 a Their sides are all the same length, so any scale factor will be constant on all dimensions.
   b The scale factor on one dimension could be different from that on another.
   ci $1:49$ **cii** $1:343$
5 a $1:1.5$ or $2:3$ **b** $34\ cm^2$ **c** $81\ cm^3$
6 Deal A

S3 Exam review

1 $(-\frac{1}{2}, -\frac{1}{2})$, $(-\frac{1}{2}, -2)$, $(-1\frac{1}{2}, -\frac{1}{2})$
2 a $1440\ cm^2$ **b** 100 ml

N5 Before you start …

1 a 2^7 **b** 3^3 **c** 5^7 **d** 6^2 **e** 7^3 **f** 4^4
2 a 1 **b** 1 **c** 5 **d** $\frac{1}{2}$
3 a 7π **b** 13π **c** 36π **d** 4 **e** $3\sqrt{2}$ **f** $\sqrt{3}$

N5.1

1 a 64 **b** 625 **c** 64 **d** 27
   e 81 **f** 2401 **g** 1 **h** 1
2 a 256 **b** 5 **c** 9
   d $x = 4$, $y = 3$
3 a 3^4 **b** 5^5 **c** 6^7 **d** 7^9
   e 2^{15} **f** 4^{10} **g** 10^{14} **h** 9^{13}

4

$\times$	x	x^3	x^4	x^9
x^2	x^3	x^5	x^6	x^{11}
x^6	x^7	x^9	x^{10}	x^{15}
x^3	x^4	x^6	x^7	x^{12}
x^5	x^6	x^8	x^9	x^{14}

5 a 4^2 **b** 5^3 **c** 9^3 **d** 6^4
   e 7^4 **f** 8^2 **g** 9^4 **h** 3^5
6 a 3^2 **b** 4^5 **c** 7^3 **d** 6^2
   e 5^4 **f** 2^5 **g** 6^1 **h** 4^3
7 a 2^6 **b** 4^{10} **c** 7^4 **d** 5^{15}
   e 3^{16} **f** 6^4 **g** 5^{21} **h** 10^{16}
8 a 4096 **b** 2 **c** 2 **d** 793
9 a 4^4 **b** 3^4 **c** 6^7 **d** 5^{12}
   e 2^{10} **f** 7^{12} **g** 3^3 **h** 9^5

N5.2

1 a 4 **b** 3 **c** 3 **d** 0 **e** 10
2 a $\frac{1}{2}$ **b** $\frac{1}{3}$ **c** $\frac{1}{2}$ **d** $\frac{1}{4}$
3 a 5 **b** $\frac{1}{5}$ **c** 1 **d** 125 **e** $\frac{1}{25}$
4 a 8 **b** 4 **c** 243 **d** $\frac{1}{10}$ **e** $\frac{1}{64}$
   f 100 **g** $\frac{1}{20}$ **h** 2197 **i** $\frac{1}{729}$ **j** $\frac{1}{8}$
5 a 4^{-1} **b** 4^2 **c** 4^{-2} **d** $4^{\frac{3}{2}}$ **e** $4^{\frac{5}{2}}$
6 a $16^{-\frac{1}{2}}$ **b** 16^1 **c** 16^{-1} **d** $16^{\frac{3}{4}}$ **e** $16^{\frac{5}{4}}$
7 a 10^2 **b** 10^{-1} **c** $10^{\frac{1}{2}}$ **d** $10^{\frac{3}{2}}$ **e** $10^{-\frac{5}{2}}$
8 a 5^{-2} **b** $5^{-\frac{1}{3}}$ **c** $5^{\frac{2}{3}}$ **d** $5^{-\frac{3}{2}}$ **e** $5^{-\frac{4}{3}}$
9 a $\frac{1}{9}, \frac{1}{3}, 1, 3, 9$
   b, c
   The graphs both pass through $(0, 1)$ and have a similar shape, $y = 9^x$ is steeper than $y = 4^x$.

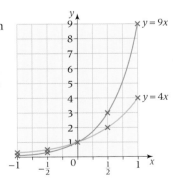

N5.3

1 a Rational **b** Irrational **c** Rational
   d Irrational **e** Rational **f** Irrational
2 All fractions are rational. The calculator does not have enough digits to show the repeating pattern.

3 a e.g. $\sqrt{3}$ **b** e.g. 25 **c** e.g. $\sqrt[3]{3}$ **d** e.g. π

4 64

5 Karla cannot be correct. If the number is rational, it can be written as a fraction $\frac{a}{b}$, and half of it can be written as the fraction $\frac{a}{2b}$.

6 Jim cannot be correct. If the final number is $\frac{a}{b}$, then the original number would have been $\frac{a-b}{b}$, which is rational.

7 Javier is correct.

8 a $\frac{\sqrt{2}}{2}$ **b** $\frac{\sqrt{3}}{3}$ **c** $\frac{\sqrt{7}}{7}$ **d** $\frac{\sqrt{6}}{6}$ **e** $\frac{\sqrt{5}}{5}$

9 a $\frac{\sqrt{8}}{4}$ **b** $\frac{\sqrt{10}}{5}$ **c** $\frac{\sqrt{12}}{4}$ **d** $\frac{\sqrt{30}}{6}$ **e** $\frac{\sqrt{40}}{5}$

10 Simone's number is rational, but Lisa's is irrational. For a repeating decimal to be rational, there must be a pattern that repeats exactly.

N5.4

1 a $\sqrt{6}$ **b** $\sqrt{15}$ **c** $\sqrt{231}$

2 a $\sqrt{2}\sqrt{7}$ **b** $\sqrt{3}\sqrt{11}$ **c** $\sqrt{3}\sqrt{7}$

3 a $2\sqrt{5}$ **b** $3\sqrt{3}$ **c** $7\sqrt{2}$

4 a $\sqrt{48}$ **b** $\sqrt{50}$ **c** $\sqrt{80}$

5 a 6 **b** 12 **c** 66

6 a $5\sqrt{5}$ **b** $7\sqrt{7}$ **c** $8\sqrt{3}$

7 a 11.180, 18.520, 13.856
  b 11.180, 18.520, 13.856
  c They are the same.

8 a $\sqrt{5}+5$ **b** $2\sqrt{3}-3$ **c** $7+3\sqrt{5}$
  d $1+\sqrt{3}$ **e** $38-14\sqrt{7}$ **f** 23
  g 16 **h** -38 **i** -26

9 a $\frac{\sqrt{11}}{11}$ **b** $\sqrt{2}-1$

  c $\frac{-1-2\sqrt{3}}{11}$ **d** $\frac{5-\sqrt{5}}{4}$

  e $2\sqrt{2}-3$ **f** $16+11\sqrt{2}$

N5.5

1 Distributive law

2 $24 \times 19 = 24 \times (20-1) = 24 \times 20 - 24$
  $= 480 - 24 = 456$

3 She has done $36 \div 6$ not $36 \div 5$. Division is not distributive over addition.

4 a e.g. $1-2 = -1$, $2-1 = 1$
  b e.g. $1 \div 2 = 0.5$, $2 \div 1 = 2$
  c e.g. $1-(2-3) = 2$, $(1-2)-3 = -4$
  d e.g. $1 \div (2 \div 3) = 1\frac{1}{2}$, $(1 \div 2) \div 3 = \frac{1}{6}$

5 No, for example, $\sqrt{4} = \sqrt{(2+2)} \neq \sqrt{2}+\sqrt{2}$.

6 a 4600 **b** 28 **c** 36 400

7 a 1225 **b** 841 **c** 16.81

8 $5.7^2 = (4.7+1)^2 = 4.7^2 + 2 \times 4.7 + 1$
  $= 22.09 + 9.4 + 1 = 32.49$

9 $14.6^2 = (15.6-1)^2 = 15.6^2 - 2 \times 15.6 + 1$
  $= 243.36 - 31.2 + 1 = 213.16$

10 $(x+0.5)^2 = x^2 + x + 0.25 = x(x+1) + 0.25$

N5 Exam review

1 ai -1 **aii** $x+\sqrt{x}-2$ **b** $\frac{2\sqrt{5}}{5}$

2 ai 1 **aii** 8 **aiii** $\frac{1}{16}$ **b** $\frac{5}{2}$

A5 Before you start ...

1 a £31.50 **b** 54 minutes

2 a $x = 15$ **b** $x = -2\frac{1}{2}$ **c** $x = \frac{5}{17}$ **d** $x = \frac{3}{7}$

3 a $A = \pi r^2$ **b** 4 cm **c** $A = d\frac{c-d}{2} + \frac{\pi d^2}{8}$

4 a Any three even numbers
  b e.g. 4
  c 0
  d e.g. non-isosceles trapezium

A5.1

1 a True for two values of x, equation
  b True for all values of x, identity
  c Not true for any value of x
  d True for one value of x, equation

2 a 6.78 cm^2 **b** 113 mm^3
  c 15.9 m **d** 1240 cm^3

3 a $A = p^2 - \frac{1}{4}\pi p^2 = p^2(1-\frac{\pi}{4})$

  b $A = (x-2)(x+3) + \frac{1}{2}(3+x+3)(x+6-(x-2))$
  $= x^2 + 5x + 18$

A5.2

1 a $m = \frac{y-c}{x}$ **b** $m = k+wt$

  c $m = \sqrt{p-kt}$ **d** $m = (l+k)^3$

2 a $x = 2(y-kw)$ **b** $x = \frac{4m+t^2}{a}$

  c $x = 4y^2$ **d** $x = \frac{k-y^2}{l}$

  e $x = \frac{(t-kh)^2}{a^2}$ **f** $x = \frac{p}{m+t}$

  g $x = \frac{p}{w-c}$ **h** $x = \frac{y}{a(b+j)}$

3 Clare: In the second line the RHS should be $-ax$. Isla: In the second line the RHS should be $px - k$.

4 $a = b - \frac{c}{def}$, $a + \frac{c}{def} = b$, $\frac{c}{def} = b - a$, $\frac{c}{df(b-a)} = e$,
  $c = def(b-a)$

5 a $\frac{p(c-qt)}{m-x^2} = wr$, $pc - pqt = mwr - wrx^2$,
  $x^2 = \frac{mwr + pqt - pc}{wr}$, $x = \sqrt{\frac{mwr + pqt - pc}{wr}}$

  b $\frac{1}{a} + \frac{1}{b} = \frac{1}{c}$, $\frac{1}{b} = \frac{1}{c} - \frac{1}{a}$, $\frac{1}{b} = \frac{a-c}{ac}$, $b = \frac{ac}{a-c}$

  c $\sqrt[4]{t} - qx = 2p$, $t - qx = 16p^4$, $t = qx + 16p^4$,
  $x = \frac{t - 16p^4}{q}$

6 False, the last formula has no t.

7 a $f = \dfrac{uv}{u+v}$ **b** $v = \dfrac{fu}{u-f}$

A5.3

1 a $x = \sqrt{\dfrac{y-b}{a}}$ **b** $x = c + 2t$

c $x = \dfrac{c-k}{b}$ **d** $x = \dfrac{b}{p+q}$

e $x = \dfrac{b+9t^2}{az}$ **f** $x = \sqrt[3]{k - t^2}$

2 Collecting the x-terms on one side gives $x^2 + 5x = -6$, so it is not possible to get x on its own.

3 a $w = \dfrac{t+r}{q-p}$ **b** $w = \dfrac{a-k}{c-l}$ **c** $w = \dfrac{py+qt}{p+q}$

d $w = \dfrac{t+r}{1+t}$ **e** $w = \dfrac{c-rg}{r-1}$ **f** $w = \dfrac{t+5r}{x+r}$

g $w = \dfrac{t+5k}{k-1}$ **h** $w = \dfrac{3q-4p}{7}$ **i** $w = \dfrac{-t-25q}{24}$

4 a $A = 2\pi r^2 + 2\pi rh$ **b** $r = \dfrac{-2\pi h \pm \sqrt{4\pi^2 h^2 + 8\pi A}}{4\pi}$

5 a $A = \pi r^2 + \pi rl$ **b** $r = \dfrac{-\pi l \pm \sqrt{\pi^2 l^2 + 4\pi A}}{2\pi}$

6 a $x = 2.098$ or -1.431 **b** $c = -ax^2 - bx$

c $b = \dfrac{-ax^2 - c}{x}$

d Because there are different powers of a.

7 $\sin(A+B) \neq \sin A + \sin B$

A5.4

1 a e.g. take 3, 4, 5 then
$(4+5) - (3+4) = 9 - 7 = 2$
bi $2n+1$ **bii** $2n+3$
c $(2n+3) - (2n+1) = 2$

2 a e.g. 1 is odd
b e.g. $2^3 - 1^3 = 8 - 1 = 7$, which is odd
c e.g. $1^2 = 1$, which is not greater than 1
d e.g. $(0+1)^2 = 0^2 + 1^2$
e e.g. $\sin(0° + 90°) = \sin 0° + \sin 90°$
f e.g. $x = 11$

3 a n, $n+1$ and $n+2$ are consecutive integers,
$(n+1)^2 = n^2 + 2n + 1 = n(n+2) + 1$
b $2n-1$ and $2n+1$ are consecutive odd numbers, $(2n-1)(2n+1) = 4n^2 - 1$
c n, $n+1$ and $n+2$ are consecutive integers,
$(n+1)^2 - n(n+2) = n^2 + 2n + 1 - n^2 - 2n = 1$

4 a Let n and m be the numbers on the top faces, then $7-n$ and $7-m$ are the numbers on the bottom faces, so sum $= nm + (7-n)(7-m) + n(7-m) + m(7-n) = 49$

5 a $n-2$, $n-1$, n, $n+1$ and $n+2$ are five consecutive integers, square of the mean $= n^2$, mean of squares $= n^2 + 2$

A5.5

1 Pythagoras: $(x+1)^2 + (x+3)^2$
$= (x+7)^2 \Rightarrow x^2 - 6x - 39 = 0$

2 $12n + (2n+1)^2 + 4n^2 + 3 = 8n^2 + 20n + 4 =$ which is even

3 $AB = \sqrt{53}$, $AC = \sqrt{29}$ and $BC = \sqrt{34}$, which are all different, so the triangle is scalene.

4 a $A = x(x+10)$
b $x(x+10) = 24 \Rightarrow (x+12)(x-2) = 0$ so $x = 2$ (reject $x = -12$ as lengths must be positive)

5 a $\dfrac{m}{m+6}$
b $\dfrac{6}{m+5}$
c P(both red) $= \dfrac{m}{m+6} \times \dfrac{m-1}{m+5}$
$= \dfrac{2}{11} \Rightarrow 3m^2 - 11m = 20$
d 5

6 $(\sqrt{3})^2 + (\sqrt{2} + 3)^2 = 14 + 6\sqrt{2} \neq (2 + \sqrt{10})^2$ so Pythagoras' theorem does not hold, and the triangle is not right-angled.

A5 Exam review

1 a $P = 6x + 4$
b $(x+1)(x-1) + (x+2)(x-1) = 52$
$\Rightarrow 2x^2 + x - 55 = 0$

2 a 24.8 **b** $r = \dfrac{P - 2a}{2 + \pi}$

S4 Before you start …

1 a 49 **b** 52 **c** 34 **d** 48 **e** 45

2 a $x = 6y$ **b** $x = 5y$ **c** $x = 10y$

d $x = \dfrac{2}{y}$ **e** $x = \dfrac{5}{y}$ **f** $x = \dfrac{8}{y}$

3 a $x = 5$ **b** $y = 77$

4 a $x = 4.93$ **b** $y = 2\frac{2}{3}$

5 a $x = 22.8$ **b** $y = 0.8$

S4.1

1 22.5 cm^2
2 a Square with sides 7.1 cm **b** 50 cm^2
3 a All 42 cm^2
b Halve the product of the diagonals.
4 60 cm^2
5 Both shapes have been split into congruent triangles; all sides equal in both cases.
6 a True; all squares have 2 pairs of sides of equal length and 4 equal angles.
b False; not all kites have diagonals which bisect each other.
c False; not all rhombuses have 4 equal angles.

S4.2

1 a (1, 4.5) **b** (2.5, 2)
2 5.83 units
3 a 8.06 units **b** 10.30 units
4 14.14 cm
5 a 7.07 cm **b** 50 cm^2
6 12.9 cm
7 16.2 cm
8 19.5 cm

S4.3

1 a 3.75 cm **b** 8.63 cm **c** 11.3 cm
d 2.04 cm **e** 2.42 cm **f** 4.73 cm
g 2.40 cm **h** 8.48 cm **i** 68.4 cm
j 4.73 cm **k** 4 cm **l** 16.3 cm

2 Right-angled isosceles triangles. The missing side is equal to the given shorter side.

S4.4

1 5.14 cm	**2** 10.4 cm	**3** 5.26 cm
4 4.75 cm	**5** 9.50 cm	**6** 11.1 cm
7 8.20 cm	**8** 8.30 cm	**9** 51.8 cm
10 10.9 cm	**11** 8.27 cm	**12** 7.65 cm
13 13.3 cm	**14** 5.89 cm	**15** 15.1 cm

S4.5

1 15.8°	**2** 18.3°	**3** 65.3°
4 67.1°	**5** 24.7°	**6** 20.5°
7 55.7°	**8** 69.3°	**9** 22.2°
10 39.2°	**11** 69.8°	**12** 25.9°
13 74.5°	**14** 35.8°	**15** 45°
16 30°	**17** 60°	

S4.6

1 a 18.9 cm   **b** 20.6 cm   **c** 4.54 cm
2 a 30.5°   **b** 29.1°   **c** 63.9°
3 31.8 cm^2
4 36.7 cm^2
5 89.8°
6 9.46 cm
7 50°, 50° and 80°
8 Edina's measurements make the height of the tree 25.6 m. Patsy's measurements make the height of the tree 27.0 m. They could both be correct as this type of measuring can be inaccurate (they may have taken different points as the top of the tree, the ground may not be level, ...).

S4 Exam review

1 529.8 cm^2
2 a 62.0°   **b** 19.1 m

N6 Before you start ...

1 a 40   **b** 20   **c** 100
   d 0.08   **e** 0.4
2 a $20 \times 50 = 1000$ **b** $4 \times 90 = 360$
   c $4000 \div 80 = 50$ **d** $600 \div 30 = 20$
3 a 0.091 152 815   **b** −0.657 142 857
   c 2.632 489 316
4 a 1.666×10^7 **b** 7.27×10^8

N6.1

1 a 8   **b** 500   **c** 0.008   **d** 10
   e 4   **f** 50
2 a 1.4   **b** 2.8   **c** 3.2   **d** 3.9
   e 4.5   **f** 5.1   **g** 5.7   **h** 6.7
   i 8.4   **j** 9.2
3 a The denominator would become zero, so the result would become undefined.
   b The approximation would be 1^8 instead of 1.49^8, and the large power would mean this was very inaccurate.

c This would simply give zero, which would not be very helpful.

4 a $\frac{300 \times 4}{0.2} = 6000$ **b** $\frac{4 \times 700}{0.8} = 3500$
   c $\frac{5 \times 100}{500 \times 0.4} = 2.5$ **d** $5^2 - 9^2 = 44$
   e $\frac{9 - 4}{0.04 - 0.02} = 250$ **f** $8 \times \frac{0.2 \times 200}{50 - 25} = 12.8$

5 a $\frac{50 \times 5}{100} = 2.5$ **b** $\frac{\sqrt{300^2}}{900} = 10$
   c $\frac{\sqrt{100}}{0.1} = 100$ **d** $\frac{49 + 144}{\sqrt{50^2}} \approx \frac{200}{50} = 4$
   e $\frac{(23 - 18)^2}{3 + 2} = 5$ **f** $\sqrt{\frac{3 + 6}{0.2^2}} = \frac{3}{0.2} = 15$

6 a $\frac{5 \times 10^3}{5 \times 10^2} = 10$ **b** $\frac{7 \times 10^3}{7 \times 10^{-2}} = 10^5$
   c $\frac{8 \times 10^4 \times 4 \times 10^{-2}}{4 \times 10^3} = 0.8$ **d** $\frac{7 \times 10^2 \times 4 \times 10^{-1}}{7 \times 10^{-2} \times 4 \times 10^{-1}} = 10^4$
   e $\frac{7 \times 10^2 + 9 \times 10^2}{6 \times 10^{-2} \times 8 \times 10^{-1}} = \frac{16}{48} \times 10^5 \approx 3 \times 10^4$
   f $\frac{3 \times 10^3 \times 5 \times 10^2}{(30 + 40)^2} \approx \frac{15 \times 10^5}{5 \times 10^3} = 3 \times 10^2$

N6.2

1 a $\frac{8}{15}$ **b** $\frac{29}{35}$ **c** $\frac{5}{56}$ **d** $\frac{44}{45}$
   e $4\frac{7}{8}$ **f** $\frac{59}{60}$ **g** $\frac{19}{40}$ **h** $9\frac{17}{36}$
2 a $\frac{1}{2}$ **b** $1\frac{2}{3}$ **c** $1\frac{9}{16}$ **d** $\frac{8}{15}$
   e $1\frac{5}{9}$ **f** $14\frac{3}{10}$ **g** $2\frac{37}{55}$ **h** $26\frac{4}{9}$
3 Exact decimal representations are possible for:
   1e 4.875   **1g** 0.475   **1j** 0.8
   2a 0.5   **2c** 1.5625   **2f** 14.3
   All of the other answers have recurring decimal representations, as the denominators have prime factors other than 2 or 5.
4 a $\frac{17}{30}$ **b** $1\frac{67}{87}$ **c** $\frac{1}{3}$
   d $\frac{207}{1715}$ **e** $1\frac{133}{324}$ **f** $-\frac{8}{21}$
5 a $5\sqrt{5}$ **b** $3 - \sqrt{3}$ **c** $\frac{3}{2} + 3\sqrt{\frac{3}{4}}$
   d $\frac{5}{3} - \sqrt{7}$ **e** $\frac{10}{3} + 10\sqrt{\frac{2}{9}}$ **f** $\frac{20}{9} + 7\sqrt{\frac{5}{9}}$
6 a $5 + \sqrt{5}$ **b** $5\sqrt{3} - 3$ **c** $9 + 5\sqrt{3}$
   d $11 - 6\sqrt{2}$ **e** 18   **f** $9 + 14\sqrt{5}$
7 a $\frac{\sqrt{2}}{2}$ **b** $1 + \frac{\sqrt{3}}{3}$ **c** $\sqrt{5} - 2$ **d** $\frac{\sqrt{7} - 1}{2}$
   e $\frac{40 + 16\sqrt{3}}{13}$ **f** $\frac{(3 - \sqrt{3})(4 - \sqrt{5})}{11}$
8 a $\frac{-(1 + \sqrt{5})(1 - \sqrt{3})}{2}$ **b** $(2 - \sqrt{3})(2 + \sqrt{7})$
   c $-3 + \sqrt{11}$ **d** $8 + 5\sqrt{3}$
   e $-9 - \frac{16\sqrt{3}}{3}$ **f** $\frac{18 - \sqrt{5}}{22}$

N6.3

1 0.455 m, 0.445 m
2 a 6.45 mm, 6.35 mm   **b** 4.725 m, 4.715 m
   c 18.5 s, 17.5 s   **d** 0.3885 kg, 0.3875 kg
   e 6.55 volts, 6.45 volts
3 20.968 m/s, 19.091 m/s

4 a 14.6 m **b** 13.7025 m²

5 14 boxes

6 7 crates

7 No, because 5 × 87.5 kg = 437.5 kg < 440 kg

8 21.71 cm

9 a 86.5 cm, 85.5 cm

b Lower bound of distance travelled = 27.36 m. This is shorter than the upper bound of the length of the path, which is 28.5 m.

N6.4

1 a 38 456 **b** 29 412 **c** 67 887 **d** 159 803

2 a 156 **b** 177 **c** 209 **d** 319

3 a 90.78 **b** −5.797 **c** −10.6 **d** −10.67

4 a 2.55 **b** 33 **c** 2.50 **d** 4.09

5 a 216.32 **b** 37.24 **c** 555.03

d 91.63 **e** 146.775 **f** 26.0925

g 1.7556 **h** 0.0595 **i** 0.216 65

6 a $\frac{13}{21}$ **b** $\frac{7}{20}$ **c** $\frac{1}{6}$ **d** $1\frac{1}{6}$

e $2\frac{1}{24}$ **f** $23\frac{11}{36}$ **g** $43\frac{7}{8}$ **h** $1\frac{13}{21}$

7 a 6.5×10^{3} **b** 2.07×10^{5}

c 1.441×10^{-7} **d** 6.721×10^{8}

e 1.09×10^{2} **f** 2.35×10^{-6}

8 a 1.62×10^{8} **b** 6.85×10^{5} **c** 1.17×10^{10}

d 4.84×10^{14} **e** 2.29×10^{9} **f** 2.41×10^{12}

N6.5

1 a 7.74×10^{-3} **b** 9.63×10^{5}

c 4.38×10^{-5} **d** 2.55×10^{2}

e 3.4×10^{5} **f** 4.47×10^{-2}

2 a 2.173×10^{-10} **b** 1.74×10^{-7}

c 1.37×10^{-2} **d** 4.32×10^{-9}

e 6.46×10^{8} **f** 6.32×10^{3}

3 a 11 **b** $9\frac{5}{28}$ **c** 2.92×10^{3}

4 a 317.869 **b** 497.372 **c** 15 625

d 162.37 **e** 34.29 **f** 6.82

5 a 30.7 **b** 7.72 **c** 8.10

d 4.19 **e** 0.0633 **f** 0.486

6 a $\frac{22}{15}$ **b** $\frac{19}{12}$ **c** $\frac{1}{9}$ **d** $\frac{1}{12}$ **e** $\frac{6}{5}$

f $\frac{15}{8}$ **g** $\frac{97}{12}$ **h** $\frac{27}{10}$ **i** $\frac{15}{16}$ **j** $\frac{273}{20}$

7 a 0.322 **b** 18.4 **c** 0.312 **d** 0.0303

N6 Exam review

1 142.7645

2 a 9.622 m/s², 11.827 m/s²

b 10.7 m/s² (halfway between lower and upper bounds)

A6 Before you start …

1 a $x = 3$ or 4 **b** $y = 0$ or 8

c $x = 0.56$ or −3.56 **d** $x = -0.5$ or −3

e $y = 12.87$ or 1.87 **f** $x = 1$ or $-\frac{1}{3}$

2 a 5, 8 **b** 2, 7 **c** 3, 4

3 a 18 **b** 4.5 **c** 90 **d** 63

4 a

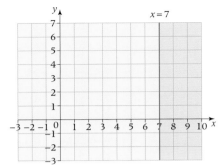

b

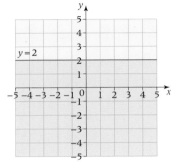

c

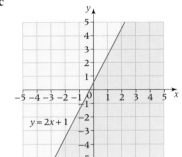

A6.1

1 a $x = 3$ or 6 **b** $x = 10$ or −10 **c** $x = 0$ or 7

2 a $x = 8.24$ or −0.243 **b** $x = 1.42$ or −0.587

c $x = 1$ or −1.43

3 ai $x^2 - 4x - 117 = 0$

aii 9 and 13 or −9 and −13

bi $x^2 - 4x - 357 = 0$ **bii** 17 cm by 21 cm

ci $x^2 - 6x - 1 = 0$ **cii** 6.16 or −0.162

di $10x^2 - x - 3 = 0$ **dii** 0.6 or −0.5

ei $x^2 + 7x - 120 = 0$ **eii** 8 mm by 15 mm

fi $2x^2 - 9x + 4 = 0$ **fii** 10 units

5 a $2\pi r^2 + 2\pi r \times 5 = 100$, so $\pi r^2 + 5\pi r - 50 = 0$

b 4.42 cm

A6.2

1 a $x = 7, y = 4$ **b** $x = -4, y = -5$

c $c = 2, d = -1$ **d** $x = 2, y = 0.5$

e $a = -\frac{2}{3}, b = \frac{3}{4}$ **f** $p = 2.5, q = -3$

2 a 9, 14 **b** −2, 4

3 The star is worth 29 and the moon 17.

4 $y = -\frac{8}{9}x + 9\frac{2}{3}$

5 73

6 $y = \frac{8}{3x^2} - \frac{19}{3x} + 7$

A6.3

1 a $x = 7, y = 6$ or $x = -7, y = 6$
 b $x = 7, y = 5$ or $x = 7, y = -5$
 c $x = 10, y = 0$ or $x = -10, y = 9$
2 a $x = 4, y = 8$ or $x = -1, y = 3$
 b $x = 1, y = 0$
 c $x = 32, y = 4$ or $x = \frac{1}{2}, y = \frac{1}{2}$
3 a $x = 3, y = 7$ or $x = 5, y = 17$
 b $p = 2, q = 1$ or $p = -2, q = -1$
4 a $y = x^2 + x - 6$
 b $y = x + 10$
 c (4, 14) and (−4, 6)
5 $x = -3, y = 2$ or $x = 3, y = -2$
6 $p = -1, q = 6$

A6.4

1 a $x^2 + y^2 = 36$ **b** $4x^2 + 4y^2 = 1$
 c $100x^2 + 100y^2 = 16$ **d** $x^2 + y^2 = 5$
2 a $x^2 + y^2 = 49$ **b** $x^2 + y^2 = 7$
3 (5, 12) and $(-\frac{33}{5}, -\frac{56}{5})$
4 They do not intersect
5 ai Once **aii** Twice **aiii** Once **b** Tangent
6 a $(\sqrt{5}, 2\sqrt{5})$ and $(-\sqrt{5}, -2\sqrt{5})$
 b (1.38, 3.76) and (−2.18, −3.36)
7 a $(x - 3)^2 + (y - 5)^2 = 36$ **b** (6, 0)

A6.5

1 a 31 **b** 10
2 False, $0 \leqslant x \leqslant 1$
3 a $x > 7\frac{2}{3}$ **b** $p \leqslant -32$ **c** $x \leqslant -2\frac{3}{4}$ **d** $y < -4$
 e $q \leqslant 10$ **f** $z \leqslant -3$ **g** $y > -10\frac{4}{5}$
4 a $x > 7\frac{2}{3}$ **b** 8
5 a $-1 < x < 4$ **b** $-5 < x < 18$
6 These give $y \leqslant 6$ and $y > 6$, which do not intersect when combined.
7 a $-1 \leqslant y \leqslant 6$ **b** $-2 < z < 6$ **c** $\frac{1}{3} < p < 5$
8 a The solution is $-5 \leqslant x \leqslant 5$
 bi $x < -4, x > 4$ **bii** $p < -2, p > 2$
9 a $30 < x < 150$
 b $0 < x < 45, 135 < x < 225, 315 < x < 360$
 c $0 < x < 30, 330 < x < 360$

A6.6

1 a $-13 < x < 13$ **b** $x \geqslant 13, x \leqslant -13$
 c $-\sqrt{5} < x < \sqrt{5}$ **d** $-3 \leqslant x \leqslant 3$
2 a $-10 < x < 16$ **b** $x \leqslant -10, x \geqslant 16$
 c $3 - \sqrt{5} < x < 3 + \sqrt{5}$ **d** $0 \leqslant x \leqslant 6$
3 a $0 \leqslant x \leqslant 10$ **b** $x \geqslant 6.5, x \leqslant 0.5$

4 a $-2 \leqslant x \leqslant 3$ ` **b** $x < -8, x > -2.5$
5 a $-5 < x < 1$ **b** $x \geqslant \frac{1}{3}, x \leqslant -\frac{3}{2}$
 c $-5 < x < 0.5$ **d** $-1.5 < x < 4$
6 a $-4, -3, -2, -1, 0$ **b** $-1, 0, 1, 2, 3, 4, 5$
 c $0, 1, 2, 3, 4, 5, 6, 7$
7 $0 < x < 1$

A6 Exam review

1 $-7 < x \leqslant \frac{1}{2}$
2 a $x^2 = -9$, which is not possible
 b $x = 3, y = 4$ or $x = -1\frac{2}{5}, y = -4\frac{4}{5}$

S5 Before you start ...

1 a $a = 8.1$ units **b** $b = 9.8$ units
2 a $x = 2\frac{6}{7}$ **b** $x = 1\frac{7}{11}$ **c** $x = 5\frac{3}{5}$
3 a 5 **b** −38 **c** 6
4 d Diagram(s) should show 5 planes of symmetry

S5.1

1 a 050° **b** 320°
2 a Bearing of 070° **b** Bearing of 155°
 c Bearing of 340° **d** Bearing of 260°
3 a 316° **b** 265° **c** 068°
4 a 284° **b** 263° **c** 117°
5 b 170°
6 aii 328° **bii** 357°

S5.2

1 a Triangle with sides 8 cm, 4 cm, 7 cm
 b Triangle with sides 3 cm and 4 cm, and 30° angle
 c Triangle with sides 10 cm, 7.5 cm, 6 cm
 d Triangle with sides 8 cm and 2 cm, and 90° angle
 e Triangle with sides 6 cm, 9 cm, 5 cm
 f Triangle with two 45° angles and a 4 cm side
2 a $4 + 3 < 9$, two short sides will never meet.
 b $4 + 5 = 9$ so triangle is a straight line.
3 a Yes **b** Yes **c** Yes **d** No
 e No **f** No **g** Yes **h** Yes
4 a Triangle with sides 7 cm, 7 cm, 5 cm
 b Triangle with sides 5 cm, 5 cm, 7 cm
5 a Equilateral triangle with sides 5 cm
 b Rhombus with sides 5 cm
 c Rhombus with sides 3.5 cm
6 a Triangle with sides 5 cm, 12 cm, 13 cm
 b Right-angled triangle
7 a Triangle with sides 3 cm, 4 cm, 5 cm
 b Rectangle with sides 3 cm, 4 cm
 c Rectangle

8 a, b

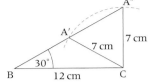

c No, SSA triangles are not unique since you can construct two different triangles with the given side and angle measurements.

S5.3

1 Angle bisectors

2 a Equilateral triangle with sides 5 cm
  b Angles bisectors of triangle in part **a**
  c Bisectors meet at a point and cut the midpoints of the opposite sides.

3 a Equilateral triangle with sides 4 cm
  b Angles bisector of triangle in part **a**

4 a Perpendicular bisector of 6 cm line
  b Perpendicular bisector of 9 cm line
  c Perpendicular bisector of 5.6 cm line
  d Perpendicular bisector of 10 cm line
  e Perpendicular bisector of 11.2 cm line

5 a Equilateral triangle with sides 5 cm
  b Perpendicular bisectors of sides of triangle in part **a**
  c Perpendicular bisectors and angle bisectors of equilateral triangles are the same.

6 d Both sets of lines meet at common points but not the same points.

7 c 90°, 135°, 180°, 225°, 270°, 315°

S5.4

1 Perpendiculars from points to lines
2 Perpendiculars from points on lines

S5.5

These sketches not drawn to scale.

1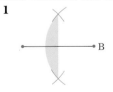

2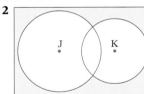

3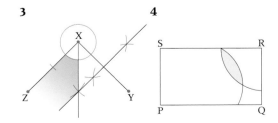

4

S5.6

1 a $a = 4.5$ cm   **b** $b = 5.4$ cm   **c** $c = 3.7$ cm
  d $d = 10.2$ cm   **e** $e = 15.1$ cm   **f** $f = 9.6$ cm

2 a $p = 32.6°$ **b** $q = 53.7°$ **c** $r = 55.6°$
  d $s = 31.6°$ **e** $t = 42.5°$

3 a $AD^2 = 3^2 + 12^2 + 4^2$, so AD = 13 cm
  b 18.2 cm

4 11.5 cm

S5.7

1 a 6.5 cm   **b** 14.3 cm   **c** 5.7 cm   **d** 10.3 cm
  e 9.5 cm   **f** 18.1 cm   **g** 28.9 cm   **h** 9.8 cm

2 a 33.4°   **b** 26.1°   **c** 13.0°   **d** 44.8°
  e 18.3°   **f** 59.8°   **g** 35.6°   **h** 25.2°

S5.8

1 a 6.2 cm   **b** 13.9 cm   **c** 12.0 cm   **d** 6.6 cm
  e 14 cm   **f** 11.3 cm   **g** 12.0 cm   **h** 18.8 cm

2 a 33.0°   **b** 34.8°   **c** 85.3°   **d** 55.8°
  e 19.9°   **f** 63.1°   **g** 54.7°   **h** 111°

3 B = 59.0°, C = 41.0°, x = 10.4 cm

S5.9

1 11.1 km, 191°   **2** 18.1 km, 261°
3 47.7 cm   **4** 41.3 cm
5 47.3 cm   **6 a** 13.4 cm   **b** 10.5 cm
7 43.7 cm   **8** 12.7 cm
9 11.8 cm

S5.10

1 a 5.9   **b** 6.6   **c** 9.7   **d** 4.9
2 a 15.3 cm   **b** 11.7 cm   **c** 12.3 cm
3 a 51.8°   **b** 43.1°   **c** 18.9°
4 a 35.3°   **b** 35.3°
5 12°
6 a 10.4 cm   **b** 16.7°

S5 Exam review

1 a 5.7 units   **b** 5.7 units
2 a 9.11 cm   **b** 19.2°

A7 Before you start …

1 ai $x = 3$ or 11   **aii** $x = 0.372$ or -5.372
  b In part **i** $b^2 - 4ac$ is a square number, in part **ii** it is not.
  c Equation **i**

2 $b^2 - 4ac$ is negative so you cannot take the square root.

5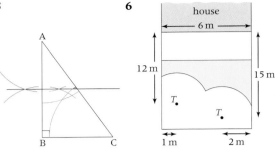

6

3 a When $x = 0$, $y = 6$; when $y = 0$, $x = -2$ or -3.
  b When $x = 0$, $y = 6$; when $y = 0$, $x = 2$ or 1.5.
4 $(3x + 2)(x - 1) = 40$ so $x = 3.912$ (accept) or -3.579 (reject); the rectangle is 13.7 units by 2.9 units

A7.1

1 a $x = -2$ or -5 **b** $x = 2$ or -6
  c $x = -7$ or 7 **d** $x = 0$ or 8
  e $x = 2$ or -6 **f** $x = \frac{1}{3}$ or 2

2 a $x = -0.838$ or -7.16 **b** $x = -0.227$ or -0.631
  c $x = 2.41$ or -0.414 **d** $x = 0.657$ or -0.457
  e $x = 0.425$ or -1.18 **f** $x = 2.65$ or 0.849

3 ai Factorises   **aii** Two solutions
  bi Factorises   **bii** One solution
  ci Does not factorise   **cii** Two solutions
  di Factorises   **dii** Two solutions
  ei Factorises   **eii** Two solutions
  fi Factorises   **fii** Two solutions

4 a $y = 8$ or -11 **b** $x = -2.5$ or 2.5
  c $x = -3$ or -0.5 **d** No solutions
  e $y = 3$ **f** $p = 0.662$ or -22.7
  g $x = 0.587$ or -0.730

5 False, as the discriminant $= -15 < 0$ there are no solutions.

6 8 cm by 15 cm

7 a $x = 0$, 2 or 3
  b $x = 0$, 0.186 or -2.69
  c $x = -3$, 0 or 2

A7.2

1 a $(0, 12)$ **b** $(0, 15)$ **c** $(0, 5)$
  d $(0, 6)$ **e** $(0, 25)$

2 a $(-6, 0)$, $(-2, 0)$ **b** $(3, 0)$, $(5, 0)$
  c $(-1, 0)$, $(-5, 0)$ **d** $(-2, 0)$, $(-3, 0)$
  e $(-5, 0)$

3 a $(-4, -4)$ **b** $(4, -1)$ **c** $(-3, -4)$
  d $(-2\frac{1}{2}, -\frac{1}{4})$ **e** $(-5, 0)$

4 $y = (x + 4)^2 + 4 \geqslant 4$, so y never intersects the x-axis.

5

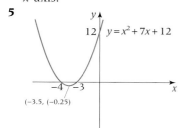

$y = x^2 + 7x + 12$, 12, -4, -3, $(-3.5, (-0.25)$

6 a ii $y = x^2 - x - 20$
  b iv $y = x^2 + 2x - 24$
  c i $y = (x - 4)^2$; graph of
    iii $y = (x - 6)(x + 4)$, has y-intercept $(0, -24)$
    and intercepts the x-axis at $(6, 0)$ and $(-4, 0)$

7 a $y = x(x + 5)$
  b $y = (x - 6)^2$

c $y = (x - 4)^2 - 2$

8 a e.g. $y = x^2 + 5$
  b e.g. $y = (x + 12)^2$
  c e.g. $y = 9 - (x - 4)^2$

9 a It is a cubic equation, not a quadratic equation, so the graph will be S-shaped.

  b

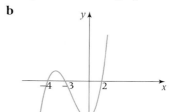

A7.3

1 a $x = 2.28$ or 0.219 **b** $y = 9.47$ or 0.528
  c $p = -3$ or -7 **d** $x = -1$ or 0.3
  e $x = 0.762$ or -2.36

2 a 34 cm   **b** 27 cm   **c** $(3, 4)$, $(-1.4, -4.8)$
  d Volume of sphere $= 19.0$ units^3, volume of cylinder $= 51.7$ units^3, so cylinder has greater volume

3 a

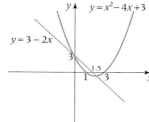

  b $(0, 3)$ **c** $(2, -1)$

4 2.8

A7 Exam review

1 a $y = 4$, 0, -2, -2, 4, 10
  b

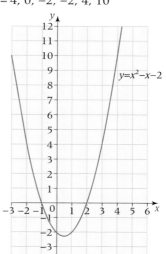

2 a $k = m^2$ **bi** $-m^2$ **bii** m

1 a 22 **b** 45

2 a 1.3 **b** 2.4 **c** 0.36 **d** 30

D4.1

1 a Class width: 2, 1, 1, 1, 3;
Frequency density: 6, 17, 19, 11, 6

b

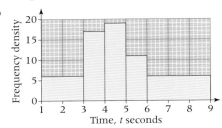

2 a Class width: 5, 5, 10, 20, 20, 40;
Frequency density: 1.2, 2, 2.3, 1.45, 1.2, 0.2

b

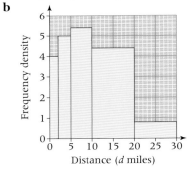

3 a Class width: 2, 3, 5, 10, 10;
Frequency density: 4, 5, 5.4, 4.4, 0.6

b

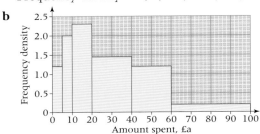

4 a Class width: 10, 30, 5, 10, 30, 5;
Frequency density: 1.2, 1.6, 3.6, 1.1, 1.0, 4.4

b

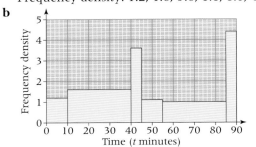

5 a Class width: 0.1, 0.3, 0.5, 1, 3;
Frequency density: 30, 40, 44, 25, 6

b

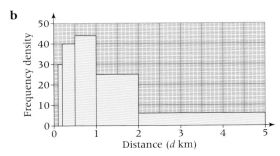

6 a Class width: 10, 20, 20, 30, 30;
Frequency density: 0.8, 0.8, 1.4, 1.3, 0.3

b

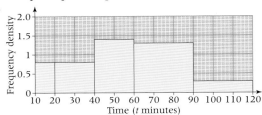

D4.2

1 a 10
b 4, 18, 53, 20, 10
c 105

2 a 48
b 3, 7, 28, 20, 8
c 66

3 a

Height (h cm)	Frequency
$110 \leqslant h < 115$	3
$115 \leqslant h < 120$	7
$120 \leqslant h < 125$	13
$125 \leqslant h < 135$	29
$135 \leqslant h < 145$	21
$145 \leqslant h < 150$	5
$150 \leqslant h < 170$	6

Total = 85

b

Height (h cm)	Frequency
$130 \leqslant h < 140$	1
$140 \leqslant h < 150$	9
$150 \leqslant h < 155$	9
$155 \leqslant h < 160$	12
$160 \leqslant h < 165$	14
$165 \leqslant h < 170$	13
$170 \leqslant h < 180$	15
$180 \leqslant h < 190$	1

Total = 74

D4.3

1 a 5, 8, 11
b

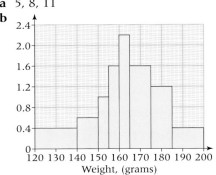

2 a 36, 42 **b**

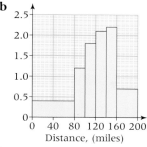

Distance, (miles)

b

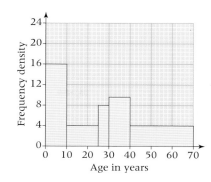

Age in years

3 a Bar height $= \frac{2f}{13}$ cm

b Bar height $= f$
Frequency density $= 0.4$ and $\frac{f}{5}$
Frequency density $= \frac{f}{30}$

D4.4

1 The range for boys aged 11 is the same as the range for boys aged 16 (60 cm). The modal class for boys aged 11 is 125–135 cm, which is smaller than for 16-year-old boys, who have a modal class of 160–165 cm. The distribution for boys aged 11 is positively skewed, whereas for boys aged 16, the data is slightly negatively skewed.

2 The range for apples (80 g) is greater than the range for pears (70 g). The modal class for apples is 160–165 g, which is larger than for pears, having a modal class of 115–120 g. The distribution for apples is roughly symmetrical with no skew, whereas for pears, the data is slightly negatively skewed.

3 The range for girls (8 s) is greater than the range for boys (6 s). The modal class for girls is 6–6.5 s, which is the same as that for boys. Both distributions are slightly negatively skewed.

D4 Exam review

1 a

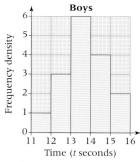

Boys — Time (t seconds)

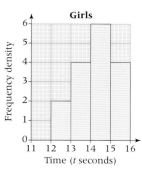

Girls — Time (t seconds)

b The range of times for the boys is larger than the range for the girls. The modal class is lower for the boys. The girl's times are more negatively skewed than the boy's, indicating that the boys were faster.

2 a 60, 40

S6 Before you start ...

1 a $x = 8$ **b** $x = 7.5$ **c** $x = 2$ **d** $x = 4$
2 ai 6.3 cm   **aii** 38.1 cm^2
   bi 8.9 cm   **bii** 57.6 cm^2

S6.1

1 a 32 m^2 **b** 4 mm   **c** 20 000 cm^3
   d 1.9 litres   **e** 5.1 km   **f** 0.58 m
   g 6.3 km   **h** 24 000 mm^3 **i** 630 000 cm^3
   j 90 000 m^2 **k** 700 m^2 **l** 1000 mm^2
2 a Length + Length + Length = Length
   b Length × Length × Length = Volume
3 There are four lengths multiplied together when there should only be three.
4 a Volume **b** Area   **c** Area   **d** Area
   e Length **f** Length   **g** Area
   h None of these   **i** None of these
   j Volume   **k** None of these
   l None of these
5 a $2n + 2m\pi$ **b** $2mn + \pi m^2$
6 a $2ab + 2ac + 2bc$ **b** abc
7 $2.4ab + 0.6b^2$ cm^3
8 $\frac{lwh}{10}$ cm^3

S6.2

1 a 25.4 cm^2 **b** 36.1 cm^2 **c** 25.9 cm^2
   d 12.4 cm^2 **e** 40.7 cm^2 **f** 21.8 cm^2
   g 26.5 cm^2 **h** 37.1 cm^2 **i** 17.7 cm^2
   j 31.6 cm^2
2 a 55.0 cm^2 **b** 34.9 cm^2 **c** 61.0 cm^2
   d 114 cm^2 **e** 49.0 cm^2 **f** 72.2 cm^2
   g 80.1 cm^2 **h** 74.8 cm^2

S6.3

1 a 2.61 cm^2 **b** 0.916 cm^2 **c** 7.50 cm^2
   d 18.2 cm^2 **e** 17.6 cm^2 **f** 14.6 cm^2
   g 90.0 cm^2 **h** 1.70 cm^2
2 a 39.1°   **b** 60.1°   **c** 17.0 cm^2
3 a 22.1 cm^2 **b** 9.25 cm^2 **c** 18.0 cm^2
4 a 56.4°   **b** 10.2 cm^2
5 a 19.9 cm^2 **b** 29.8 cm^2 **c** 36.9 cm^2
6 1356 cm^2

S6.4

1 ai 490 cm^3 **aii** 659 cm^2
   bi 1186 cm^3 **bii** 707 cm^2

ci	1057 cm^3	**cii** 704 cm^2
di	1982 cm^3	**dii** 1152 cm^2
ei	638 cm^3	**eii** 467 cm^2
fi	159 cm^3	**fii** 252 cm^2
gi	650 cm^3	**gii** 510 cm^2
hi	1985 cm^3	**hii** 1020 cm^2
i i	3146 cm^3	**i ii** 1528 cm^2
ji	7242 cm^3	**jii** 3339 cm^2
ki	3398 cm^3	**kii** 4610 cm^2
li	16 040 cm^3	**li** 4682 cm^2

2 a $\frac{7\pi r^2 h}{3}$ cm^3 **b** $5\pi r^2 + 3\pi r\sqrt{r^2 + h^2}$ cm^2

S6.5

1 $r = 10$ cm and $h = 2\sqrt{11}$ cm so $V = \frac{200}{3}\pi\sqrt{11}$ cm^3

2 $r = 5$ cm and $h = 10\sqrt{2}$ cm so $V \frac{250}{3}\pi\sqrt{2}$ cm^3

3 $4\pi r^2 = 12\pi \Rightarrow r = \sqrt{3}$ so $V = \frac{4}{3}\pi(\sqrt{3})^3 = 4\pi\sqrt{3}$

4 103.1π cm^2

5 48.17π cm^2

6 $2\pi(2r)h = \pi rl \Rightarrow h = \frac{1}{4}l$

7 4 cm

S6 Exam review

1 a ii   **b** iii   **c** i   **d** iv

2 a 56.4 cm^2

  b 7.84 cm

A8 Before you start ...

1 a Horizontal   **b** Diagonal   **c** Vertical

  d Horizontal   **e** Diagonal

2 a 40   **b** 6   **c** −80

3 ai 5   **aii** 11

  bi $x^2 + y^2 = 4$ **bii** $x^2 + y^2 = \frac{9}{64}$

A8.1

1 a $y = x^2 + 2x + 1$ **b** $g(x) = x^2 - x^3$

  c $f(x) = x^3 + x - 2$ **d** $y = 4 - x^2$

2 a

x	−4	−3	−2	−1	0	1	2	3	4
$2x^2$	32	18	8	2	0	2	8	18	32
$3x$	−12	−9	−6	−3	0	3	6	9	12
−6	−6	−6	−6	−6	−6	−6	−6	−6	−6
y	14	3	−4	−7	−6	−1	8	21	38

  b

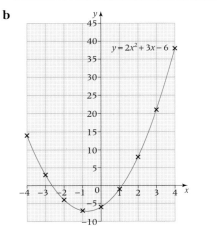

c $\left(-\frac{3}{4}, -7\frac{1}{8}\right)$

3 a

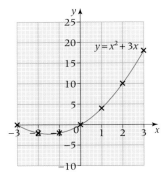

b

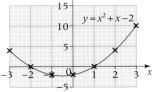

c

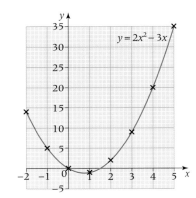

d

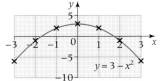

e   **f**

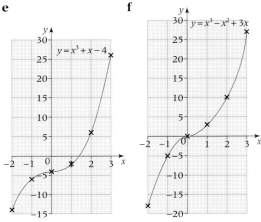

4 a

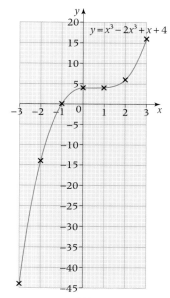

$y = x^3 - 2x^3 + x + 4$

b

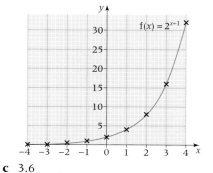

$f(x) = 2^{x+1}$

c 3.6

4

x	-4	-3	-2	-1	0	1	2	3	4	5
$\frac{20}{x}$	-5	$-6\frac{2}{3}$	-10	-20	$-$	20	10	$6\frac{2}{3}$	5	4
-5	-5	-5	-5	-5	-5	-5	-5	-5	-5	-5
y	-14	$-14\frac{2}{3}$	-17	-26	$-$	16	7	$4\frac{2}{3}$	4	4

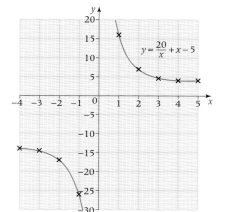

$y = \frac{20}{x} + x - 5$

bi -2.3 **bii** 4.8

5 ai $l = 100 - 2w$ **aii** $A = w(100 - 2w)$

b 25 m by 50 m

A8.2

1 a $y = 2^{-x}$ **b** $y = \frac{4}{x}$ **c** $y = 3^x$ **d** $y = \frac{10}{x-3}$

2 a

x	-6	-5	-4	-3	-2	-1	0	1	2	3	4	5	6
$f(x)$	-2	-2.4	-3	-4	-6	-12	Asymptote	12	6	4	3	2.4	2

b

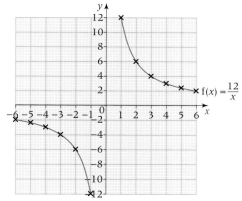

$f(x) = \frac{12}{x}$

c The graph is a hyperbola. The axes are the asymptotes.

d 4.8

3 a

x	-4	-3	-2	-1	0	1	2	3	4
$g(x)$	$\frac{1}{8}$	$\frac{1}{4}$	$\frac{1}{2}$	1	2	4	8	16	32

5 a

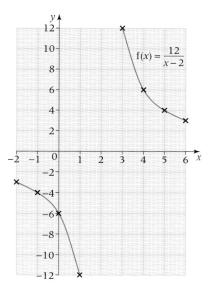

$f(x) = \frac{12}{x-2}$

b

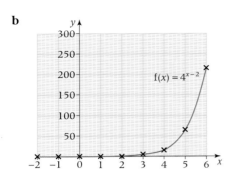

c

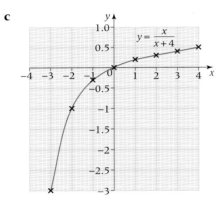

d

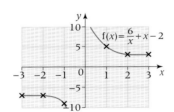

e

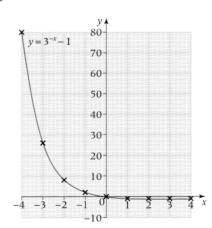

A8.3

1 **a** $x = 3$, $y = 7$ **b** $x = 1$, $y = 1$
   c $x = 3$, $y = -1$

2 **ai** $x = -1.6$ or 2.6
   aii $x = -0.6$ or 1.6
   aiii $x = -1$ or 3
   b On the x-axis; $x = -1$ or 2
   c $y = 2 - x$

3 **a** $x = -0.7$ or 2.7 **b** $x = -1.2$ or 3.2
   c $x = -0.8$ or 3.8 **d** $x = -1.8$ or 2.8
   e $x = -0.9$ or 3.4 **f** $x = 1$, $y = 2$

4 **a** $x = 4.8$ **b** $x = 2.3$
   c No solutions   **d** $x = 3.2$

5 **a** None   **b** None   **c** One   **d** One

A8.4

1 **a** $x = -3$ or 1 **b** $x = -2.6$ or 1.6
   c $x = -3.4$ or 1.4 **d** $x = -1$ or 0

2 **a** $y = 0$ **b** $y = x + 2$
   c $y = 2x - 3$, $y = 0$

3 **a** $y = 3$ **b** $y = 0$ **c** $y = 2x + 1$
   d $y = 4$ **e** $y = 2$ **f** $y = 8x + 1$

4 **a** $y = 2x - 1$ **b** $y = 5x - 3$ **c** $y = 2x + 3$
   d $y = 0$ **e** $y = x^2 - 1$ **f** $y = x^3 + x^2 - 1$

5 **bi** $x = 1.6$ **bii** $x = 0.2$ **biii** $x = -0.8$, 2, 4

6 **bi** $x = 8$ **bii** $x = 2.9$ **c** $y = x + 1$

A8.5

1 $x = 1.6$, $y = 2.6$ and $x = -1.6$, $y = -2.6$

2 **a** Two   **b** Two   **c** One   **d** Two
   e None   **f** Two   **g** Two   **h** Three

3 **a** $x = 4$, $y = 3$ and $x = -4$, $y = 3$
   b $x = 1.4$, $y = 1.4$ and $x = -1.4$, $y = -1.4$
   c $x = 2.2$, $y = 5.6$ and $x = -1.6$, $y = -5.8$
   d $x = 0.8$, $y = 0.6$ and $x = -0.8$, $y = 0.6$

4 **a** $x = 1.8$, $y = 3.6$ and $x = -1.8$, $y = -3.6$
   b $x = 3.5$, $y = 3.5$ and $x = -3.5$, $y = -3.5$
   c $x = 1.2$, $y = 2.7$ and $x = -0.6$, $y = -2.9$
   d $x = 2.6$, $y = 6.5$ and $x = -2.6$, $y = 6.5$

5 Yes, as in question **2h**.

A8 Exam review

1 **a** ii   **b** iv   **c** iii   **d** i

2 **a** $y = -\frac{1}{2}x + 3$ **b** $(0.5, 0.5)$

A9 Before you start …

1 **a** £16.71   **b** $15\frac{3}{4}$ hours

2 **a** 6 days   **b** 4.8 minutes

3 **ai** 10   **aii** 5.5
   bi 48   **bii** 5 or -5
   ci 1.5   **cii** 1.2

4 **a**   **b**

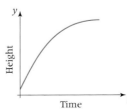

A9.1

1 a 60 km   **b** 1 hours 30 minutes
  c 60 km/h   **d** 11:30 am and 12 noon
  e 30 km/h

2 Claire and Christina: Claire's vertical line means
she travelled a distance in no time, and
Christina's line sloping backwards means she
has travelled backwards in time.

3 a

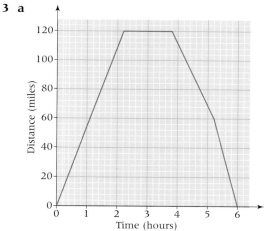

b

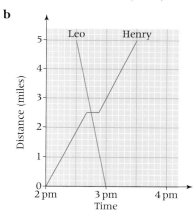

4

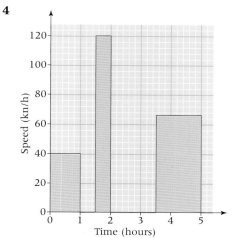

A9.2

1 A3, B1, C4, D2
2 Growth spurt at around 5, steadies down until
next growth spurt at about 13–17 years
(puberty) then steadies down again.

3

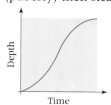

4 a

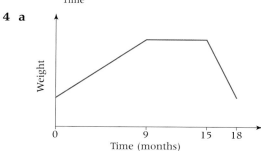

b

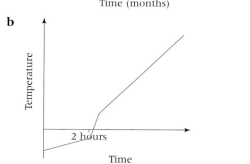

5 a   **b**

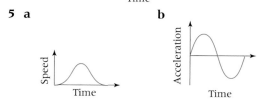

A9.3

1 a 5, 10, 15, 20, 25, 30
  b Cost, C is directly proportional to hours spent
on the phone, h. This means $C = kh$ for some
constant k. In Paul's case, $C = 5h$.
  ci £87.50   **cii** $10\frac{1}{4}$ hours   **d** Graph ii

2 a $A = 3.5b$
  bi 45.5   **bii** 20
3 a $q = 1\frac{5}{12}t$
  bi 5.1   **bii** $9\frac{3}{17}$
4 a $d = 50t$ **b** 200 miles   **c** 2 hours

d

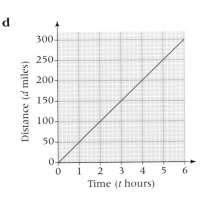

Distance (*d* miles) vs Time (*t* hours)

5 $V = 0.7R$, 7.14 volts

A9.4

1 a

x	1	2	3	4	5	6
x^2	1	4	9	16	25	36
y	4	16	36	64	100	144

  b y is proportional to x^2. This means $y = kx^2$ for some constant, k. In this case, $k = 4$.

  c $y = 4x^2$, $x = 14$ **d** Graph ii

2 a $R = 100s^2$ **b** 64   **c** 1.41

3 $A = \pi r^2$ so the area is directly proportional to the square of the radius, with constant of proportionality π. Also, $C = 2\pi r$ so the circumference is directly proportional to the radius, with constant of proportionality 2π.

4 $615\frac{2}{9}$ cm^2

5 a $y = 2x^3$ **b** 4

6 50

A9.5

1 a As speed increases, time decreases, so the quicker you drive the sooner you arrive.

  b Speed: 100, 50, 25, 20, 12.5

  c $S = \frac{100}{t}$, $28\frac{4}{7}$ mph   **d** Graph ii

2 a Time taken: 4, 2, $1\frac{1}{3}$, 1, 0.8, $\frac{2}{3}$

  b

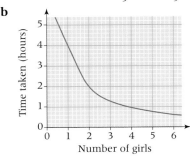

Time taken (hours) vs Number of girls

  c As the number of girls increases, the time taken to eat the chocolate decreases.

3 a $y = \frac{30}{x}$

  bi 6   **bii** 0.5

4 a $A = \frac{100}{c}$

  bi 2   **bii** $\frac{1}{4}$

A9.6

1 a $F = 100/d^2$; when $d = 3$, $F = 11\frac{1}{9}$, when $d = 4$, $F = 6\frac{1}{4}$

  b

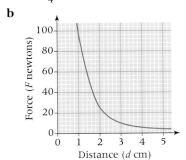

Force (*F* newtons) vs Distance (*d* cm)

2 a ii $y \propto \frac{1}{x}$ **b iii** $y \propto x$

  c i $y \propto \frac{1}{x^2}$ **d iv** $y \propto x^2$

3 1 hour

4 a $y = \frac{10}{\sqrt{x}}$ **bi** 1   **bii** $\frac{25}{36}$

5 a $P = \frac{32}{t^3}$ **bi** $1\frac{5}{27}$ **bii** 4

A9.7

1 a $y = \frac{1}{x-2}$ **b** $f(x) = x(1 - x^2)$

  c $y = 4^x$ **d** $y = x^3 + x^2 - 6x$

  e $f(x) = x^2 + x - 2$

2

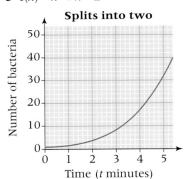

Splits into two

Number of bacteria vs Time (*t* minutes)

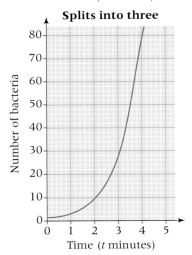

Splits into three

Number of bacteria vs Time (*t* minutes)

3 a 5 elephants

  bi 8 elephants   **bii** 11 elephants

c 4 years

d
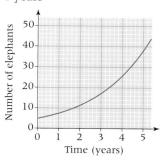

4 a $p = 3, q = 5$
  b False, because $3 \times 5^2 \neq 100$.

5 a $m = 2000, n = 1.1$
  b 10 years

A9 Exam review

1 $93\frac{3}{4}$ seconds

2 i H
  ii D
  iii A

S7 Before you start …

1

	Square	Rhombus	Rectangle	Parallelogram	Trapezium	Kite
1 pair opposite sides parallel	✓	✓	✓	✓	✓	
2 pair opposite sides parallel	✓	✓	✓	✓		
Opposite sides equal	✓	✓	✓	✓		
All sides equal	✓	✓				
All angles equal	✓		✓			
Opposite angles equal	✓	✓	✓	✓		One pair
Diagonals equal	✓		✓			
Diagonals perpendicular	✓	✓		✓		✓
Diagonals bisect each other	✓	✓	✓			One bisects the other
Diagonals bisect the angle	✓	✓				One diagonal

2

S7.1

1
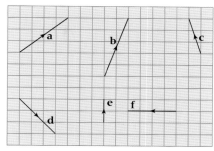

2 a
$\begin{pmatrix} 1 \\ -2 \end{pmatrix}$ 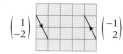 $\begin{pmatrix} -1 \\ 2 \end{pmatrix}$

b They are parallel but in opposite directions.

3 $\overrightarrow{AB} = \overrightarrow{BC} = \overrightarrow{FH}$, $\overrightarrow{FG} = \overrightarrow{DE} = \overrightarrow{GH} = \overrightarrow{IJ}$,
$\overrightarrow{JH} = \overrightarrow{IG} = \overrightarrow{GE} = \overrightarrow{FD} = \overrightarrow{DB}$, $\overrightarrow{HC} = \overrightarrow{IE} = \overrightarrow{FB}$,
$\overrightarrow{IF} = \overrightarrow{JG} = \overrightarrow{GD} = \overrightarrow{HE} = \overrightarrow{EB}$, $\overrightarrow{FA} = \overrightarrow{JD} = \overrightarrow{HB}$,
$\overrightarrow{BA} = \overrightarrow{CB} = \overrightarrow{HF}$, $\overrightarrow{GF} = \overrightarrow{ED} = \overrightarrow{HG} = \overrightarrow{JI}$,
$\overrightarrow{HJ} = \overrightarrow{GI} = \overrightarrow{EG} = \overrightarrow{DF} = \overrightarrow{BD}$, $\overrightarrow{CH} = \overrightarrow{EI} = \overrightarrow{BF}$,
$\overrightarrow{FI} = \overrightarrow{GJ} = \overrightarrow{DG} = \overrightarrow{EH} = \overrightarrow{BE}$, $\overrightarrow{AF} = \overrightarrow{DJ} = \overrightarrow{BH}$

4 ai $\overrightarrow{XA}, \overrightarrow{CB}, \overrightarrow{DX}, \overrightarrow{EF}$ **aii** $\overrightarrow{AB}, \overrightarrow{FX}, \overrightarrow{XC}, \overrightarrow{ED}$
  bi $\overrightarrow{AX}, \overrightarrow{BC}, \overrightarrow{XD}, \overrightarrow{FE}$ **bii** $\overrightarrow{BA}, \overrightarrow{XF}, \overrightarrow{CX}, \overrightarrow{DE}$

5 ai $\overrightarrow{OJ}, \overrightarrow{NK}$ **aii** $\overrightarrow{OM}, \overrightarrow{QK}$ **aiii** $\overrightarrow{OP}, \overrightarrow{LK}$
  bi $\overrightarrow{JO}, \overrightarrow{KN}$ **bii** $\overrightarrow{MO}, \overrightarrow{KQ}$ **biii** $\overrightarrow{PO}, \overrightarrow{KL}$

S7.2

1 a 13   **b** 13

2 a $\begin{pmatrix} -8 \\ -2 \end{pmatrix}$ **b** $\begin{pmatrix} 8 \\ 2 \end{pmatrix}$ **c** (0,0)   **d** (10,3)

3 a $\begin{pmatrix} 0 \\ 0 \end{pmatrix}$ **b** $\begin{pmatrix} 6 \\ 8 \end{pmatrix}$ **c** $\begin{pmatrix} 30 \\ 40 \end{pmatrix}$

  d $\begin{pmatrix} -70 \\ 70 \end{pmatrix}$ **e** $\begin{pmatrix} -3 \\ -4 \end{pmatrix}$ **f** $\begin{pmatrix} 13 \\ 1 \end{pmatrix}$

4 a $p = -8, q = 8$ **b** $p = 4, q = 6.5$

5 a $\begin{pmatrix} -6 \\ -8 \end{pmatrix}$ or $\begin{pmatrix} 6 \\ 8 \end{pmatrix}$ **b** $\begin{pmatrix} 7k \\ -7k \end{pmatrix}$

6 a $\begin{pmatrix} 10 \\ -3 \end{pmatrix}$ **b** $\begin{pmatrix} 2 \\ -5.5 \end{pmatrix}$ **c** $p = -\frac{6}{7}$

S7.3

1

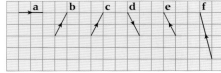

2

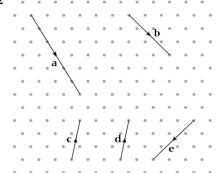

3 a r **b** $p + r$ **c** $-p - r$ **d** $p - r$

4 a $a - b$ **b** $b - a$ **c** $-a$ **d** $b - a$

5 a l **b** $l - j$ **c** $-j - l$ **d** $-j$

S7.4

1 a $3p$ **b** $5p$ **c** $5p + 2q$

   d $5p + 6q$ **e** $4q$ **f** $p + 3q$

   g $p + 6q$ **h** $3p + 3q$ **i** $p + 2q$

   j $p + 6q$ **k** $p + q$ **l** $q - p$

   m $-4q$ **n** $-p - 4q$ **o** $-3p - 4q$

   p $3q - p$ **q** $p - 3q$ **r** $p - 3q$

   s $-4p - 4q$ **t** $-p - 4q$ **u** $-3p + 4q$

2

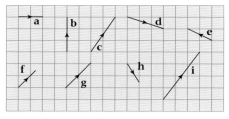

3 a $a + d$ **b** $4d - a$

4 a $1.5j$ **b** $-1.5j$

S7.5

1 a z **b** $w + z$ **c** $\frac{1}{2}(w + z)$

   d $\frac{1}{2}(w + z)$ **e** $\frac{1}{2}(w - z)$ **f** $\frac{1}{2}(z - w)$

2 a d **b** a **c** $a - d$ **d** $a - d$

3 a $2c$ **b** $-2b$ **c** $c - b$ **d** $2(c - b)$

4 a t **b** $r + t$ **c** $\frac{3}{4}(r + t)$

   d $-\frac{1}{4}(r + t)$ **e** $\frac{3}{4}r - \frac{1}{4}t$

5 a $b - a$ **b** $\frac{4}{5}a + \frac{1}{5}b$

S7.6

1 $\overrightarrow{AB} = 2(b - a)$,
$\overrightarrow{CD} = 3(b - a) \Rightarrow \overrightarrow{CD} = 1.5\overrightarrow{AB} \Rightarrow \overrightarrow{AB}$ is parallel to $\overrightarrow{CD}$.

2 ai $2r + 2a$ **aii** r **aiii** a **aiv** $r + a$

   b $\overrightarrow{RT} = 2\overrightarrow{MN}$ so $\overrightarrow{MN}$ is parallel to $\overrightarrow{RT}$.

3 ai $3p$ **aii** $r + 3p$ **aiii** $p - r$ **aiv** $2p + r$

   b $\overrightarrow{OX} = \frac{1}{4}(3p + r) = \frac{1}{4}\overrightarrow{OQ}$, so $\overrightarrow{OX}$ and $\overrightarrow{OQ}$ are parallel. $\overrightarrow{OX}$ and $\overrightarrow{OQ}$ have the point O in common. So O, X and Q lie on the same straight line.

4 ai $j + k$ **aii** k **aiii** $k - j$

   b $\overrightarrow{JX} = -j\frac{1}{3}k$ and $\overrightarrow{JM} = k - 3j = 3\overrightarrow{JX}$, so $\overrightarrow{JX}$ and $\overrightarrow{JM}$ are parallel. $\overrightarrow{JX}$ and $\overrightarrow{JM}$ have the point J in common. So J, X and M lie on the same straight line.

S7 Exam review

1 a $\begin{pmatrix} 3 \\ -1 \end{pmatrix}$ **b** $\begin{pmatrix} -1 \\ 7 \end{pmatrix}$ **c** $\begin{pmatrix} 0 \\ -10 \end{pmatrix}$

2 a $\frac{1}{2}p + \frac{1}{2}q$

   b $\overrightarrow{RS} = \frac{1}{2}q = \frac{1}{2}\overrightarrow{OQ}$, so $\overrightarrow{RS}$ is parallel to $\overrightarrow{OQ}$.

D5 Before you start …

1 a 0.55 **b** 0.04 **c** 0.72 **d** 0.625

   e 0.6 **f** 0.34 **g** 0.9 **h** 0.18

   i 0.17 **j** 0.192

2 a $\frac{1}{6}$ **b** $\frac{4}{5}$ **c** $\frac{2}{9}$ **d** $\frac{13}{15}$

   e $\frac{11}{12}$ **f** $\frac{5}{9}$ **g** $\frac{8}{45}$

D5.1

1

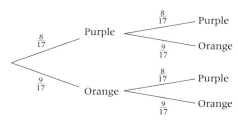

2

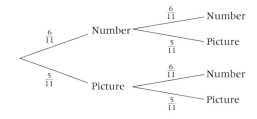

3

50p coin	20p coin

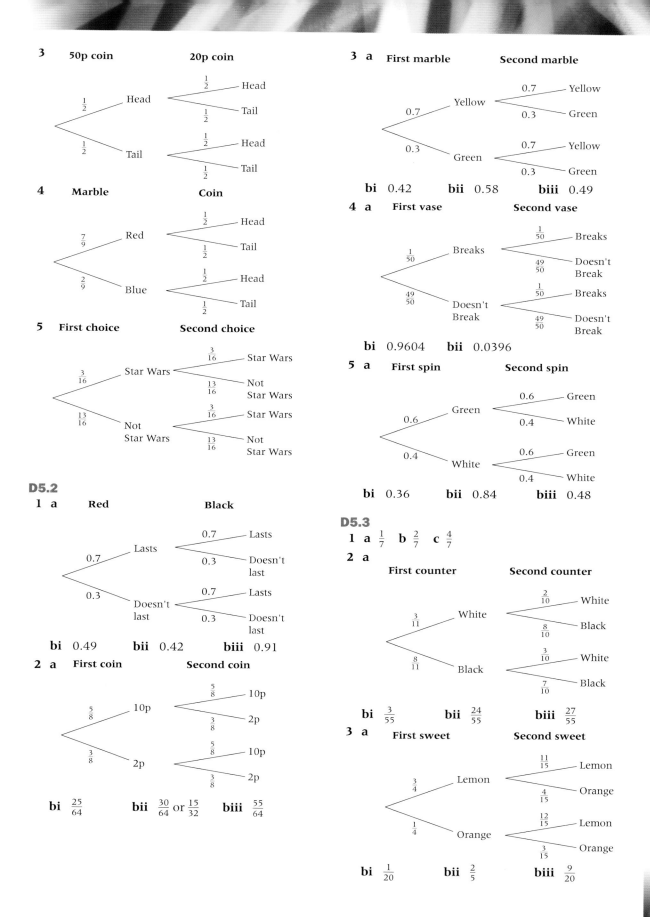

$\frac{1}{2}$ Head $\frac{1}{2}$ Head / $\frac{1}{2}$ Tail

$\frac{1}{2}$ Tail $\frac{1}{2}$ Head / $\frac{1}{2}$ Tail

4 Marble Coin

$\frac{7}{9}$ Red $\frac{1}{2}$ Head / $\frac{1}{2}$ Tail

$\frac{2}{9}$ Blue $\frac{1}{2}$ Head / $\frac{1}{2}$ Tail

5 First choice Second choice

$\frac{3}{16}$ Star Wars $\frac{3}{16}$ Star Wars / $\frac{13}{16}$ Not Star Wars

$\frac{13}{16}$ Not Star Wars $\frac{3}{16}$ Star Wars / $\frac{13}{16}$ Not Star Wars

D5.2

1 a Red Black

0.7 Lasts 0.7 Lasts / 0.3 Doesn't last

0.3 Doesn't last 0.7 Lasts / 0.3 Doesn't last

bi 0.49 **bii** 0.42 **biii** 0.91

2 a First coin Second coin

$\frac{5}{8}$ 10p $\frac{5}{8}$ 10p / $\frac{3}{8}$ 2p

$\frac{3}{8}$ 2p $\frac{5}{8}$ 10p / $\frac{3}{8}$ 2p

bi $\frac{25}{64}$ **bii** $\frac{30}{64}$ or $\frac{15}{32}$ **biii** $\frac{55}{64}$

3 a First marble Second marble

0.7 Yellow 0.7 Yellow / 0.3 Green

0.3 Green 0.7 Yellow / 0.3 Green

bi 0.42 **bii** 0.58 **biii** 0.49

4 a First vase Second vase

$\frac{1}{50}$ Breaks $\frac{1}{50}$ Breaks / $\frac{49}{50}$ Doesn't Break

$\frac{49}{50}$ Doesn't Break $\frac{1}{50}$ Breaks / $\frac{49}{50}$ Doesn't Break

bi 0.9604 **bii** 0.0396

5 a First spin Second spin

0.6 Green 0.6 Green / 0.4 White

0.4 White 0.6 Green / 0.4 White

bi 0.36 **bii** 0.84 **biii** 0.48

D5.3

1 a $\frac{1}{7}$ **b** $\frac{2}{7}$ **c** $\frac{4}{7}$

2 a First counter Second counter

$\frac{3}{11}$ White $\frac{2}{10}$ White / $\frac{8}{10}$ Black

$\frac{8}{11}$ Black $\frac{3}{10}$ White / $\frac{7}{10}$ Black

bi $\frac{3}{55}$ **bii** $\frac{24}{55}$ **biii** $\frac{27}{55}$

3 a First sweet Second sweet

$\frac{3}{4}$ Lemon $\frac{11}{15}$ Lemon / $\frac{4}{15}$ Orange

$\frac{1}{4}$ Orange $\frac{12}{15}$ Lemon / $\frac{3}{15}$ Orange

bi $\frac{1}{20}$ **bii** $\frac{2}{5}$ **biii** $\frac{9}{20}$

D5.4

1 a   Transport   Late?

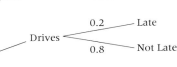

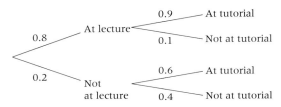

- Drives 0.2 — Late
- Drives 0.8 — Not Late
- Cycles 0.4 — Late
- Cycles 0.6 — Not Late

(0.25 Drives, 0.75 Cycles)

b 0.65

2 a   Lecture   Tutorial

- At lecture 0.9 — At tutorial
- At lecture 0.1 — Not at tutorial
- Not at lecture 0.6 — At tutorial
- Not at lecture 0.4 — Not at tutorial

(0.8 At lecture, 0.2 Not at lecture)

bi 0.12   **bii** 0.84

3 a   Gender   Completed

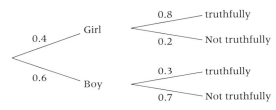

- Girl 0.8 — truthfully
- Girl 0.2 — Not truthfully
- Boy 0.3 — truthfully
- Boy 0.7 — Not truthfully

(0.4 Girl, 0.6 Boy)

bi 0.5   **bii** 0.32

D5.5

1 a $\frac{103}{220}$ **b** $\frac{51}{103}$ **c** $\frac{17}{36}$

2 a $\frac{9}{25}$ **b** $\frac{4}{9}$ **c** $\frac{4}{11}$

3 a $\frac{1}{12}$ **b** $\frac{1}{17}$

4 a $\frac{2}{3}$ **b** $\frac{6}{7}$ **c** $\frac{3}{4}$

D5 Exam review

1 a   Gavin   Alan

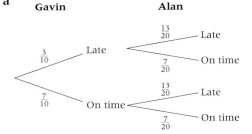

- Late $\frac{13}{20}$ — Late
- Late $\frac{7}{20}$ — On time
- On time $\frac{13}{20}$ — Late
- On time $\frac{7}{20}$ — On time

($\frac{3}{10}$ Late, $\frac{7}{10}$ On time)

b $\frac{39}{200}$

2 a Snooker   Billiards

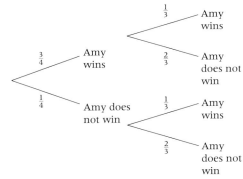

- Amy wins $\frac{1}{3}$ — Amy wins
- Amy wins $\frac{2}{3}$ — Amy does not win
- Amy does not win $\frac{1}{3}$ — Amy wins
- Amy does not win $\frac{2}{3}$ — Amy does not win

($\frac{3}{4}$ Amy wins, $\frac{1}{4}$ Amy does not win)

b $\frac{7}{12}$ **c** 14 Fridays

S8 Before you start ...

1 ai 0.5   **aii** 0.5
   bi 0.643   **bii** 0.643
   c In a right-angled triangle, the sine of one of the smaller angles is equal to the cosine of the other smaller angle.

S8.1

1 a $x = 44°$ or $136°$ **b** $x = 53°$ or $127°$
   c $x = 233°$ or $307°$ **d** $x = 210°$ or $330°$
   e $x = 192°$ or $348°$ **f** $x = 24°$ or $156°$
   g $x = 217°$ or $323°$ **h** $x = 17°$ or $163°$
   i $x = 197°$ or $343°$ **j** $x = 270°$
   k $x = 0°$ or $360°$ **l** $x = 30°$ or $150°$

2 a 0.5   **b** 0.5   **c** 0.71   **d** −0.71
   e −0.71   **f** 1   **g** −1   **h** 0
   i 0.97   **j** −0.87   **k** 0.98   **l** −0.87

3 a 130°   **b** 112°   **c** 40°   **d** 323°
   e 278°   **f** 82°   **g** 210°   **h** 297°
   i 168°   **j** 231°   **k** 8°   **l** 253°

4 ai 72°   **aii** 108°
   bi 34.6°   **bii** No second value possible
   ci 60.3°   **cii** No second value possible

S8.2

1 a $x = 46°$ or $314°$ **b** $x = 37°$ or $323°$
   c $x = 143°$ or $217°$ **d** $x = 120°$ or $240°$
   e $x = 102°$ or $258°$ **f** $x = 66°$ or $294°$
   g $x = 127°$ or $233°$ **h** $x = 73°$ or $287°$
   i $x = 107°$ or $253°$ **j** $x = 0°$ or $360°$
   k $x = 90°$ or $270°$ **l** $x = 60°$ or $300°$

2 a 0.5   **b** −0.77   **c** 0.71   **d** −0.71
   e 0.71   **f** 0   **g** 0   **h** −1
   i 0.57   **j** −0.34   **k** 0.17   **l** 0.87

3 a 330°   **b** 294°   **c** 210°   **d** 123°
   e 196°   **f** 282°   **g** 235°   **h** 126°
   i 342°   **j** 161°   **k** 58°   **l** 92°

4 b $\cos\theta = \frac{a}{b} = \sin(90° - \theta)$

S8.3

1 a $x = 35°$ or $215°$ **b** $x = 39°$ or $219°$
c $x = 141°$ or $321°$ **d** $x = 124°$ or $304°$
e $x = 62°$ or $242°$ **f** $x = 67°$ or $247°$
g $x = 149°$ or $329°$ **h** $x = 48°$ or $228°$
i $x = 132°$ or $312°$ **j** $x = 114°$ or $294°$
k $x = 0°, 180°$ or $360°$ **l** $x = 45°$ or $225°$

2 a 0.6 **b** 1.7 **c** 1 **d** 1
e -1 **f** -0.6 **g** 1.7 **h** 0
i 3.7 **j** 0.4 **k** -1 **l** -1.7

3 a $250°$ **b** $237°$ **c** $10°$ **d** $68°$
e $152°$ **f** $227°$ **g** $132°$ **h** $313°$
i $196°$ **j** $209°$ **k** $291°$ **l** $107°$

4 a Height of triangle $= \sqrt{3}$ so

$\tan 30° = \frac{1}{\sqrt{3}} = \frac{\sqrt{3}}{3}$

b $\sqrt{3}$

ci 1 **cii** $\frac{1}{\sqrt{2}}$ or $\frac{\sqrt{2}}{2}$ **ciii** $\frac{1}{\sqrt{2}}$ or $\frac{\sqrt{2}}{2}$

S8.4

1 $x = 32.3°$ or $147.7°$
2 $x = 69.1°$ or $110.9°$
3 $x = 265.6°$ or $274.4°$
4 $x = 58.6°$ or $301.4°$
5 $x = 82.9°$ or $277.1°$
6 $x = 7.1°$ or $172.9°$
7 $x = 51.3°$ or $231.3°$
8 $x = 69.0°$ or $249.0°$
9 $x = 120.9°$ or $300.9°$
10 $x = 38.3°$ or $141.7°$
11 $x = 120.5°$ or $239.5°$
12 $x = 62.1°$ or $242.1°$
13 $x = 117.9°$ or $297.9°$
14 $x = 60.0°$ or $120.0°$
15 $x = 150.0°$ or $210.0°$
16 $x = 47.1°$ or $132.9°$
17 $x = 115.5°$ or $244.5°$
18 $x = 77.5°$ or $257.5°$
19 $x = 102.5°$ or $282.5°$
20 $x = 38.9°$ or $218.9°$
21 $x = 136.5°$ or $223.5°$
22 $x = 207.8°$ or $332.2°$
23 $x = 18.7°$ or $161.3°$
24 $x = 120.0°$ or $240.0°$
25 $x = 45.0°$ or $225.0°$
26 $x = 30.0°$ or $150.0°$
27 $x = 83.4°$ or $263.4°$
28 $x = 80.4$ or $260.4°$
29 $x = 66.3°$ or $293.7°$
30 $x = 134.0°$ or $226.0°$

S8 Exam review

1 a 0.42 **b** $295°$
2 ai $y = \sin x + 1$ **aii** $y = 2 \sin x$
b Stretch with scale factor $\frac{1}{2}$ parallel to the x-axis, stretch with scale factor 3 parallel to the y-axis

N7 Before you start ...

1 a $x > 1$ **b** $x \leqslant 2$ **c** $x \leqslant \frac{7}{3}$
d $5 < x < 8$ **e** $4 < x < 6$
2 a 4 **b** $-1, 0, 1$

N7.1

1 a $\subset$ **b** $\in$
c $\cap$ **d** $\cup$
2 $A = \{17, 19\}$ $n(A) = 2$
3 $Q = \{1, 2, 3, 4, 5\}$ $n(Q) = 5$
4 a $\{2, 3, 5\}$ **b** $\{1, 2, 3, 4, 5, 6, 7, 11\}$ **c** 8
5 a $\{1, 3, 5, 7, 9\}$ **b** $\{3, 5, 7\}$
c $\{3, 5, 7\}$ **d** $\{1, 9\}$
e 5 **f** 3
6 $R \cap S = S$
7 a no hatchbacks are made in Japan
b All hatchbacks are diesels
c $J \cap D = J$ or $J \subset D$
8 a C **b** $\varnothing$ **c** $\mathscr{E}$
9 No, elements can be different
10 $Q \subset P$

N7.2

1 a $(P \cup Q)'$ **b** $(P \cup Q)' \cap R$

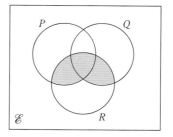

2

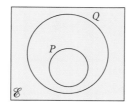 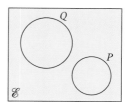

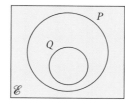 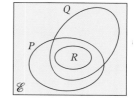

3

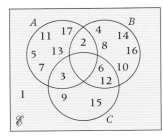

4 a ∅

   b {2,3,4,5,6,7,8,9,10,,11,12,13,14,15,16,17}

   c {2,3,4,6,8,9,10,12,14,15,16}

   d {6,12}

   e {1,2,3,5,7,9,11,13,15,17}

   f {3,5,7,11,13,17}

   g 0   **h** 11   **i** 17

N7.3

1 3

2 13

3 15

4 a 36   **b** 36   **c** 11

5 3

6

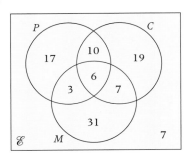

7 a

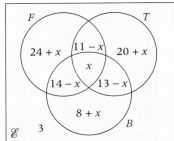

   b 5

8 a

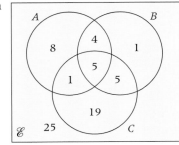

   b 8

1 a 1, 4, 6, 8   **b** 2, 3, 5 or 2, 3, 7 or 2, 5, 7 or 3, 5, 7

2 26

3 2

A10 Before you start ...

1 $50, 0, -\frac{9}{4}, -\frac{26}{9}$

2 a 1.6   **b** ±6   **c** 40   **b** −6, 7

A10.1

1 a 15   **b** 0   **c** −5

2 a 0   **b** 0   **c** $\frac{3}{8}$

3 a 8   **b** −8

4 $\frac{2}{3}$

5 a 7   **b** $x = 5, x = 1$

6 a $-\frac{1}{2}$ **b** 5

7 a $d = -5$

   b $-a + b - c + d = -5$

   c $8a + 4b + 2c + d = 13$

8 –

9 $2\sqrt{3}, -2\sqrt{3}$

10 $(x + 1)^2$

A10.2

1

$$-3 \longrightarrow -7$$
$$-2 \longrightarrow -5$$
$$-1 \longrightarrow -3$$
$$0 \longrightarrow -1$$
$$1 \longrightarrow 1$$
$$2 \longrightarrow 3$$
$$3 \longrightarrow 5$$

   Range {−7, −5, −3, −1, 1, 3, 5}

2 a $-1 \leqslant f(x) \leqslant 1$

   b $0 \leqslant f(x) \leqslant 1$

3 $-10 \leqslant f(x) \leqslant 10$

4 a

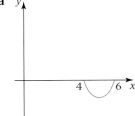

   b –   **c** $-1 \leqslant f(x) \leqslant 0$

5 a $x \geqslant 0$ **b** any value except 0

   c any value except −2

6 a $x \neq -5$ **b** $x \neq 2.5$ **c** $x \neq -2$ **d** $x < 0$

A10.3

1 **a** 10   **b** 50   **c** 6   **d** 17

2 **a** $fg(x) = x$, $gf(x) = x$

3 $\dfrac{1}{(x-1)^2} - 4$

4 –

5 15

6 $a = 3, b = -10$

7 2

8 **a** $\dfrac{3x+6}{-x-6}$ **b** $\dfrac{x+2}{4x-4}$ **c** $\dfrac{3x-2}{6-x}$

9 **a** 0   **b** 4

A10.4

1 **a** $\dfrac{4x+2}{3}$ **b** $\dfrac{1}{1-x}$ **c** $\dfrac{5+5x}{x-1}$ **d** $x^2 + 4$

2 **a** 4.5

3 **a** $f^{-1}(x) \geqslant 3$ **b** $3 + \sqrt{x}$

4 $\dfrac{x-3}{2}$

5 **a** x **b** inverse function   **c** $\sqrt{5}$

6 **a** $x = 1, x = -1.5$

7 **a** –   **b** $\dfrac{1+x}{x}$ **c** $f^{-1}(x) > 1$

8 –

A10 Exam review

1 **a** -1 **b** x^2 **c** 23

2 $8x^2 + 8x + 1$

3 **a** 0.5   **b** $\dfrac{1}{3}$ **c** $\dfrac{1-2x}{x}$ **d** -2 **e** -3

4 $x \geqslant 0$ and $x \neq \dfrac{3}{2}$ **5** $\dfrac{x-3}{2}$ **6** $0 \leqslant f(x) \leqslant 1$

A11 Before you start ...

1 **a** m   **b** c

2 **a** $\dfrac{8}{3}$ **b** -3 **c** $-\dfrac{a}{b}$

3 **a** $-\dfrac{a}{b}$ **b** $-\dfrac{c}{b}$

A11.1

1 **a** -2 **b** -2 **c** -2

2 **a** -1.5 **b** -1.5 **c** -1.5

3 **a** $\frac{1}{3}$ **b** $\frac{1}{3}$

4 $y = x + c$

5 **a** 7   **b** $C = 7t + 13$

6 **a** 2π **b** 2π

7 **a** $C = 2\pi(5t + 4)$ **b** 10π

A11.2

1 **a** 2   **b** $4x + 5$ **c** $2x - 1$ **d** $6x + 5$

     e $50x + 12x^3 + x$ **f** $x^3 - x^2 + 2x$

2 **a** -3 **b** 0   **c** $18x^2 - 14x + 1$

     d $-x^2$ **e** $2 - 8x^{-3}$ **f** $-x^{-2} + 3x^{-4}$

3 **a** $2t$ **b** t **c** $5 - 8t$

4 **a** $-\dfrac{5}{33} + \dfrac{v}{50}$

5 **a** $2r$ **b** $2\pi r$ **c** $8\pi r$ **d** $10\pi + 4\pi r$

6 **a** $12\pi r$ **b** $4\pi r^2$

7 **a** $2x - 1$ **b** $4x + 1$ **c** $4x^3 + 30x^2 + 50x$

8 **a** $5x$ **b** 1   **c** $1 - x^{-2}$

     d $-3x^{-4} - 4x^{-3}$ **e** $-2x^{-3} + 2x$ **f** $4 - \dfrac{1}{2}x^{-2}$

A11.3

1 4

2 5

3 -2.25

4 $(1, -2)$ and $\left(-\frac{1}{3}, \frac{14}{27}\right)$

5 $\frac{5}{3}$

6 0

7 5

8 0.75

9 $(2.5, -1.25)$

10 **a** $3x^2 - 4x - 4$ **b** -5 **c** $x = -\frac{2}{3}, x = 2$

11 $p = 2, q = 16$

12 $p = 3, q = 5$

A11.4

1 $y = 3$

2 $y = 5x + 1$

3 $y = -2x + 3$

4 $y = -7x + 8$

5 $y = -10x - 13$

6 $y = 65x - 48$

7 $y = 3.5x - 2$

8 $y = -11x - 19$

9 $(2x - 1)^2 = -2$ has no solutions since it requires $\sqrt{-2}$

10 $y = 3x - 4$

11 a $(2,8)$ $(-2,16)$ **b** – **c** $y = -10x - 4$

12 $x = 3$

A11.5

1 a $(-1.5, 2.5)$ **b** $(1.5, 11.5)$

c $(3, -44)$ and $(-3, 64)$

2 $(3.5, -12.25)$ minimum because graph $\cup$

3 $(2, -1)$ maximum because graph $\cap$

4 $(2, 12)$

5 $(2, -11)$ Minimum, $(-2, 21)$ Maximum

6 a $x = 4$ **b** $x = 4$ **c** $x = 2$

7 Minimum 0, maximum $\frac{500}{27}$

8 Maximum $v = 4$ when $t = 2$. Maximum because graph $\cap$

A11.6

1 a $100 - 2x$ **b** $x(100 - 2x)$

c $100 - 4x$ **d** 1250 m^2

2 a – **b** 40 cm

3 a – **b** – **c** 47.5 m^2

4 a –

b $a = 50$, $b = 50$ maximum because graph $\cap$

A11.7

1 a $-10t + 3t^2$ **b** $-10 + 6t$ **c** 100 m

d 0 m s^{-1} **e** -10 m s^{-2}

2 5 m s^{-1} **3** -7.5 m s^{-2}

4 a 413 m **b** 55 m s^{-1}

5 a $2t - 3t^2$ **b** $2 - 6t$ **c** $\frac{1}{3} \text{ m s}^{-1}$

6 a $-14 + 18t^2$ **b** $36t$ **c** 7 m

d 4 m s^{-1} **e** 36 m s^{-2} **f** 0.882 seconds

g 6.77 m

A11 Exam review

1 a $-2x - 4$ **b** $2x - 3x^2 + 3$

c $-\dfrac{2}{x^3}$ **d** $-\dfrac{42}{x^7}$

2 -885

3 15

5 $(1, 10)$. It is a minimum because the graph is $\cup$ shaped.